Alkaline Fuel Cells

Fundamentals, Design, and Applications

Renewable Energy Fundamentals Series

The *Elsevier Renewable Energy Fundamentals Series* brings together introductory titles on a range of topics related to renewable energy, covering solar energy, wind power, bioenergy and biofuels, hydrogen and fuel cells, marine energy, geothermal energy, and hydropower, as well as technologies, models, and methods for the utilization of those resources, energy storage, hybrid energy systems, carbon neutral energy, energy efficiency and energy management, sustainability, and policy.

The Series aims to support senior undergraduate, graduate, and PhD students studying the development and implementation of renewable energy and its related disciplines, as well as researchers, academics, and engineers.

Series Editors

Jingzheng Ren, Associate Professor, The Hong Kong Polytechnic University
Chang He, Associate Professor, Sun Yat-Sen University

Volumes in the Series

Grajales Agudelo & Wang (Eds.), Fundamentals and Advances in Sustainable Energy Processes and Applications, 9780443415753, 2026
Liang An, Zhefei Pan & Rong Chen (Authors), Alkaline Fuel Cells: Fundamentals, Design, and Applications, 9780443366383, 2026

Renewable Energy Fundamentals

Alkaline Fuel Cells

Fundamentals, Design, and Applications

Liang An
Department of Mechanical Engineering,
The Hong Kong Polytechnic University,
Hung Hom, Hong Kong SAR, China

Zhefei Pan
Key Laboratory of Low-grade Energy Utilization
Technologies and Systems, Ministry of Education

School of Energy and Power Engineering,
Institute of Engineering Thermophysics,
Chongqing University, Chongqing, China

Rong Chen
School of Energy and Power Engineering,
Institute of Engineering Thermophysics,
Chongqing University, Chongqing, China

ELSEVIER

ACADEMIC PRESS
An imprint of Elsevier

Academic Press is an imprint of Elsevier
125 London Wall, London EC2Y 5AS, United Kingdom
50 Hampshire Street, 5th Floor, Cambridge, MA 02139, United States

ISBN: 978-0-443-36638-3

For information on all Academic Press publications visit our website at https://www.elsevier.com/books-and-journals

Publisher: Priscilla Braglia
Acquisitions Editor: Ambika Kapoor
Editorial Project Manager: Debarati Roy
Production Project Manager: Fizza Fathima
Cover Designer: Arumugam Kothandan

Typeset by Integra Software Services

Contents

3 Catalysts for hydrogen oxidation in alkaline media: Mechanisms, classifications, designs and challenges

Wenzhi Li, Yun Liu, Xiaohong Zou, and Liang An

4 Anion exchange membranes for alkaline fuel cells: Synthesis, properties and perspectives

Anastasiia Konovalova, David Aili, and Dirk Henkensmeier

5 Alkaline oxygen reduction reaction catalysts: Mechanisms, classification, design and challenges

Aoxin Ran and Linhao Fan

6 Cell design and stack engineering: Theory, integration and perspectives

Haoran Du and Zhiming Bao

7 Performance testing and characterization techniques

Xingyi Shi, Dixian Wu, and Qiang Zheng

Contributors

David Aili Department of Energy Conversion and Storage, Technical University of Denmark, Kongens Lyngby, Denmark

Liang An Department of Mechanical Engineering, The Hong Kong Polytechnic University, Hung Hom, Hong Kong SAR, China

Zhiming Bao Tianjin University, Tianjin, China

Haoran Du Tianjin University, Tianjin, China

Liuke Duan Key Laboratory of Low-grade Energy Utilization Technologies and Systems, Ministry of Education; School of Energy and Power Engineering, Institute of Engineering Thermophysics, Chongqing University, Chongqing, China

Linhao Fan State Key Laboratory of Engines; School of Energy (National Industry-Education Platform for Energy Storage), Tianjin University, Tianjin, China

Dirk Henkensmeier KIST School, Energy and Environment Technology, University of Science and Technology; Center for Hydrogen and Fuel Cells, Korea Institute of Science and Technology; Graduate School of Energy and Environment, Korea University, Seoul, South Korea

Anastasiia Konovalova Department of Energy Conversion and Storage, Technical University of Denmark, Kongens Lyngby, Denmark

Qiao Lan School of Petrochemical Engineering, Lanzhou University of Technology, Lanzhou, Gansu, China

Dennis Y.C. Leung Department of Mechanical Engineering, The University of Hong Kong, Hong Kong, China

Michael K.H. Leung School of Energy and Environment, City University of Hong Kong, Hong Kong, China

Haozhen Li School of Energy and Power Engineering, Institute of Engineering Thermophysics, Chongqing University, Chongqing, China

Wenzhi Li Department of Mechanical Engineering, The Hong Kong Polytechnic University, Hung Hom, Hong Kong SAR, China

Jian Liu School of Petrochemical Engineering, Lanzhou University of Technology, Lanzhou, Gansu, China

Yun Liu Department of Mechanical Engineering, The Hong Kong Polytechnic University, Hung Hom, Hong Kong SAR, China

Wang Ming School of Energy and Power Engineering, Institute of Engineering Thermophysics, Chongqing University, Chongqing, China

Yutong Mu School of Human Settlements and Civil Engineering, Xi'an Jiaotong University, Xi'an, Shaanxi, China

Muhammad Mubeen The Islamia University of Bahawalpur, Bahawalpur, Pakistan

Meng Ni Department of Building Environment and Energy Engineering, The Hong Kong Polytechnic University, Hong Kong, China

Zhefei Pan School of Energy and Power Engineering, Institute of Engineering Thermophysics; Key Laboratory of Low-grade Energy Utilization Technologies and Systems, Ministry of Education, Chongqing University, Chongqing, China

Zhiguo Qu School of Energy and Power Engineering, Xi'an Jiaotong University, Xi'an, Shaanxi, China

Aoxin Ran State Key Laboratory of Engines, Tianjin University, Tianjin, China

Xingyi Shi City University of Hong Kong (Dongguan), Dongguan, China

Yifei Wang School of Robotics and Advanced Manufacture, Harbin Institute of Technology, Shenzhen, China

Dixian Wu City University of Hong Kong (Dongguan), Dongguan, China

Xinhai Xu School of Robotics and Advanced Manufacture, Harbin Institute of Technology, Shenzhen, China

Qiang Zheng Eastern Institute of Technology, Zhejiang Key Laboratory of Industrial Intelligence and Digital Twin, Ningbo, China

Yao Zhu Department of Mechanical Engineering, The Hong Kong Polytechnic University, Hung Hom, Hong Kong SAR, China

Xiaohong Zou Department of Mechanical Engineering, The Hong Kong Polytechnic University, Hung Hom, Hong Kong SAR, China

About the authors

Professor Liang An received his bachelor degree in thermal and power engineering from Harbin Institute of Technology and his PhD in mechanical engineering from the Hong Kong University of Science and Technology (HKUST). After graduation, he worked as a visiting scholar and a postdoctoral fellow in the same research group. His research interests are in the areas of clean energy-conversion technologies and renewable energy storage technologies. His research interests include fuel cells and batteries.

Professor Zhefei Pan received his BS degree (2015) and MS degree (2017) from the School of Energy Science and Engineering at the Harbin Institute of Technology, and his PhD (2021) from the Department of Mechanical Engineering at the Hong Kong Polytechnic University. After working as a post-doctoral fellow in the same department for three years, he became a Professor at Chongqing University. Professor Pan has nearly 10 years of experience in experimental and numerical investigations of fuel cells and electrolyzers. His scientific interests include direct liquid fuel cells fueled by ammonia and ethylene glycol and the electrochemical production of chemicals such as ammonia and hydrogen peroxide. He has published more than 40 publications, including works in *Progress in Energy and Combustion Science* and *ACS Energy Letters*.

Professor Rong Chen received his BS degree (2000) and MS degree (2003) from School of Energy and Power Engineering from Chongqing University, and PhD degree (2007) in mechanical engineering from the Hong Kong University of Science and Technology. He completed his postdoctoral research in the same place and was appointed as a visiting assistant professor at the Hong Kong University of Science and Technology from 2008–2009. He joined the School of Energy and Power Engineering at Chongqing University in 2010. He has published more than 200 peer-reviewed papers in the international journals. His research interests mainly cover the solar utilization by photochemistry, optofluidics, new energy technologies, micro-scale transport, and interfacial phenomena, as well as the heat/mass transport in porous media.

Preface

The global energy system is accelerating toward a renewable and zero-carbon future. Climate change, the depletion of fossil fuels, and fluctuations in energy supply are driving the development of clean secondary energy. Hydrogen energy, with its zero emissions, high energy density, and ease of storage and transportation, is gradually becoming an important carrier connecting renewable energy with diverse end-use sectors such as transportation, industry, and buildings. Alkaline fuel cells (AFCs), with their cost advantages and excellent performance, are emerging as a key technology pathway for the efficient conversion and utilization of hydrogen energy. As the first electrochemical energy conversion device to be engineered for application in the aerospace domain, AFCs have recently become a research hotspot in the energy transition. Their main advantages include: achieving efficient oxygen reduction reactions without the need for precious metal catalysts, significantly reducing the cost of core materials; having low-temperature start-up capabilities, meeting the application requirements of multiple scenarios; and generating only water during power generation, achieving zero carbon emissions. These characteristics give AFCs unique competitiveness in the conversion and utilization of hydrogen energy. However, before AFCs can be scaled up for commercial application, they still face three major bottlenecks: performance, lifespan, and cost. Trace impurities in the fuel gas can cause catalyst poisoning and deactivation, and CO_2 in the air can react with the liquid KOH electrolyte to form carbonate precipitates, blocking the electrode interface; the strong coupling of multiphase flow, heat transfer, and electrochemical reactions within the system increases the difficulty of simulation and optimization, limiting the prediction and control of full life-cycle performance. To address these challenges, multidisciplinary collaborative optimization of materials, interfaces, and systems is needed. This includes developing CO_2-resistant high-performance electrolytes and highly active non-precious metal catalysts, optimizing the microstructure of electrodes and the wettability of electrolytes, and combining multiphysics field simulation with experimental verification to achieve the large-scale application and long-term stable operation of AFC technology.

To present the overall development and cutting-edge trends of the AFC technology system, this book comprehensively reviews AFC technology from the dimensions of technological evolution, core scientific issues, key material development, system design, performance characterization, modeling methods, application scenarios, and industrial ecology. The core content and positioning of each chapter are as follows:

Chapter 1 systematically reviews the development history and technical breakthroughs of AFCs from its early space applications to multiscenario exploration, presenting a complete framework from theoretical exploration to engineering application, laying the foundation for the subsequent chapters.

Chapter 2 introduces the mass transfer phenomena in AFCs and analyzes the transfer processes of gas, charge, and water, and their effect on the fuel cell in detail and the optimization directions involved.

Chapter 3 introduces the reaction mechanisms, catalyst development, and existing challenges for the alkaline hydrogen oxidation reaction in fuel cells.

Chapter 4 provides a comprehensive introduction to the synthesis methods, performance characterization, and structure-activity relationships of anion exchange membranes (AEMs), and looks forward to its application and improvement directions in high-performance AFCs.

Chapter 5 describes the mechanism of the oxygen reduction reaction, summarizes state-of-the-art catalysts for alkaline exchange membrane fuel cells, discusses the catalyst design strategies such as atomic-scale regulation and nanoscale optimization, and proposes the future directions of cross-scale simulation and artificial intelligence-aided design.

Chapter 6 focuses on the development and integration methods of key components such as electrodes, electrolytes, and flow field plates. It assesses the impact of different designs on mass transfer efficiency, heat transfer, and electrochemical performance, and proposes strategies to address technical bottlenecks.

Chapter 7 provides a comprehensive overview of common performance testing and characterization methods for AFCs, including electrochemical measurements and physical and compositional analysis techniques.

Chapter 8 systematically reviews mainstream modeling methodologies for AFC and alkaline AEM fuel cells, with emphasis on multiscale modeling approaches that optimize anion transport in membranes to achieve high ion conductivity and cell performance.

Chapter 9 provides a comprehensive overview of the advancements in alkaline liquid fuel cells, including the physicochemical properties of diverse liquid fuels and the working principles underlying alkaline liquid fuel cell systems. The research advancements across material innovation, reactor design, and energy system integration, as well as the prospective application scenarios, are highlighted.

Chapter 10 introduces the application of AFCs in various sectors, including its glorious history in aerospace, military, and civil transportation, its recent transformation into stationary power generation, as well as the new generation of AEM fuel cell that is coming back for the transportation sector.

Chapter 11 analyzes the global market prospects, cost composition, policy barriers, and regional differences, and explores the supporting role of hydrogen infrastructure in promoting the industry and the potential of emerging markets.

Chapter 12 provides a comprehensive overview of the AFC industrial policy frameworks and strategic plans of major economies, presenting the tripartite interaction mechanism among policies, technologies, and the market. It also explores the development paths and opportunities for industries under policy guidance.

Chapter 13 dissects the problems faced by the AFC technology system, including dynamic lag, material stability, system cost, and environmental adaptability. It then looks forward to the strategic value and development potential in the future from the dimensions of material innovation, system optimization, application expansion, and policy-driven factors.

This book systematically reviews the technical framework and cutting-edge developments of AFCs, aiming to provide researchers, engineers, and policymakers in the field of fuel cells with a reference that combines theoretical depth with practical guidance value. It is intended to promote the innovative development and industrialization implementation of AFC technology in the global clean energy transition, and provide technical support for the sustainable development of the energy system.

Acknowledgements

This book focuses on the cutting-edge advancements and practical experiences in the field of alkaline fuel cells (AFC), serving as an indispensable reference resource for researchers in the academic community, technicians in the industrial sector, and policymakers in government departments worldwide. The successful completion of this work was made possible by support from the Research Grants Council of the Hong Kong Special Administrative Region (N_PolyU559/21, 15308024), the National Key Research and Development Program (2024YFB4006500), the National Natural Science Foundation of China (52161160333, 52576192), and the Chongqing Talent Project (CSTB2024YCJH-KYXM0082). We sincerely thank all the authors who contributed their articles, integrating their valuable professional knowledge, technical experience, and research insights in the field of AFCs into this book, laying a solid foundation for its academic value and practical significance. At the same time, we express our heartfelt gratitude to every expert and scholar who participated in the review process, who meticulously reviewed the content and provided meticulous guidance with a rigorous professional attitude, effectively contributing to the improvement of the content and quality. Finally, we would like to extend our special thanks to Ming Wang for his diligent assistance and strong support during the completion of this book.

Chapter 1

Introduction to alkaline fuel cells

Wang Ming[a] **and Zhefei Pan**[a,b]
[a]*School of Energy and Power Engineering, Institute of Engineering Thermophysics, Chongqing University, Chongqing, China*
[b]*Key Laboratory of Low-grade Energy Utilization Technologies and Systems, Ministry of Education, Chongqing University, Chongqing, China*

1 The birth of alkaline fuel cells

At a time when the global energy landscape is undergoing profound changes, energy and environmental issues have become key factors restricting the sustainable development of human society. The extensive consumption of traditional fossil fuels not only leads to the depletion of resources but also triggers severe environmental pollution and climate change problems. According to data from the International Energy Agency, the annual carbon dioxide emissions from the combustion of fossil fuels worldwide amount to hundreds of billions of tons, posing a significant threat to the ecological balance of the Earth (Staffell et al., 2019). Against this backdrop, fuel cells, as an efficient and clean energy conversion technology, have emerged, bringing new hope for addressing energy and environmental issues (Walsh, 2001). The working principle of fuel cells is based on electrochemical reactions, which can directly convert the chemical energy of fuel and oxidant into electrical energy, avoiding energy loss and pollutant emissions in the traditional combustion process (Carrette et al., 2001). Compared with traditional internal combustion engines, the energy conversion efficiency of fuel cells can be increased by 2 to 3 times, and they produce almost no harmful gases such as nitrogen oxides and sulfur oxides, earning them the title of "the new energy star of the 21st century" (O'hayre et al., 2016; Springer et al., 1991). Among various types of fuel cells, alkaline fuel cells (AFCs) have become an important research object in the field of fuel cells due to their unique advantages and development history.

The origin of AFCs can be traced back to the early 19th century when scientists' exploration in the field of electrochemistry laid the theoretical foundation for their birth. In 1838, German chemist Christian Friedrich Schönbein first proposed the principle of fuel cells, suggesting that the redox reaction between hydrogen and oxygen could be used to generate electricity. This theoretical

Alkaline Fuel Cells. https://doi.org/10.1016/B978-0-443-36638-3.00009-7

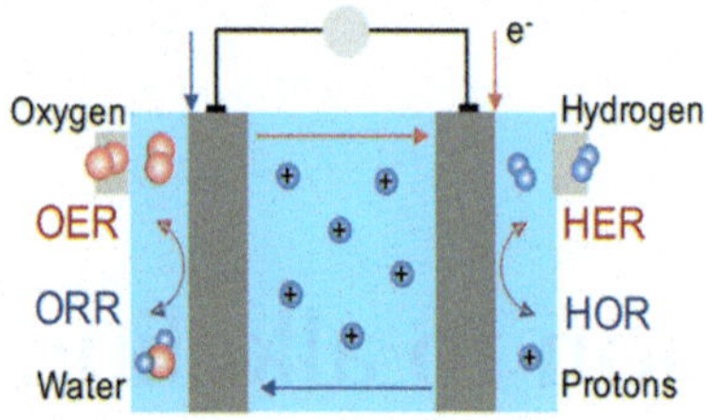

FIG. 1.1 Hydrogen oxygen fuel cell. (Wei et al., 2019). Schematic diagram of a two-electrode fuel cell. *(From Wei, C., Rao, R.R., Peng, J., Huang, B., Stephens, I.E.L., Risch, M., Xu, Z.J., Shao-Horn, Y., 2019. Recommended Practices and Benchmark Activity for Hydrogen and Oxygen Electrocatalysis in Water Splitting and Fuel Cells. Adv. Mat., 31 (31). https://doi.org/10.1002/adma.201806296)*

proposal was like a seed, sowing hope for the future development of fuel cells. Based on Schönbein's theory, in February 1839, British physicist William Robert Grove conducted in-depth research and experiments (Grove, 1842; see Fig. 1.1). He designed a device that successfully converted the chemical energy of hydrogen and oxygen into electrical energy, which was the initial prototype of a fuel cell. Grove published his theoretical proof in the Philosophical Magazine and Journal of Science, attracting widespread attention from the scientific community. His discovery not only provided an important theoretical basis for the development of fuel cells but also laid the foundation for the later birth of AFCs. Although Grove's fuel cell design was simple, it contained the basic principles of AFCs. In his design, sulfuric acid was used as the electrolyte, and current was generated through the reaction of hydrogen and oxygen on the electrodes (Wisniak, 2015). Although this design was far from perfect at the time, it provided valuable research ideas for later scientists.

Since then, scientists have conducted in-depth research on the principles and applications of fuel cells, constantly exploring methods to improve their performance. The related paper, "*From Sir William Grove to today: fuel cells and the future*" provides a detailed review and analysis of this history. It points out that Grove's discovery opened up a new path for the development of fuel cells, and subsequent researchers have continuously innovated on this basis, promoting the progress of fuel cell technology (Appleby, 1990). After Grove's research, although some progress was made in the development of fuel cells, for a long time, their performance and efficiency could not meet the demands of practical applications. It was not until the early 20th century, with the continuous advancement of materials science and engineering technology, that the development of fuel cells encountered a new opportunity. During this period, scientists began to experiment with different electrolytes and electrode materials to enhance the performance of fuel cells. Among these efforts, the use of alkaline electrolytes became an important direction in the development of fuel cells.

2 Early technical exploration of AFC

In the 1940s and 1950s, Francis Thomas Bacon of Cambridge, United Kingdom, initiated the crucial transition of AFCs from theory to practice. Bacon was passionate and persistent in his research on fuel cells, devoting a great deal of time and energy to solving numerous key problems in fuel cell technology. At that time, fuel cells faced multiple challenges, including the selection of electrode materials, the stability of electrolytes, and the design of battery structures. To find suitable electrode materials, Bacon conducted countless experiments and studied various metals and alloys. He discovered that nonprecious metals such as Raney Nickel (Fig. 1.2) had excellent catalytic activity for the oxidation of hydrogen in an alkaline environment, while silver-based catalysts performed well in the reduction of oxygen (Kiros and Schwartz, 2000). This discovery provided an important direction for the selection of electrode materials in AFCs, reducing the reliance on expensive precious metals like platinum and lowering costs while enhancing the practicality of the cell (Brouzgou, Song and Tsiakaras, 2012). The book "*Recent Advances in Fuel Cell Technology and Its Applications*" points out that Bacon's experiments with different electrode materials confirmed the feasibility of using Raney Nickel and silver-based catalysts in AFCs, laying the foundation for subsequent research (Gasteiger et al., 2005). Regarding the electrolyte, Bacon chose potassium hydroxide (KOH) solution as the electrolyte for AFCs. KOH solution has good ionic conductivity and can effectively transfer ions during battery operation, promoting electrochemical reactions. However, KOH solution is prone to reacting with carbon dioxide (CO_2) during use, forming potassium carbonate (K_2CO_3), which degrades the performance of the electrolyte. To address this issue, Bacon proposed a series of measures, such as strictly purifying the fuel and oxidant to reduce the CO_2 content, and designing a special battery structure to allow the electrolyte to circulate within the battery and promptly remove the generated K_2CO_3, maintaining the stability of the electrolyte. After years of relentless efforts, in 1959, Bacon

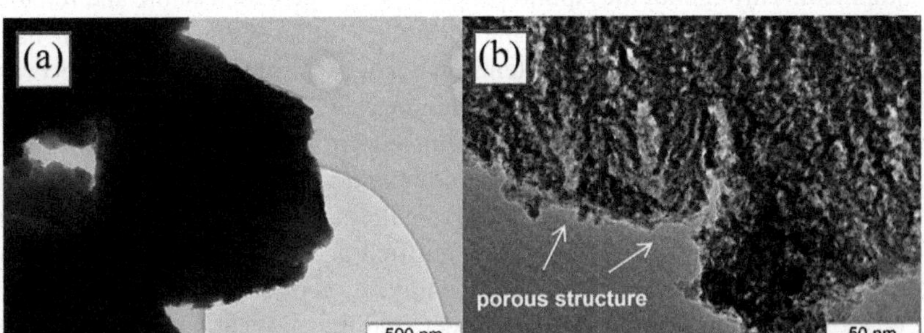

FIG. 1.2 TEM images: (a) nickel agglomerates; (b) a sponge-like structure (Philippov et al., 2020). TEM images of the freshly prepared Raney nickel. *(From Philippov, A.A, Chibiryaev A.M, Prosvirin, I.P, et al. Some like it weak: different activity of Raney® nickel in transfer hydrogenation under air and inert atmosphere[J]. App. Cat. A: General, 2020, 605:117788)*

successfully developed a 5 kW AFC system. This achievement was of milestone significance, marking the transition of AFC technology from theoretical research to practical application. Bacon's AFC system demonstrated numerous advantages at that time, with an energy conversion efficiency of 60–70%, far exceeding that of traditional power generation equipment. For instance, compared to the common internal combustion engines of the time, which typically had an energy conversion efficiency of around 30–40%, Bacon's AFC system could more efficiently convert the chemical energy of fuel into electrical energy, significantly improving energy utilization efficiency. Additionally, the battery system operated stably, with low noise and environmental friendliness, generating almost no pollutant emissions. This technological breakthrough attracted the attention of numerous research institutions and enterprises worldwide, providing valuable experience and a technical foundation for the subsequent development of fuel cells. Many researchers built upon this foundation to further optimize and improve the performance of AFCs, driving the continuous development of fuel cell technology.

3 Applications of AFC in the aerospace field

Entering the 1960s, with the rapid development of aerospace technology, extremely high demands were placed on efficient and reliable energy supply systems. Among various energy technologies, National Aeronautics and Space Administration (NASA) chose AFCs as the main power source for the Apollo spacecraft, a decision that completely transformed the fate of AFCs and made them a star technology in the aerospace field (Halpert et al., 1999). During the Apollo moon landing program, NASA faced the arduous task of providing stable power and water resources for the spacecraft in the extreme space environment. AFCs stood out due to their unique advantages. They use hydrogen and oxygen as fuel and directly convert chemical energy into electrical energy through electrochemical reactions (Fig. 1.3), providing power for various spacecraft equipment. Throughout the Apollo program, AFCs provided stable and reliable power for the spacecraft, ensuring the normal operation of communication, navigation, life support, and other systems on board. According to relevant records, during the Apollo 11 moon landing mission, AFCs worked continuously and stably, providing sufficient power support for astronauts' activities on the lunar surface, enabling them to successfully complete various scientific exploration tasks and return to Earth (Launius, 1994). At the same time, the only by-product of the fuel cell reaction—water—was purified and treated to meet the astronauts' daily water needs, maximizing the utilization of energy. The paper "*Fuel Cells for Space Science Applications*" detailed the application of AFCs in the aerospace field, pointing out that their high energy conversion efficiency and reliability are key factors in meeting the demands of space missions (Burke, 2003). The AFCs used in the Apollo spacecraft adopted a highly integrated and compact design. The battery system was composed of multiple single cells,

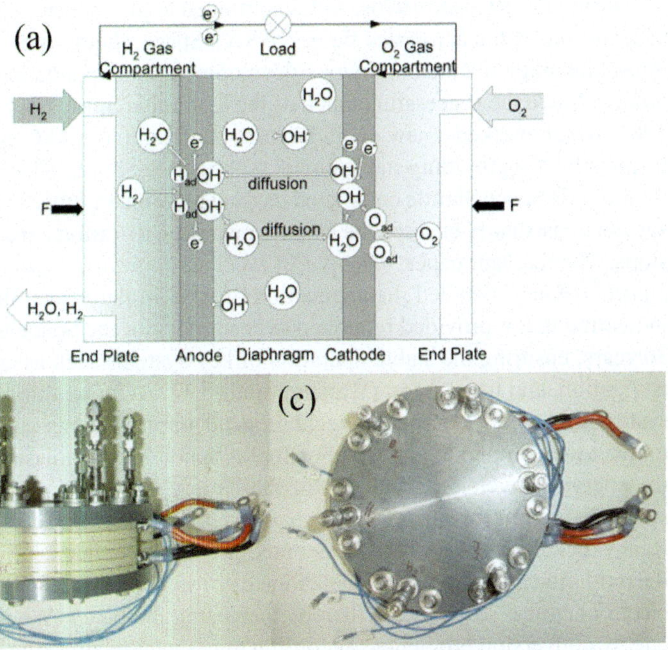

FIG. 1.3 (a) An alkaline fuel cell operates inside the Apollo spacecraft; (b-c) Side and top views of the AFC (Markgraf et al., 2012). The basic assembly and working principle of a fixed AFC. *(From Markgraf, S., Hörenz, M., Schmiel, T., Jehle, W., Lucas, J., Henn, N., 2012. Alkaline fuel cells running at elevated temperature for regenerative fuel cell system applications in spacecrafts. J. Power Sources. 03787753. 201, 236–242. doi:10.1016/j.jpowsour.2011.10.118.)*

which were arranged and connected in a reasonable manner to form an efficient power generation unit. To adapt to the extreme conditions of the space environment, such as high vacuum, strong radiation, and drastic temperature changes, the battery casing was made of special materials and designed with a special structure, effectively resisting the influence of the external environment and ensuring the normal operation of the battery. In addition, NASA equipped the battery system with an advanced control system that could monitor the battery's working status in real time, such as voltage, current, and temperature, and adjust and optimize it according to actual needs, ensuring stable operation of the battery in various complex situations. The successful application of AFCs in the Apollo spacecraft not only made significant contributions to the development of the aerospace industry but also sparked global attention and in-depth research on AFC technology. It proved the feasibility and superiority of AFCs in extreme conditions, laying a solid foundation for the subsequent application of fuel cells in other fields. Many countries and enterprises began to increase their investment in fuel cell technology research and development, hoping to apply this advanced technology to a wider range of fields and promote the transformation of the energy sector.

During the 1970s to 1990s, AFCs continued to play an irreplaceable and significant role in the aerospace field. NASA's Space Shuttle program became an important stage for the continuous development and application of AFC technology. Since the successful launch of the Columbia Space Shuttle on April 12, 1981, which marked a new era in human spaceflight, the AFC system has been responsible for providing stable power to the Space Shuttle. This continued until the Atlantis Space Shuttle completed its final mission on July 21, 2011. The AFC system consistently ensured the smooth operation of various Space Shuttle missions. The relevant paper, "*The NASA fuel cell upgrade program for the Space Shuttle orbiter*" detailed the application of AFC in the Space Shuttle program, indicating that it provided reliable power support for the Space Shuttle for over 30 years, ensuring the stable operation of key systems such as communication, navigation, and life support (Warshay et al., 1997). During this period, the AFC system underwent continuous technological improvements and optimizations to meet the complex demands of different Space Shuttle missions. To enhance the energy conversion efficiency of the battery, scientists conducted in-depth research and improvements on electrode materials and catalysts. A new type of Raney nickel-based catalyst was adopted, which has higher catalytic activity for the oxidation of hydrogen in an alkaline environment, effectively promoting the electrochemical reaction and thereby increasing the battery's output power and energy conversion efficiency. According to relevant records, the energy conversion efficiency of the improved AFC system increased by 10–15% compared to before, significantly enhancing the energy utilization of the Space Shuttle. At the same time, to enhance the stability and reliability of the battery, NASA comprehensively upgraded the battery's structural design and control system. In terms of battery structure, a more compact and robust design was adopted, which could better withstand the various mechanical environments and extreme temperature changes that the Space Shuttle faced during launch, flight, and return. For example, by optimizing the battery's casing material and structure, it could withstand higher pressure and impact forces, ensuring normal operation of the battery in harsh environments. In terms of the control system, advanced sensors and intelligent control algorithms were introduced, which could monitor the battery's working status in real time and automatically adjust the battery's operating parameters according to actual needs, ensuring stable operation of the battery under various working conditions. The relevant research results were detailed in the paper, "Advances in Alkaline Fuel Cell Technology for Space applications", which mentioned that through improvements in battery structure and control systems, the reliability of the AFC system was significantly enhanced, and the failure rate was significantly reduced (Kordesch et al., 2000). The outstanding performance of the AFC system in the Space Shuttle program further verified the feasibility and superiority of AFC technology in extreme environments. It not only made significant contributions to the development of the aerospace industry but also provided valuable experience and technical support for the subsequent application of fuel cell technology in other fields.

The accumulation of these experiences and technologies laid a solid foundation for the further development of fuel cell technology and promoted global enthusiasm for research and application of fuel cell technology.

4 The application of AFC in the civil field

During the 1970s to 1990s, with the successful application of AFCs (AFC) in the aerospace field, people began to turn their attention to the civilian sector, attempting to apply this advanced technology to more practical scenarios (Carrette et al., 2000). During this period, the application of AFC in the civilian sector was initially expanded. Although it faced many challenges, it also accumulated valuable experience for subsequent development. In the field of ground transportation, AFC was applied to some special vehicles. In 1971, Kordesh et al. attempted to use AFCs in cars to provide power (Larminie and McDonald, 2003). However, due to the technological limitations at that time, the AFC system had problems such as high cost, large volume, extremely high requirements for fuel purity, and vulnerability to carbon dioxide contamination, which greatly restricted its application in cars (Abdelkareem et al., 2021). According to relevant research data, the cost of the fuel cell system in AFC cars at that time was several times that of traditional internal combustion engines. Moreover, due to the large volume of the system, the interior space of the vehicle was severely compressed, greatly reducing its practicality. In addition, AFC was extremely sensitive to carbon dioxide in the fuel. Even a trace amount of carbon dioxide could cause a sharp decline in battery performance, which required strict purification of the fuel, further increasing the cost and complexity of use. The relevant paper "Intermittent Use of a Low-Cost Alkaline Fuel Cell-Hybrid System for Electric" analyzed in detail the problems faced by AFC in automotive applications, pointing out that these problems severely restricted the development of AFC in the automotive field. Besides cars, AFC was also applied to some other ground transportation vehicles. For example, in 1960, Ihrig attempted to use AFC as the power source for driving agricultural tractors (Kordesch et al., 1999). In the field of construction machinery, such as cranes, some researchers also attempted to apply AFC technology. However, these applications also faced similar problems as those in automotive applications, such as high cost, poor operability, insufficient durability, and safety. In actual use, the maintenance cost of the AFC system was high, and due to its inferior stability and reliability compared to traditional power systems, the failure rate of the equipment was high, affecting normal operational efficiency (Ma et al., 2023). In the field of ships, AFC also showed certain application potential. Some researchers attempted to use AFC to provide power for ships and supply power to marine navigation equipment. Compared with traditional ship power systems, AFC had the advantages of low noise and low emissions, which could reduce pollution to the marine environment. However, AFC also faced many challenges in ship applications. Due to the complex working environment of ships, extremely

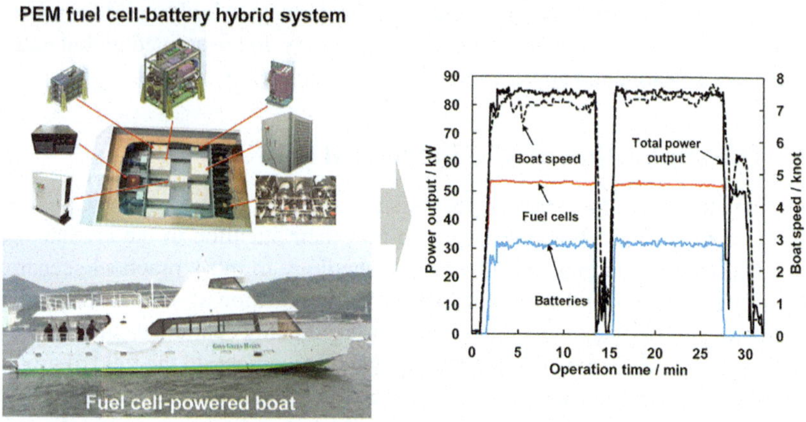

FIG. 1.4 Alkaline fuel cell sightseeing boat (Choi et al., 2016): Installation layout of the system components, along with power-output and boat-speed profiles during the test operation of the fuel-cell-powered boat. *(From Choi, C.H., Yu, S., Han, I.S., Kho, B.K., Kang, D.G., Lee, H.Y., Seo, M.S., Kong, J.W., Kim, G., Ahn, J.W., Park, S.K., Jang, D.W., Lee, J.H., Kim, M., 2016. Development and demonstration of PEM fuel-cell-battery hybrid system for propulsion of tourist boat. Int. J. Hydrogen Energy. 03603199. 41 (5), 3591–3599. Available from: http://www.journals.elsevier.com/international-journal-of-hydrogen-energy/. doi:10.1016/j.ijhydene.2015.12.186.)*

high requirements were placed on the reliability and stability of the equipment, and the AFC technology at that time still had certain deficiencies in this regard (Fig. 1.4). Ships needed to navigate at sea for long periods and had high requirements for fuel storage and supply. The strict requirements of AFC for fuel purity made fuel storage and supply a difficult problem. The relevant paper "*The Future of Fuel Cell Technology in Maritime Applications*" discussed the prospects and challenges of AFC in ship applications, pointing out that although AFC had certain advantages in the ship field, to achieve large-scale application, many technical problems still needed to be overcome.

Although the application of AFC in the civilian sector has achieved some progress during this period, its development has been significantly impeded by technical limitations and high costs. Nevertheless, these pioneering efforts provide valuable references and insights for subsequent applied research in the civilian domain, encourage researchers to continuously explore methods to enhance AFC technology, and establish a solid foundation for the future advancement of AFC in the civilian field.

5 The development of AFC technology in China

In the 1970s, during the global upsurge in fuel cell research, China also actively engaged in the research and development of AFC technology (余沛亮, 我国燃料电池技术发展现状概述, 2009). During this period, relevant Chinese institutions conducted in-depth studies on AFC technology, aiming to apply it to microwave communication relay power supply to meet the urgent need for stable and efficient power sources in the communication industry at that time. Wuhan University achieved remarkable results in this field. They developed a 200-watt AFC fueled by ammonia decomposition gas, which demonstrated unique advantages in the communication field at that time. Ammonia decomposition gas, as a fuel, has the characteristics of wide availability and relatively low cost, providing a feasible fuel option for AFC applications. In 1976, a powerful earthquake of magnitude 7.8 on the Richter scale struck Tangshan, causing devastating damage to local communication facilities. In the emergency situation where microwave communication relays between Beijing and Tianjin were on the verge of collapse, the 200-watt AFC developed by Wuhan University stepped forward to serve as a critical communication relay power source. With its stable power output, it ensured the smooth operation of microwave communication between Beijing and Tianjin, providing crucial communication support for the smooth progress of earthquake relief efforts. Relevant reports detailed this event, highlighting that in extreme situations such as earthquakes, the stable operation of AFC demonstrated its significant potential in emergency communication. However, with the rapid development of technology, profound changes occurred in the field of communication technology (Barbir, 2012). Satellite communication technology gradually emerged and rapidly developed, featuring wide coverage, large communication capacity, high transmission quality, and no geographical restrictions, gradually replacing microwave communication as the dominant technology in long-distance communication. Against this backdrop, the demand for AFC as a power source for microwave communication relays significantly decreased. At the same time, research on proton exchange membrane fuel cells (PEMFC) gained momentum worldwide. PEMFC has the advantages of low operating temperature, fast start-up, high power density, and simple structure, attracting the attention of many researchers and the investment of research resources (Wang et al., 2011; Sharaf and Orhan, 2014). In contrast, AFC gradually lost its competitive edge in terms of technological development and application prospects, and the research interest in it gradually waned. In China, research on AFC became less active. Although AFC research in China gradually quieted down during this period, the previous research achievements and practical experience accumulated valuable knowledge and technical reserves for the development of fuel cell technology in China (衣宝廉 邵志刚, 2019). These experiences and lessons provided important references for subsequent research and development in the field of fuel cells in China, promoting continuous exploration and innovation in other types of fuel cell technologies (Zhang and Cooke, 2010).

6 Recent research progress of AFC

Entering the 21st century, with the global demand for clean energy constantly increasing, research on AFC technology has witnessed a new upsurge. In terms of material and catalyst innovation, researchers have achieved a series of remarkable results, laying a solid foundation for improving the performance and reducing the cost of AFCs. In the development of nonprecious metal catalysts, many research teams have devoted considerable efforts and made breakthroughs (Li et al., 2010). The team from Chongqing University has made a key breakthrough in the nonprecious metalization of the anode in alkaline membrane fuel cells (Zhou et al., 2024). By constructing a crystalline doped carbon-molybdenum layer at the nickel catalytic interface, they successfully achieved gated electrochemical reactions. This unique design confines the electrons of nickel metal within the low-energy valley of the quasiquantum well catalytic structure, effectively suppressing the electrochemical oxidation of nickel. When hydrogen atoms are adsorbed, the charge charging degree of the quasiquantum well catalytic structure changes, opening the charge transfer channel and enabling selective catalytic oxidation of hydrogen. Based on this, the researchers assembled the first all-nonprecious metal AFC stack and achieved small power output, significantly promoting the development of hydrogen energy technology and solving the technical difficulty of nonprecious metalization of the anode in alkaline membrane fuel cells. The related paper in *Nature Energy* elaborates on this research achievement, pointing out that the quantum confinement charge selective transfer theory and the high antielectrochemical oxidation characteristic theory proposed by the team decouple the corrosion process from the catalytic process, providing new ideas for the design of nonprecious metal catalysts. Professor Zhuang Zhongbin's team from Beijing University of Chemical Technology, addressing the issues of low activity and high cost of anode catalysts in alkaline membrane fuel cells, has achieved significant results by designing highly active and stable hydrogen oxidation catalytic materials with low precious metal content and nonprecious metal bases (Wang et al., 2024). The NiCuCr/C catalyst they prepared, which is surface-modified with Cr_2O_3, has achieved high HOR activity and stability through a dual strategy of electrode component regulation and optimization of the electrode interface water structure. The specific exchange current density of NiCuCr/C catalyzing HOR reached 0.136 mA cm^{-2} ECSA, surpassing all previously reported nonprecious metal-based HOR catalysts (Fig. 1.5). The peak power density of the alkaline membrane fuel cell assembled with this catalyst can reach 577 MW cm^{-2}, significantly higher than other comparison catalysts, and it can stably operate for 156 hours at a constant potential of 0.7 V, with an average degradation rate of only 0.6 mA cm^{-2} h^{-1}, outperforming all Ni-based catalysts reported in the literature. The related research results were published in Nature Communications, where the paper detailed the mechanism of the catalyst, indicating that the formation of an alloy between Cu and Ni weakens the hydrogen binding energy of Ni, and the introduction of Cr enriches the adsorption of hydroxyl groups at the interface and enhances the binding of interface water, jointly optimizing the apparent hydrogen binding energy of the catalyst and thereby enhancing

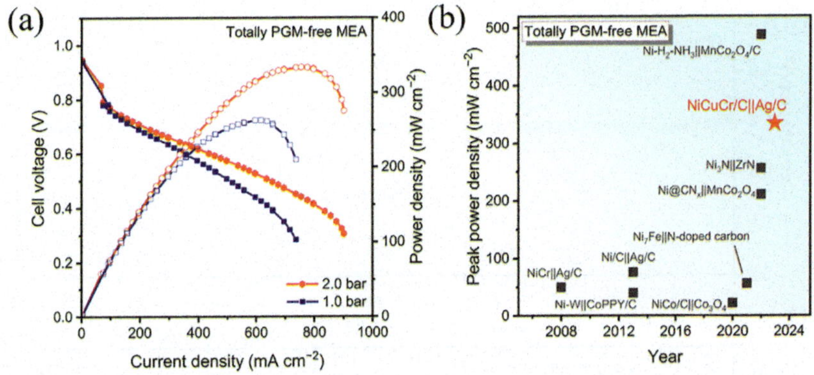

FIG. 1.5 (a) H_2-2HEMFC polarization and power density curves of the MEA using NiCuCr/C anode and Ag/C cathode. (b) Summary of the peak power density of reported totally PGM-free MEA. HEMFC performance and durability. *(From Wang X, Liu X, Fang J, et al., 2024. Tuning the apparent hydrogen binding energy to achieve high-performance Ni-based hydrogen oxidation reaction catalyst. Nat. Commun. 15(1), 1137.)*

the HOR catalytic performance, providing a highly promising solution for reducing the cost of fuel cells.

A team led by Liu Jing and Jiang Luhua from the School of Materials Science and Engineering has developed a new type of nonprecious metal catalyst (Yu et al., 2024). See Fig. 1.6. This catalyst is based on transition metal nitrides and, through a unique preparation process, enhances catalytic activity while reducing costs. Experimental data show that this new catalyst performs exceptionally well in AFCs, with catalytic activity comparable to that of traditional precious metal catalysts but at a fraction of the cost. In a comparative experiment, an AFC using the new nonprecious metal catalyst achieved an output power similar to that of a cell using a precious metal catalyst under the same conditions, but at a significantly lower cost. The research results were published in *Advanced Functional Materials*, where the preparation method, structural features, and application performance of the new catalyst in fuel cells were comprehensively analyzed, providing an important reference for the further development of nonprecious metal catalysts. The development of these new nonprecious metal catalysts not only reduces the cost of AFCs but also improves their performance and stability. Their emergence brings new hope for the large-scale commercial application of AFCs and is expected to break the limitations of precious metal catalysts on the development of fuel cells, promoting the application of fuel cell technology in a wider range of fields. In addition to nonprecious metal catalysts, the exploration of new electrolyte materials has also become a research hotspot in recent years. The new generation of AFC technology aims to replace the traditional KOH aqueous solution with solid-state alkaline polymer electrolytes (APE), an innovative approach designed to address the numerous issues of traditional electrolytes and open up new paths

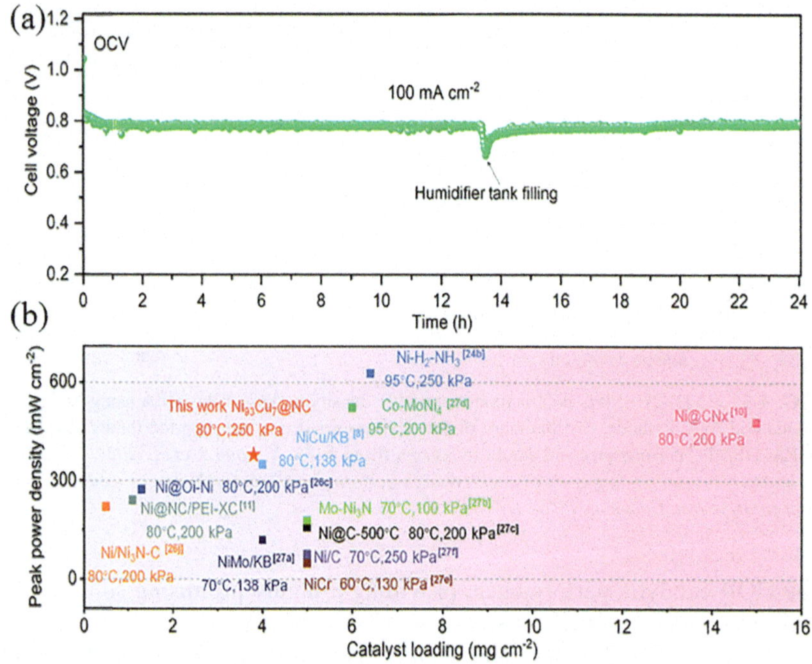

FIG. 1.6 (a) Galvanic discharging at current densities of 100 mA·cm^{-2} for 24 h; (b) Comparison of the performance of $Ni_{93}Cu_7$@NC-based AEMFC with literature with Ni-based catalysts. AEMFC performance. *(From Yu, W., Liu, J., Hu, H., et al., 2024. Optimizing Nickel orbital occupancy via co-modulation of core and shell structures to accelerate alkaline hydrogen electro-oxidation reaction. Adv. Funct. Mat. 34(22), 2315062.)*

for the development of AFCs (Wang et al., 2013; Pan et al., 2010). The working principle of APE is based on its unique molecular structure, which enables the conduction of hydroxide ions (OH^-) in a solid state, facilitating ion transport within the battery. Compared with the traditional KOH aqueous solution, APE has significant advantages. It eliminates the risk of electrolyte leakage, greatly enhancing the safety and reliability of the battery and making it more suitable for use in various complex environments. APE also has a higher ion conductivity, effectively reducing the internal resistance of the battery and improving its energy conversion efficiency. In some experimental tests, AFCs using APE have shown an energy conversion efficiency 10–20% higher than those using KOH aqueous solution, demonstrating promising application prospects. The paper "*Recent Research Progress in Alkaline Polymer Electrolyte Membranes for Alkaline Solid Fuel Cells*" provides a comprehensive review of the research progress of APE, highlighting its great potential in improving battery performance and stability (Hong-Ying, 2014).

However, APE still faces some challenges in practical applications, among which the issue of chemical stability is one of the key factors restricting its development. During the operation of the battery, APE may be attacked by free radicals and other substances produced by electrode reactions, leading to the destruction of its molecular structure and thus affecting its ionic conductivity and the battery's service life. To address this problem, researchers have conducted extensive research. They have modified and optimized the molecular structure of APE, introducing stable functional groups and enhancing the interaction between molecular chains to improve its chemical stability. Introducing fluorine atoms into the APE molecule can enhance its antioxidant properties and stability in the battery operating environment. Researchers have also explored new preparation processes to improve the quality and performance of APE. When preparing APE membranes using the solution casting method, by precisely controlling parameters such as solution concentration, temperature, and drying conditions, uniform and high-performance electrolyte membranes can be fabricated. Relevant research results have been published in journals such as Journal of Membrane Science, providing effective solutions to the chemical stability problem of APE. Research on the compatibility between APE and electrode materials is also deepening (Steele and Heinzel, 2001). Good compatibility can ensure good interface contact between the electrolyte and the electrode, reduce interface resistance, and improve the overall performance of the battery. Researchers have optimized the surface properties of electrode materials through surface modification and other methods to better match them with APE. Coating the electrode surface with a polymer film that has good compatibility with APE can effectively improve the interface performance between the electrode and the electrolyte, enhancing the battery's performance and stability. With the continuous deepening of research, it is believed that new electrolyte materials will bring greater breakthroughs to the development of AFCs and promote their wide application in the energy field.

7 A new breakthrough in AFC technology

In the research of AFCs, enhancing energy conversion efficiency and reducing system cost have consistently been critical objectives pursued by researchers (McLean et al., 2002). In recent years, substantial advancements have been achieved in these two pivotal areas through the optimization of electrode structure, electrolyte formulation, and catalyst design. To improve energy conversion efficiency, numerous studies have focused on optimizing electrode structures. Traditional AFC electrodes encounter challenges during reactions, such as significant mass transfer resistance and inadequate utilization of reactive sites, which constrain the battery's energy conversion efficiency. To address these issues, researchers have proposed various innovative electrode structure designs. Among them, the three-dimensional porous electrode structure has garnered considerable attention. This electrode structure features a high specific surface area and excellent porosity, providing more active sites for electrochemical reactions while facilitating the transport of reactants and products.

By leveraging nanotechnology, three-dimensional porous electrodes with nanoscale pores have been fabricated, significantly increasing the electrode's specific surface area and the number of reactive sites (Rolison et al., 2009). Experimental data indicate that the energy conversion efficiency of AFCs with three-dimensional porous electrodes is 15–25% higher compared to traditional electrode structures. In certain practical application scenarios, such as distributed power generation systems, this high-efficiency AFC can more effectively convert chemical energy into electricity, offering users a more stable and efficient power supply. Relevant papers, such as "Structural Analysis of Alkaline Fuel Cell Electrodes and Electrode Materials," have elaborated on the design principles and preparation methods of three-dimensional porous electrodes (Tomantschger and Kordesch, 1989).

The optimization of electrolyte formulation remains a critical factor in enhancing energy conversion efficiency. Researchers continue to refine the electrolyte composition through in-depth investigations into the proportions of various components, types and concentrations of additives, and other relevant parameters. Incorporating an appropriate amount of lithium salt into the conventional KOH electrolyte can significantly improve its ionic conductivity, thereby enhancing battery performance (Tang et al., 2021). The addition of lithium salt modifies the ion transport mechanism within the electrolyte, reduces ion migration resistance, and consequently boosts the energy conversion efficiency of the battery. Experimental findings indicate that the ionic conductivity of the electrolyte increases by 20–30% upon lithium salt incorporation, leading to a marked improvement in the energy conversion efficiency of corresponding AFCs. Furthermore, researchers have explored novel electrolyte additives, such as ionic liquids. Ionic liquids possess unique physical and chemical properties that not only enhance ionic conductivity but also improve the stability of the electrolyte. In one study, the inclusion of ionic liquid in an alkaline electrolyte resulted in sustained high ionic conductivity even at elevated temperatures, effectively improving the energy conversion efficiency of AFCs under high-temperature conditions. These research outcomes, published in the journal Electrochimica Acta, offer innovative insights and methodologies for optimizing electrolyte formulations.

The advancement in catalyst design plays a pivotal role in enhancing energy conversion efficiency (Qiu et al., 2021). With the ongoing progress in materials science, novel catalyst materials and preparation technologies have been emerging continuously. Some studies have employed nanostructured catalysts to improve catalytic activity and selectivity by precisely controlling the particle size and morphology of the catalysts. A prepared nanoscale platinum-based catalyst, with a particle size of only a few nanometers, exhibits exceptionally high catalytic activity. In the cathode reaction of AFCs, nanostructured catalysts can more effectively promote the oxygen reduction reaction, thereby enhancing battery performance. Theoretical calculations and experimental investigations indicate that nanostructured catalysts increase the exposure of active sites, improve catalyst utilization, and significantly enhance the energy conversion efficiency of batteries. Researchers have further optimized catalyst performance through surface modification and composite structure design. By coating a layer of transition metal oxide on the surface of platinum-based catalysts, the electronic structure of the catalyst can be altered, enhancing its adsorption capacity for reactants and improving catalytic activity. The application of surface-modified catalysts in AFCs has further improved the energy conversion efficiency of these batteries. Recent advancements in catalyst design were reviewed in the paper "Alkaline Fuel Cells: Status and Prospects," where the effects of various catalyst design strategies on energy conversion efficiency improvement were analyzed, providing valuable insights for future catalyst research and development (Hamada et al., 2023).

In the context of reducing system costs, researchers have achieved a series of positive outcomes from multiple perspectives. Given that material cost constitutes a significant portion of the overall cost of AFC systems, the pursuit of low-cost alternative materials has emerged as a key research focus. As previously mentioned, substantial progress has been made in the development of nonprecious metal catalysts, offering feasible solutions to decrease catalyst expenses. Beyond catalysts, reducing the costs of electrode and electrolyte materials remains a critical concern. Regarding electrode materials, some studies have investigated the substitution of traditional metal-based electrodes with carbon-based materials. Carbon-based materials exhibit advantages such as low cost, excellent electrical conductivity, and robust chemical stability, making them an ideal alternative for electrode applications. Through surface modification and doping treatments, the catalytic activity and stability of carbon-based materials can be enhanced, enabling them to satisfy the performance requirements of AFC electrodes (Fig. 1.7). In one study, AFCs fabricated using modified carbon-based materials as electrodes demonstrated performance comparable to conventional metal-electrode cells, while achieving a cost reduction of 30–40%. The paper "The Role of Carbon in Fuel Cells" provides a comprehensive summary of the research advancements in carbon-based electrode materials, highlighting their considerable potential for cost reduction in battery systems (Dicks, 2006). The optimization of the preparation process represents a critical approach to reducing costs. The traditional AFC preparation process is characterized by its complexity, low production efficiency, and high cost. To address these challenges, researchers have developed a range of innovative preparation techniques. Screen printing technology, for instance, enables large-scale and cost-effective electrode production. This method allows for precise control over electrode thickness and patterning, thereby enhancing electrode quality and performance. Compared with conventional preparation methods, screen printing technology achieves several times higher production efficiency while significantly reducing costs. Additionally, some studies have investigated integrated preparation processes, where individual battery components are fabricated within a single process, minimizing assembly-related costs and

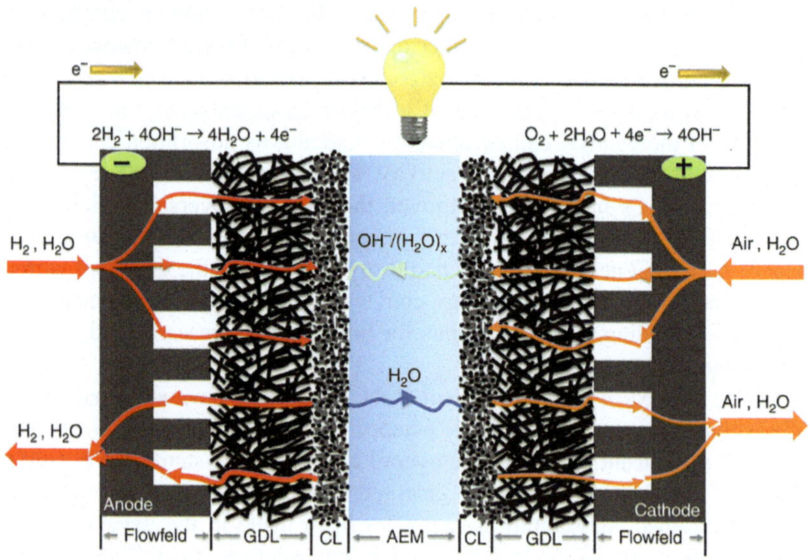

FIG. 1.7 Overall fuel cell configuration (Peng et al., 2020). Illustration of AMFC water dymamics. *(From Peng, X., Kulkarni, D., Huang, Y., et al., 2020. Using operando techniques to understand and design high performance and stable alkaline membrane fuel cells. Nat. commun 11(1), 3561.)*

time consumption. AFCs prepared via this integrated approach not only reduce costs but also enhance overall battery performance. Relevant research findings have been published in reputable journals such as the Journal of Manufacturing Processes, offering valuable insights for optimizing preparation processes. Large-scale production remains one of the pivotal factors in cost reduction.

With the growing market demand for AFCs, large-scale production is becoming increasingly feasible. Through such production, enterprises can significantly reduce per-unit production costs and enhance their market competitiveness. In the process of scaling up, companies can leverage advanced manufacturing equipment and automated production lines to improve both efficiency and product quality. Furthermore, by optimizing supply chain management and reducing raw material procurement costs, the overall cost of battery systems can be further minimized. Some enterprises have already initiated the construction of large-scale AFC production facilities, and it is anticipated that over the next few years, as large-scale production becomes more widespread, the cost of AFCs will decrease substantially. According to relevant market research reports, large-scale production could reduce the cost of AFCs by over 50% within the next 5–10 years, thereby laying a robust foundation for their extensive commercial deployment. With continuous advancements in AFC technology, its application domains are expanding. Beyond its traditional strengths in aerospace applications and initial explorations in civilian sectors, AFCs have demonstrated significant potential in recent years for distributed power generation, backup power supply, transportation, and other areas. Practical application cases have emerged, offering new solutions to address energy challenges and promote sustainable development.

In the domain of distributed power generation, AFCs have emerged as a prominent energy supply technology due to their high efficiency, cleanliness, and flexibility. A distributed power generation system strategically places power generation equipment near end-users, enabling on-site electricity production and consumption. This approach minimizes transmission losses and enhances energy utilization efficiency. AFCs can be integrated with renewable energy sources such as solar and wind power to create a complementary energy system. During periods of abundant sunlight or strong winds, electricity is generated using solar panels or wind turbines; when energy supply is insufficient, AFCs provide electricity to ensure a stable power supply. Such a distributed energy system not only improves the stability and reliability of energy usage but also reduces reliance on traditional power grids and decreases carbon emissions. In remote areas with limited grid coverage, distributed power generation systems have become a critical means for local residents to access electricity. The deployment of AFCs in these regions has provided a reliable power source, significantly improving living conditions. The related paper "*Alkaline Fuel Cells Applications*" provides an in-depth analysis of the role of AFCs in distributed power generation, highlighting their significance in enhancing energy utilization efficiency and advancing the development of renewable energy (Kordesch et al., 2000; Huang et al., 2006). In the domain of backup power supply, AFCs exhibit substantial application potential. Data centers, communication base stations, and other critical facilities impose stringent requirements on power supply stability. A power outage could result in significant losses. As a backup power solution, AFCs offer advantages such as rapid startup, stable operation, and low noise levels. In the event of grid failure, they can be promptly activated to provide reliable power support for these facilities. In certain data centers, AFCs serve as backup power sources, enabling swift startup during grid outages to ensure uninterrupted operation of servers and equipment, thereby preventing data loss and business disruptions. Compared with conventional backup power systems like diesel generators, AFCs demonstrate superior environmental performance with virtually no pollutant emissions, aligning with modern society's demand for green energy. Relevant research reports indicate that as the number of data centers and communication base stations continues to rise, so does the demand for backup power. AFCs, as an efficient and clean backup power source, possess very promising market prospects. Furthermore, in the transportation sector, the application potential of AFCs is becoming increasingly evident.

In the realm of new energy vehicles, while PEMFC currently dominate, AFCs exhibit promising application prospects in certain specific scenarios due to their unique advantages. AFCs are characterized by relatively low costs

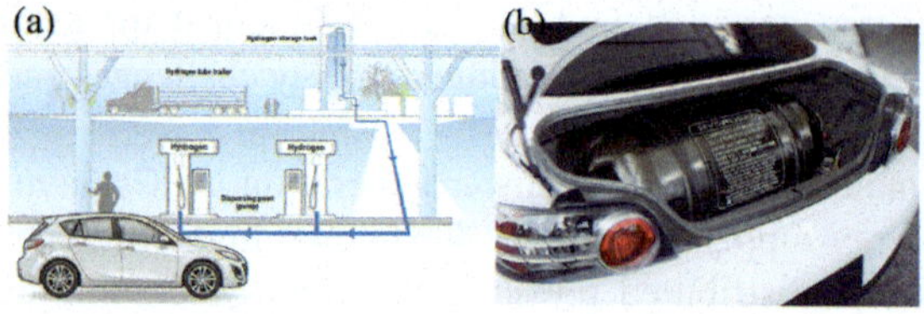

FIG. 1.8 (a) Hydrogen refueling infrastructure; (b) The hydrogen storage tank installed in a Mazda RX-8 Hydrogen car. Hydrogen fuel cell vehicles. *(From Alaswad, A., Baroutaji, A., Achour, H., et al., 2016. Developments in fuel cell technologies in the transport sector. Int. J. Hydrogen Energy 41(37), 16499–16508.)*

and reduced requirements for fuel purity, making them competitive in cost-sensitive markets (Fig. 1.8). In the commercial vehicle sector, such as buses and logistics vehicles, AFCs can serve as a power source, providing efficient and clean energy for vehicles. Some cities have initiated pilot programs utilizing AFC-powered buses. During operation, these buses achieve zero emissions, minimize urban environmental pollution, and maintain relatively low operating costs, thereby delivering significant economic and social benefits. In marine and rail transit applications, AFCs also demonstrate potential. Ships require substantial energy during navigation, and AFCs can supply power, reducing reliance on traditional fuels and decreasing carbon emissions. In rail transit, AFCs can function as auxiliary power sources, enhancing train operational efficiency and energy utilization. Relevant literature, such as "Developments in Fuel Cell Technologies in the Transport Sector: Current Challenges and Developments," has analyzed the application potential of AFCs in transportation, indicating that with ongoing technological advancements and cost reductions, AFCs are anticipated to gain broader adoption in this field (Alaswad et al., 2016).

Although AFCs exhibit significant potential in these emerging applications, they also encounter several challenges. In the realm of distributed power generation, the integration technology between AFCs and renewable energy requires further enhancement to ensure the stability and reliability of the energy system. In the context of backup power supply, the high cost of fuel cells remains a barrier, necessitating additional cost reductions to enhance market competitiveness. Within the transportation sector, the relatively low energy density of AFCs limits driving range, thus requiring advancements in energy density alongside improvements in hydrogen infrastructure construction, such as the strategic deployment of hydrogen refueling stations. Looking ahead, with ongoing technological advancements and supportive policies, it is anticipated that AFCs will surmount these challenges and achieve broader application across various domains, playing a pivotal role in facilitating global energy transformation and sustainable development.

8 Challenges and prospects of AFC

Reflecting on the evolution of AFCs, from their theoretical inception in the early 19th century, through their remarkable achievements in aerospace during the 20th century, to their current exploration and breakthroughs in civilian and emerging sectors in the 21st century, AFC technology has undergone a significant leap forward, progressing from foundational research to practical implementation. Throughout this journey, the relentless dedication and innovative spirit of numerous researchers have driven the continuous advancement of AFC technology. Looking ahead, AFCs are poised to play an increasingly critical role in the global energy transition. In terms of technological innovation, despite substantial progress, there remains considerable potential for further enhancement. Through the interdisciplinary integration of materials science, chemical engineering, and other fields, the development of novel materials and catalysts will continue to deepen, potentially leading to improvements in both performance and stability of AFC systems. In the future, it may be possible to develop highly efficient nonprecious metal catalysts with superior catalytic activity and stability compared to existing options, thereby addressing the issue of high costs associated with fuel cells. Research into new electrolyte materials will also advance, with the aim of developing electrolytes that exhibit higher ion conductivity and chemical stability, further enhancing the energy conversion efficiency and lifespan of batteries. Cost control remains one of the pivotal factors for achieving large-scale commercialization of AFCs.

With the continuous advancement of technology and the promotion of large-scale production, the cost of AFCs is expected to gradually decrease. By optimizing manufacturing processes, enhancing production efficiency, and exploring more cost-effective raw materials, the production cost of AFCs is anticipated to decline significantly. In the context of mass production, enterprises can reduce the per-unit production cost through automated production lines and advanced management strategies. As market demand increases, the procurement cost of raw materials will also decrease due to economies of scale. Once costs are reduced to a certain threshold, AFCs will become economically competitive in a broader range of applications, facilitating large-scale commercial deployment. Policy support plays a pivotal role in the development of AFCs. Governments should enhance their support for AFC-related research and development as well as industrialization efforts by establishing relevant policies and regulations. This includes encouraging collaboration between enterprises and research institutions to advance the study and application of AFCs. Additionally, governments can mitigate the R&D costs and market risks for companies by offering financial subsidies and tax incentives, thereby accelerating the adoption and dissemination of this technology. Furthermore, governments should prioritize planning and investment in hydrogen energy infrastructure, such as the strategic layout and construction of hydrogen refueling stations, to provide essential support for the practical application of AFCs. In terms of application, the potential of AFCs is extensive and promising. In the domain of distributed power generation, AFCs are expected to achieve tighter integration with renewable energy sources, such as solar and wind power. This synergy will establish a complementary energy system capable of delivering stable and reliable electricity to remote regions and islands. Within the context of backup power, AFCs' inherent efficiency, cleanliness, and reliability render them an ideal solution for critical infrastructure, including data centers and communication base stations. Their application ensures uninterrupted operation of these facilities during grid failures. In the transportation sector, ongoing technological advancements and cost reductions are anticipated to drive significant breakthroughs in the utilization of AFCs within new energy vehicles, maritime vessels, and rail transit systems. These developments will play a pivotal role in reducing carbon emissions and enhancing environmental quality.

Although AFCs encounter numerous challenges, with the continuous advancement of technology, cost reduction efforts, and supportive policies, their future development prospects remain highly promising. It is anticipated that, in the near future, AFCs will become a critical component of the global energy sector and significantly contribute to the sustainable development of human society (Fig. 1.9). This book aims to offer readers a comprehensive overview of the technical principles of AFCs, their current application status, and future developmental directions. Through this publication, we hope to provide a valuable resource for researchers, engineers, students, and practitioners, fostering collaborative advancements in the development and application of AFC technology and promoting a cleaner, more efficient, and more sustainable energy future.

FIG. 1.9 Alkaline fuel cells of the future—tubular alkaline fuel cells.

AI disclosure

During the preparation of this work, the author(s) used Fig. 1.9 in order to generate pictures of future alkaline fuel cells using AI. After using this tool/service, the author(s) reviewed and edited the content as needed and take(s) full responsibility for the content of the publication.

References

Abdelkareem, M.A., Elsaid, K., Wilberforce, T., Kamil, M., Enas Taha Sayed, A.O., 2021. Environmental aspects of fuel cells: a review. Sci. Total Environ. 00489697. 752, 141803. doi:10.1016/j.scitotenv.2020.141803.

Alaswad, A., Baroutaji, A., Achour, H., Carton, J., Ahmed Al Makky, A.G.O., 2016. Developments in fuel cell technologies in the transport sector. Int. J. Hydrogen Energy. 03603199. 41 (37), 16499–16508. doi:10.1016/j.ijhydene.2016.03.164.

Appleby, A.J., 1990. From Sir William Grove to today: fuel cells and the future. J. Power Sources. 03787753. 29 (1-2), 3–11. doi:10.1016/0378-7753(90)80002-u.

Barbir, F., 2012. PEM Fuel Cells: Theory and Practice PEM Fuel Cells: Theory and Practice. Elsevier, United States, pp. 1–518. Available from: https://www.sciencedirect.com/book/9780123877109. doi:10.1016/C2011-0-06706-6.

Brouzgou, A., Song, S.Q., Tsiakaras, P., 2012. Low and non-platinum electrocatalysts for PEMFCs: current status, challenges and prospects. Appl. Catal. B: Environ. 09263373. 127, 371–388. doi:10.1016/j.apcatb.2012.08.031.

Burke, K.A. 2003 1st International Energy Conversion Engineering Conference IECEC American Institute of Aeronautics and Astronautics Inc. United States Fuel cells for space science applications

Carrette, L., Friedrich, K.A., Stimming, U., 2000. Fuel cells: principles, types, fuels, and applications. ChemPhysChem. 14394235. 1 (4), 162–193. Available from: http://onlinelibrary.wiley.com/journal/10.1002/(ISSN)1439-7641. doi:10.1002/1439-7641(20001215)1:4<162::aid-cphc162>3.3.co;2-q.

Carrette, L., Friedrich, K.A., Stimming, U., 2001. Fuel cells-fundamentals and applications. Fuel Cells 1 (1), 5–39.

Choi, C.H., Yu, S., Han, I.S., Kho, B.K., Kang, D.G., Lee, H.Y., Seo, M.S., Kong, J.W., Kim, G., Ahn, J.W., Park, S.K., Jang, D.W., Lee, J.H., Kim, M., 2016. Development and demonstration of PEM fuel-cell-battery hybrid system for propulsion of tourist boat. Int. J. Hydrogen Energy. 03603199. 41 (5), 3591–3599. Available from: http://www.journals.elsevier.com/international-journal-of-hydrogen-energy/. doi:10.1016/j.ijhydene.2015.12.186.

Dicks, A.L., 2006. The role of carbon in fuel cells. J. Power Sources. 03787753. 156 (2), 128–141. doi:10.1016/j.jpowsour.2006.02.054.

Gasteiger, H.A., Kocha, S.S., Sompalli, B., Wagner, F.T., 2005. Activity benchmarks and requirements for Pt, Pt-alloy, and non-Pt oxygen reduction catalysts for PEMFCs. Appl. Catal. B: Environ. 09263373. 56 (1-2), 9–35. doi:10.1016/j.apcatb.2004.06.021.

Grove, W.R., 1842. LXXII, On a gaseous voltaic battery. Lond.Edinb.Dubl.Phil.Mag. 21 (140), 417–420.

Halpert, G., Frank, H., Surampudi, S., 1999. Batteries and fuel cells in space. ECS Interface. 1064-8208. 8 (3), 25–30. doi:10.1149/2.f06993if.

Hamada, A.T., Orhan, M.F., Kannan, A.M., 2023. Alkaline fuel cells: status and prospects. Energy Rep. 23524847. 9, 6396–6418. Available from: http://www.journals.elsevier.com/energy-reports/. doi:10.1016/j.egyr.2023.05.276.

Hong-Ying, H.O.U., 2014. Recent research progress in alkaline polymer electrolyte membranes for alkaline solid fuel cells. Acta Phys.-Chim. Sin. 1000-6818. 30 (8), 1393–1407. doi:10.3866/pku.whxb201406171.

Huang, X., Zhang, Z., Jiang, J. 2006 IEEE International Symposium on Industrial Electronics. doi: 10.1109/ISIE.2006.295713 1613 1618 Canada Fuel cell technology for distributed generation: An overview 2.

Kiros, Y., Schwartz, S., 2000. Long-term hydrogen oxidation catalysts in alkaline fuel cells. J. Power Sources. 03787753. 87 (1-2), 101–105. doi:10.1016/s0378-7753(99)00436-x.

Kordesch, K., Gsellmann, J., Cifrain, M., Voss, S., Hacker, V., Aronson, R.R., Fabjan, C., Hejze, T., Daniel-Ivad, J., 1999. Intermittent use of a low-cost alkaline fuel cell-hybrid system for electric vehicles. J. Power Sources. 03787753. 80 (1-2), 190–197. doi:10.1016/s0378-7753(98)00261-4.

Kordesch, K., Hacker, V., Gsellmann, J., Cifrain, M., Faleschini, G., Enzinger, P., Fankhauser, R., Ortner, M., Muhr, M., Aronson, R.R., 2000. Alkaline fuel cells applications. J. Power Sources. 03787753. 86 (1-2), 162–165. doi:10.1016/s0378-7753(99)00429-2.

Larminie, A.D., McDonald, M., (2003).

Launius, R.D., Apollo, A retrospective analysis[R]. (1994).

Li, X., Popov, B.N., Kawahara, T., Yanagi, H., 2010. Recent advances in nonprecious metal catalysts for oxygen reduction reaction in fuel cells. ECS Trans. 1938-5862. 33 (1), 1769–1776. doi:10.1149/1.3484666.

Ma, R., Chai, X., Geng, R., Liangcai, X., Xie, R., Zhou, Y., Wang, Y., Li, Q., Jiao, K., Gao, F., 2023. Recent progress and challenges of multi-stack fuel cell systems: fault detection and reconfiguration, energy management strategies, and applications. Energy Convers. Manag. 01968904. 285, 117015. doi:10.1016/j.enconman.2023.117015.

Markgraf, S., Hörenz, M., Schmiel, T., Jehle, W., Lucas, J., Henn, N., 2012. Alkaline fuel cells running at elevated temperature for regenerative fuel cell system applications in spacecrafts. J. Power Sources. 03787753. 201, 236–242. doi:10.1016/j.jpowsour.2011.10.118.

McLean, G.F., Niet, T., Prince-Richard, S., Djilali, N., 2002. An assessment of alkaline fuel cell technology. Int. J. Hydrogen Energy. 03603199. 27 (5), 507–526. doi:10.1016/S0360-3199(01)00181-1.

O'hayre, R., Cha, S.W., Colella, W., 2016. Fuel Cell fundamentals[M], John Wiley & Sons.

Pan, J., Lu, S., Li, Y., Huang, A., Zhuang, L., Lu, J., 2010. High-performance alkaline polymer electrolyte for fuel cell applications. Adv. Funct. Mater. 1616-301X. 20 (2), 312–319. doi:10.1002/adfm.200901314.

Peng, X., Kulkarni, D., Huang, Y., Omasta, T.J., Ng, B., Zheng, Y., Wang, L., LaManna, J.M., Hussey, D.S., Varcoe, J.R., Zenyuk, I.V., Mustain, W.E., 2020. Using operando techniques to understand and design high performance and stable alkaline membrane fuel cells. Nat. Commun. 2041-1723. 11 (1), doi:10.1038/s41467-020-17370-7.

Philippov, A.A., Chibiryaev, A.M., Prosvirin, I.P., Martyanov, O.N., 2020. Some like it weak: different activity of Raney® nickel in transfer hydrogenation under air and inert atmosphere. Appl. Catal. A: Gen. 0926860X. 605, 117788. doi:10.1016/j.apcata.2020.117788.

Qiu, Y., Xie, X., Li, W., Shao, Y., 2021. Electrocatalysts development for hydrogen oxidation reaction in alkaline media: from mechanism understanding to materials design. Chin. J. Catal. 18722067. 42 (12), 2094–2104. doi:10.1016/s1872-2067(21)64088-3.

Rolison, D.R., Long, J.W., Lytle, J.C., Fischer, A.E., Rhodes, C.P., Mc Evoy, T.M., Bourg, M.E., Lubers, A.M., 2009. Multifunctional 3D nanoarchitectures for energy storage and conversion. Chem. Soc. Rev. 14604744. 38 (1), 226–252. doi:10.1039/b801151f.

Sharaf, O.Z., Orhan, M.F., 2014. An overview of fuel cell technology: fundamentals and applications. Renew. Sustain. Energy Rev. 13640321. 32, 810–853. doi:10.1016/j.rser.2014.01.012.

Springer, T.E., Zawodzinski, T.A., Gottesfeld, S., 1991. Polymer electrolyte fuel cell model. J. Electrochem. Soc. 0013-4651. 138 (8), 2334–2342. doi:10.1149/1.2085971.

Staffell, I., Scamman, D., Velazquez Abad, A., Balcombe, P., Dodds, P.E., Ekins, P., Shah, N., Ward, K.R., 2019. The role of hydrogen and fuel cells in the global energy system. Energy Environ. Sci. 17545706. 12 (2), 463–491. Available from: http://pubs.rsc.org/en/journals/journal/ee. doi:10.1039/c8ee01157e.

Steele, B.C.H., Heinzel, A., 2001. Materials for fuel-cell technologies. Nature. 00280836. 414 (6861), 345–352. doi:10.1038/35104620.

Tang, S., Guo, W., Fu, Y., 2021. Advances in composite polymer electrolytes for lithium batteries and beyond. Adv. Energy Mater. 1614-6832. 11 (2), doi:10.1002/aenm.202000802.

Tomantschger, K., Kordesch, K.V., 1989. Structural analysis of alkaline fuel cell electrodes and electrode materials. J. Power Sources. 03787753. 25 (3), 195–214. doi:10.1016/0378-7753(89)80004-7.

Walsh, F.C., 2001. Electrochemical technology for environmental treatment and clean energy conversion. Pure Appl. Chem. 1365-3075. 73 (12), 1819–1837. doi:10.1351/pac200173121819.

Wang, X., Liu, X., Fang, J., Wang, H., Liu, X., Wang, H., Chen, C., Wang, Y., Zhang, X., Zhu, W., Zhuang, Z., 2024. Tuning the apparent hydrogen binding energy to achieve high-performance Ni-based hydrogen oxidation reaction catalyst. Nat. Commun. 2041-1723. 15 (1), doi:10.1038/s41467-024-45370-4.

Wang, Y., Chen, K.S., Mishler, J., Cho, S.C., Adroher, X.C., 2011. A review of polymer electrolyte membrane fuel cells: technology, applications, and needs on fundamental research. Appl. Energy. 03062619. 88 (4), 981–1007. Available from: http://www.elsevier.com/inca/publications/store/4/0/5/8/9/1/index.htt. doi:10.1016/j.apenergy.2010.09.030.

Wang, Y.J., Qiao, J., Baker, R., Zhang, J., 2013. Alkaline polymer electrolyte membranes for fuel cell applications. Chem. Soc. Rev. 14604744. 42 (13), 5768–5787. doi:10.1039/c3cs60053j.

Warshay, M., Prokopius, P., Le, M., Voecks, G., 1997 Proceedings of the Intersociety Energy Conversion Engineering Conference 228 231 IEEE United States NASA fuel cell upgrade program for the Space Shuttle Orbiter 1

Wei, C., Rao, R.R., Peng, J., Huang, B., Stephens, I.E.L., Risch, M., Xu, Z.J., Shao-Horn, Y., 2019. Recommended practices and benchmark activity for hydrogen and oxygen electrocatalysis in water splitting and fuel cells. Adv. Mater. 15214095. 31 (31). Available from: http://onlinelibrary.wiley.com/journal/10.1002/(ISSN)1521-4095. doi:10.1002/adma.201806296.

Wisniak, J., 2015. Electrochemistry and fuel cells: the contribution of William Robert Grove. Indian J. Hist. Sci. 50 (3), 476–490.

Yu, W., Liu, J., Hu, H., He, Z., Cui, X., Jiang, L., 2024. Optimizing nickel orbital occupancy via co-modulation of core and shell structures to accelerate alkaline hydrogen electro-oxidation reaction. Adv. Funct. Mater. 1616-301X. 34 (22), doi:10.1002/adfm.202315062.

Zhang, F., Cooke, P., 2010. Hydrogen and fuel cell development in China: a review. Eur. Plan. Stud. 0965-4313. 18 (7), 1153–1168. doi:10.1080/09654311003791366.

Zhou, Y., Yuan, W., Li, M., Xie, Z., Song, X., Yang, Y., Wang, J., Li, L., Ding, W., Lin, W.-F., Wei, Z., 2024. Quantum confinement-induced anti-electrooxidation of metallic nickel electrocatalysts for hydrogen oxidation. Nat. Energy. 2058-7546. 9 (10), 1297–1309. doi:10.1038/s41560-024-01604-9.

余沛亮, 我国燃料电池技术发展现状概述, (2009), 103–120.

衣宝廉 邵志刚, 2019. 氢能与燃料电池发展现状及展望. 中国科学院院刊 34 (04), 469–477.

Chapter 2

Mass transport phenomena in alkaline fuel cells

Qiao Lan and Jian Liu
School of Petrochemical Engineering, Lanzhou University of Technology, Lanzhou, Gansu, China

1 Introduction

The performance of alkaline fuel cells (AFCs) is collectively determined by three core processes: electrochemical reaction kinetics, charge transfer processes, and mass transport behavior. Among these, mass transport constitutes a multiphysics coupling process involving the transport of reactants, products, and electrolytes, which directly impacts cell efficiency, stability, and lifespan. This chapter will analyze the mass transport dynamics at the porous electrode-electrolyte interface to elucidate the migration mechanisms of reactants and products at the gas-liquid-solid three-phase interface and establish mathematical correlations between transport processes and polarization.

AFCs utilize alkaline substances as electrolyte solutions, typically employing aqueous solutions of 35% to 50% potassium hydroxide (KOH) or NaOH by mass. These fuel cells can operate at lower temperatures (<120°C), while at higher temperatures, such as 200°C, an electrolyte solution with a high mass fraction of KOH (85%) may be used. Compared with acidic electrolytes, redox reactions are more likely to occur in alkaline environments. For instance, when hydrogen is used as fuel and pure oxygen or CO_2-free air serves as the oxidant, gas diffusion electrodes prepared with electrocatalysts such as Pt–Pd/C, Pt/C, and Ni, which exhibit excellent catalytic activity for hydrogen oxidation, are commonly employed as hydrogen electrodes. Meanwhile, porous gas diffusion electrodes prepared with catalysts such as Pt/C and Ag, which demonstrate strong catalytic activity for the electrochemical reduction of oxygen, are utilized as oxygen electrodes. The working principle of AFCs is illustrated in Fig. 2.1.

In AFCs, whether they are H_2/O_2 fuel cells or other types of AFCs, the gas transport mainly involves the transport of fuel. Taking H_2/O_2 fuel cells as an example:

As shown in Fig. 2.2, hydrogen is typically stored in external storage devices in the form of high-pressure gas, liquid, or solid hydrogen storage materials.

Alkaline Fuel Cells. https://doi.org/10.1016/B978-0-443-36638-3.00006-1

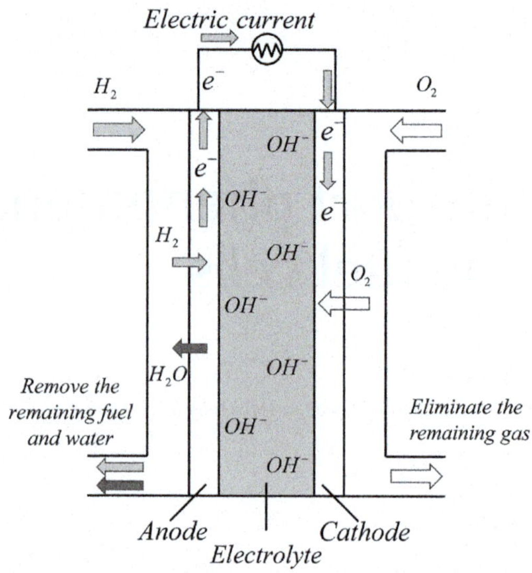

FIG. 2.1 Schematic diagram of the working principle of an alkaline fuel cell.

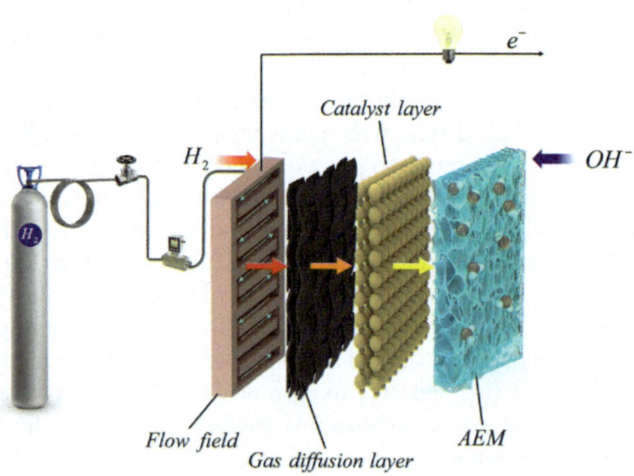

FIG. 2.2 Hydrogen Transmission Process.

When the fuel cell starts to operate, the hydrogen in the hydrogen storage device is precisely and stably delivered to the anode inlet of the fuel cell through pipelines, valves, flow control systems, etc. Upon reaching the anode inlet, the hydrogen enters the anode flow field plate, which is designed with special flow channel structures. These flow channels serve to uniformly distribute hydrogen to the entire surface of the anode catalyst layer, ensuring that the catalyst can fully contact hydrogen and promote subsequent electrochemical reactions.

After being distributed by the flow field, hydrogen reaches the gas diffusion layer (GDL), which is generally made of porous carbon materials and possesses good air permeability and electrical conductivity. In the diffusion layer, hydrogen gradually penetrates from the flow channels to the catalyst layer by diffusion, relying on the concentration gradient. During this process, the diffusion layer not only provides a transport channel for hydrogen but also supports the catalyst layer, collects electrons, and removes the water generated by the reaction, ensuring that hydrogen can smoothly reach the catalyst layer to participate in the reaction while maintaining the material transport balance inside the cell.

Once hydrogen reaches the anode catalyst layer, an oxidation reaction occurs under the action of the catalyst. Common anode catalysts for AFCs include platinum-based catalysts and nonnoble metal catalysts, such as nickel-based catalysts. On the catalyst surface, hydrogen molecules are decomposed into hydrogen ions and electrons with the participation of hydroxide ions, and the chemical reaction equation is as follows: The hydrogen ions combine with hydroxide ions to form water, while the electrons flow to the cathode through the external circuit to form an electric current, providing electrical energy for external loads and realizing the conversion of chemical energy into electrical energy. Part of the generated water is discharged in the form of water vapor together with the unreacted gas, and the other part may be absorbed by the electrolyte, which requires a reasonable water drainage design to maintain the concentration stability of the electrolyte.

At the anode of anion exchange membrane fuel cell (AEMFC), the main reaction is hydrogen oxidation reaction (HOR), as shown in Equation (1). In alkaline media, HOR has two mechanisms: 1) Tafel-Volmer step 2) Heyrovsky-Volmer step:

$$2H_2 + 4OH^- \rightarrow 4H_2O + 4e^- E_a^\theta = -0.83V vs. SHE$$

Tafel step:

$$H_2 + 2^* \rightarrow 2H_{ad}$$

Heyrovsky step:

$$H_2 + 2OH^- + ^* \rightarrow Had + H_2O + e^-$$

Volmer step:

$$H_{ad} + OH^- \rightarrow H_2O + e^- + ^*$$

Generally, hydrogen molecules are adsorbed on active sites (*) to produce adsorbed hydrogen (H_{ad} or $H_{2,ad}$), which then reacts with hydroxide anions (OH^-) to form water molecules (H_2O). For example, the HOR mainly occurs on the Pt surface via the Tafel-Volmer pathway, where the rate-determining step is the Volmer step (Strmcnik et al., 2016) When the reaction medium changes from acid to base, the activation energy of the Volmer step increases from 17.1 kJ · mol^{-1} to 34.0 kJ · mol^{-1} (Rheinländer et al., 2014). Therefore,

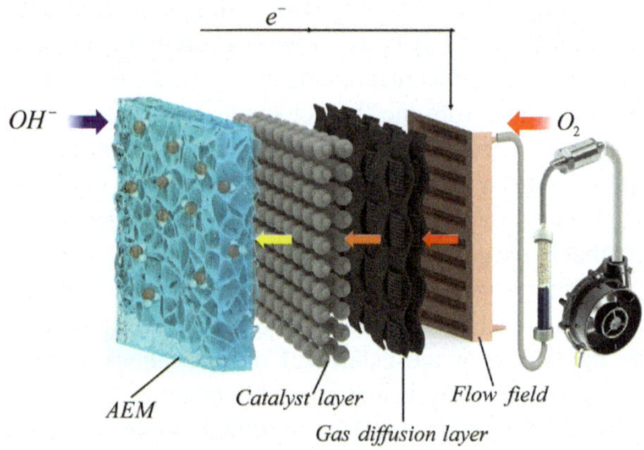

FIG. 2.3 Oxygen transport process.

compared with the acidic environment of PEM fuel cell (FC), the HOR kinetics in the alkaline environment of AEMFC is much slower, which leads to a significant decrease in cell voltage (Campos-Roldán et al., 2020).

The transport of oxidants is a key link in achieving efficient power generation. Taking oxygen as the oxidant for the analysis of the transport process in fuel cells, as shown in Fig. 2.3.

In practical applications, if fuel cells are used in special fields such as aerospace and submarines, due to the extremely high requirements for energy density and performance, high-purity canned pure oxygen is usually used as the oxidant. For commercial fuel cells, such as power sources for electric vehicles and small fuel cell power stations, considering safety and cost, oxygen is directly obtained from the air. Air is forcibly introduced into the fuel cell system through the air inlet with the help of equipment such as fans, and then passes through a series of air pretreatment devices. Since air contains not only oxygen but also nitrogen, carbon dioxide, water vapor, and various tiny particulate impurities, these components may have negative impacts on the performance and service life of the fuel cell. Therefore, the air first enters a filter to remove mechanical impurities such as dust and particulates. It then enters a carbon dioxide removal device because carbon dioxide reacts with the alkaline electrolyte KOH to form potassium carbonate, thereby reducing the electrolyte concentration and cell performance. At the same time, the air also needs to be dehumidified to control the water vapor content—excessive moisture will dilute the electrolyte and affect ion conduction, and methods such as condensation dehumidification or desiccant adsorption are generally used to precisely regulate the air humidity. The pretreated oxygen-rich air also enters the GDL through the flow field and diffuses from the diffusion layer to the cathode catalyst layer. When oxygen reaches the cathode catalyst layer, a reduction reaction occurs under the action of the catalyst. Common cathode catalysts for AFCs include platinum-based, palladium-based, silver-based, and some nonnoble metal catalysts. Taking the common four-electron reaction pathway as an example, the electrode reaction is as follows: In this reaction process, oxygen molecules receive electrons from the external circuit and combine with water to form hydroxide ions. These hydroxide ions move through the electrolyte to the anode to participate in the anode reaction, while electrons form an electric current from the anode to the cathode through the external circuit, providing electrical energy for the load. After the cathode reaction is completed, unreacted oxygen and other gases in the air are discharged from the fuel cell through the exhaust port, completing the entire cycle of oxidant transport and reaction.

As shown in Equation (2), the cathode reaction of AEMFC is the oxygen reduction reaction (ORR), which is the most challenging in the proton exchange membrane fuel cell (PEMFC) because noble metals are required to achieve acceptable kinetic activity and durability (Ma et al., 2020). Fortunately, a broader library of inexpensive materials exists as ORR catalysts in alkaline media, providing a wider range of promising options for AEMFC catalysts. This is because, when the pH changes from 0 to 14, the equilibrium potential for the reduction of O_2 to H_2O/OH^- decreases from 1.23 V to 0.40 V (i.e., 59 mV pH^{-1}) relative to the standard hydrogen electrode, and the potential required for the first electron transfer step also decreases from 1.53 V to 0.70 V, a decrease of 0.83 V (Wan et al., 2015). These are the thermodynamic reasons why ORR occurs at lower overpotentials in alkaline media compared to acidic media.

ORR can occur through the two-electron and four-electron pathways in an alkaline medium:

Four electrons pathway:

$$O_2 + 2H_2O + 4e^- \rightarrow 4OH^- E^\theta = 0.4V\ vs.SHE$$

Two-plus-two electrons pathway:

$$O_2 + H_2O + 2e^- \rightarrow HO_2^- + OH^- E^\theta = -0.076V\ vs.SHE$$

$$HO_2^- + H_2O + 2e^- \rightarrow 3OH^- E^\theta = 0.87V\ vs.SHE$$

In AFCs, the transport pathways indicate that mass transfer behaviors occur in the flow field plate, GDL, and electrolyte. Through these transport pathways, the core mechanisms and influencing factors of mass transport can be designed and optimized.

2 Gas transmission

2.1 Flow field design and gas distribution

The flow field is a key component of fuel cells, responsible for collecting electric current, providing mechanical support for the membrane electrode assembly (MEA), distributing reaction gases, and removing liquid water through flow channels. It ensures the uniform distribution of reaction gases, efficient utilization, and timely discharge of products. Flow fields typically refer to the groove structures on the surface of bipolar plates, and their design directly affects gas transport pathways, pressure drop distribution, water management, and overall performance. Therefore, their design is crucial for the initial performance and durability of fuel cells. The main goals of flow field design are to enhance water and thermal management and ensure uniform gas distribution. Currently, significant efforts have been made to optimize and design different types of flow fields, which can be broadly classified into traditional flow fields, biomimetic flow fields, and 3D flow fields. Serpentine, parallel, pin-type, interdigitated, and their variants are considered traditional flow fields, among which parallel and serpentine geometries are the most used (Bunyan et al., 2023). These designs rely on an open flow channel network to distribute gas reactants and provide pathways for electron conduction. For example, Chen et al. designed and optimized an oriented-type flow channel with porous-blocked baffles (H. Chen et al., 2021b), while Jiao et al. compared the differences between wavy flow field designs and serpentine, interdigitated, and straight parallel flow field designs (Najmi et al., 2021). Their results showed that the wavy flow field configuration helps improve cell performance due to its enhanced mass transport, reduced parasitic power, and better humidity tolerance compared to other traditional flow field designs. Inspired by nature, various biomimetic flow field designs have been developed to enhance mass and heat transport and cell performance. These designs include leaf-shaped, snowflake-shaped, lung-shaped, honeycomb-shaped, rib-shaped, fishbone-shaped structures, etc. (Li et al., 2024). Additionally, novel 3D flow fields have attracted extensive attention due to their ability to enhance cell performance by promoting forced convection.

Different flow channel structures affect the uniformity and efficiency of fuel cells by regulating the gas flow path. Li et al. [Flow channel geometry optimization and novel criterion for improved water and thermal management in anion-exchange membrane fuel cell] tested the performance of parallel flow fields with different channel shapes by constructing a three-dimensional, two-phase, nonisothermal AEMFC model (as shown in Fig. 2.4). The results showed that at low voltages, the channel shape significantly influenced cell performance in the order of inverted trapezoid > square > triangle > rectangle > trapezoid > semicircle. Compared with the cell with semicircular channels, the cell with inverted trapezoid channels increased the peak power density by 12.9% and reduced the current density nonuniformity by 83.4%.

2.2 Gas diffusion layer

The GDL, a core component of fuel cell devices, is positioned between the bipolar plate and the catalyst layer. Its primary role is to optimize the transport of reaction gases, discharge of products, and current conduction through its porous structure and material properties, thereby enhancing cell performance

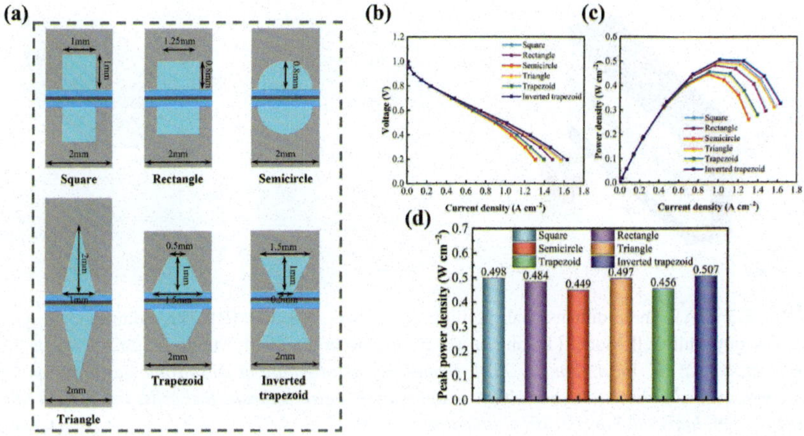

FIG. 2.4 (a) Channel cross-section of flow field performance of AEMFC with different flow channel geometries: (b) polarization curves; (c) power density curves; (d) peak power density (Li et al., 2024). *(From Li, F., Cai, S., Tu, Z., 2024. Flow channel geometry optimization and novel criterion for improved water and thermal management in anion-exchange membrane fuel cell. Int. J. Hydrogen Energy, 87, 199–213. https://doi.org/10.1016/j.ijhydene.2024.09.012)*

and stability. The thickness and porosity of the GDL significantly affect the mass transport resistance of reactants/products. A thicker GDL increases mass transfer resistance, leading to lower reactant delivery rates and product removal rates. Therefore, from a mass transfer perspective, a thinner GDL is preferred as it enables faster reactant transport and product removal (Mathematical modeling of alkaline direct ethanol fuel cells). As shown in Fig. 2.5. The GDL is typically composed of carbon fibers and hydrophobic agents, which help prevent water flooding within the GDL.Van Men Truong et al. (2021) studied the impact of GDL thickness on AEMFC performance and found that AEMFC performance is more affected by the anode-side GDL thickness than the cathode-side, indicating that water removal from the anode is more challenging than water supply to the cathode. Jiao et al. (2014) constructed a three-dimensional multiphase AEMFC model to investigate the influence of contact angles in the cathode and anode catalyst layer (CL) and GDL on AEMFC performance. Mu et al. (2024) developed an anisotropic GDL model and a multiphysics AEMFC model considering multiphase flow in anisotropic GDLs. This study examined the effects of different porosity gradients and wettability changes in GDLs on cell performance. It was found that in GDLs with different porosity gradients, reducing the porosity near the interface between the anode GDL and the CL can promote uniform distribution of reaction gases, thereby improving cell performance.

2.3 Transmission at the reaction interface

In fuel cells, anion-exchange ionomers (AEIs) are typically used as binders and ion conductors in electrodes to form the Three-Phase Boundary (TPB), i.e., the

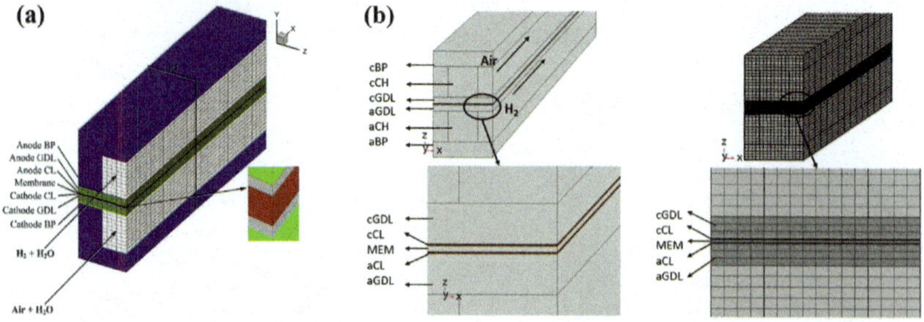

FIG. 2.5 (A) Three-dimensional multiphase AEMFC model. (B) Three-dimensional GDL Anisotropy Full-Cell Model (Truong et al., 2021; Jiao et al., 2014; Mu et al., 2024). *(From Mu, M., Liu, W., Xi, W., Yu, A., Shi, L., 2024. Numerical investigation of anisotropic gas diffusion layers with graded porosity and wettability in anion exchange membrane fuel cells. Int. J. Heat Mass Trans., 226. https://doi.org/10.1016/j.ijheatmasstransfer.2024.125493[11]K; Jiao, P. He, Q. Du, Y. Yin, 2014. Three-dimensional multiphase modeling of alkaline anion exchange membrane fuel cell, Int. J. Hydrog. Energy 39, 5981–5995. https://doi.org/10.1016/j.ijhydene.2014.01.180.[12]M; Mu, W. Liu, W. Xi, A. Yu, L. Shi, 2024. Numerical investigation of anisotropic gas diffusion layers with graded porosity and wettability in anion exchange membrane fuel cells, Int. J. Heat Mass Transf. 226, 125493. https://doi.org/10.1016/j.ijheatmasstransfer.2024.125493.)*

interface among the catalyst, membrane, and gas diffusion region (Peron et al., 2011). During the formation of TPB, the favorable porous structure in the catalyst layer requires the participation of AEIs to enhance the transport of ions and reactants to active sites. AEIs need to exhibit long-term durability, even at high temperatures and high current densities, to ensure high-performance AEMFCs.

In electrochemical systems, transport at the reaction interface is a core process determining reaction rates and device performance, involving the collaborative processes of reactant migration to the interface, product departure from the interface, and charge transfer. Oxygen molecules in the air are adsorbed onto the surface of the porous cathode. This adsorption occurs at the cathode-electrolyte-gas interface. The adsorbed oxygen molecules then react to form oxygen ions, which are subsequently transported through the electrolyte to the anode, where they react with the fuel. The capability to catalytically reduce oxygen, high electrical conductivity (>100 S/cm), ionic conductivity, suitable porosity, and chemical and thermal stability are all essential requirements for cathode materials. According to different conduction mechanisms, cathode materials can be divided into three categories: electron conductor cathodes, mixed oxygen ion-electron conductor cathodes, and triple-conductor cathodes. As shown in Fig. 2.6A, in electron conductor cathodes, only electrons are conducted within the cathode, while the cathode reaction occurs at the three-phase interface involving the electrolyte, cathode material, and oxygen. As shown in Fig. 2.6B, mixed oxygen ion-electron conductor cathodes can simultaneously conduct oxygen ions and electrons, enabling the cathode reaction to proceed through the bulk phase of the cathode. Triple-conductor cathodes are rich in

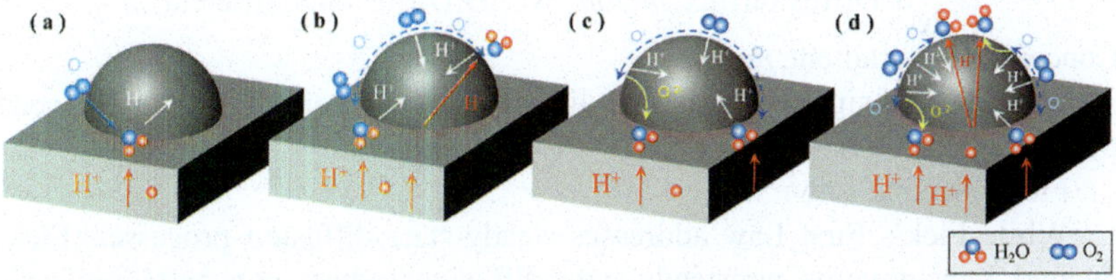

FIG. 2.6 Schematic Illustration of Particle Transport at the Three-Phase Interface: (a) Oxygen Ion Transport; (b) Proton Transport; (c) Proton-Ion Transport; (d) Electron-Proton-Ion Transport (Ya et al., 2025). *(From Ya, Y., Xu, Y.S., Elbanna, A.M., Liu, Y., Sun, B., Cheng, X., 2025. Review of direct ammonia solid oxide fuel cells: Low temperature cell structure and ammonia decomposition strategies., Renew. Sustain. Energy Rev. 213, 115350. https://doi.org/10.1016/j.rser.2025.115350.)*

transport carriers of protons, oxygen vacancies, and electron holes. As shown in Fig. 2.6C, these materials have the ability to conduct electrons, oxygen ions, and protons simultaneously. These properties significantly enhance the spatial transport of reactive species to active reaction sites within the cell, thereby improving the overall proton conductivity of the cathode. The reaction interface of triple-conductor cathodes penetrates the entire cathode layer, remarkably enhancing reaction efficiency and electrochemical activity, especially under low-temperature conditions (Adler et al., 1996).

Three-phase interface (gas-liquid-solid): Effective reactions occur at the junction of gas (gas phase), electrolyte (liquid phase), and catalyst (solid phase). Gases must dissolve into the electrolyte and come into contact with the catalyst surface through diffusion or convection to complete the ORR (cathode) or HOR (anode).

Concentration gradient-driven: Gas transport primarily relies on concentration gradients and follows Fick's diffusion law. At high reaction rates, it is necessary to prevent concentration polarization caused by local gas depletion, which leads to performance degradation. A detailed explanation of Fick's diffusion law will be provided in subsequent sections.

2.4 Fick's laws

Fick's laws are core theories describing the phenomenon of mass diffusion, widely applied in chemistry, materials science, biology, and engineering. They are divided into Fick's first law and Fick's second law, corresponding to the mathematical descriptions of steady-state diffusion and unsteady-state diffusion, respectively.

Fick's First Law:

Under steady-state diffusion conditions (where concentration does not vary with time), the diffusion flux is proportional to the concentration gradient and opposite in direction. Its mathematical expression is:

$J = -D\frac{\partial c}{\partial x}$ J:Diffusionflux, $mol{\cdot}m^{-2}{\cdot}s^{-1}$;D:Diffusioncoefficient, $m^2{\cdot}s^{-1}$; $\frac{\partial c}{\partial x}$: Concentration gradient, $mol{\cdot}m^{-4}$.

The negative sign indicates that diffusion occurs from high to low concentration regions.

Fick's Second Law:

While Fick's First Law addresses steady-state diffusion processes, Fick's Second Law governs nonsteady-state diffusion (where concentration varies with time). In such conditions, the diffusion process depends not only on the spatial concentration gradient but also on time. Its mathematical formulation is:

$\frac{\partial c}{\partial t} = D\frac{\partial^2 c}{\partial x^2}$ $\frac{\partial c}{\partial t}$: Temporal rate of concentration change, $mol{\cdot}m^{-3}{\cdot}s^{-1}$; $\frac{\partial^2 c}{\partial x^2}$:

Second spatial derivative of concentration, $mol{\cdot}m^{-5}$.

Where the temporal evolution of concentration is governed by the diffusion coefficient and the curvature of the concentration distribution, a region with a positive curvature in concentration exhibits decreasing concentration over time, whereas a negative curvature leads to increasing concentration.

Fick's laws form the theoretical foundation for understanding mass diffusion behavior.

First Law: Quantifies steady-state diffusion through linear flux-gradient relationships.

Second Law: Extends to dynamic processes by incorporating temporal dependencies.

In engineering applications, such as fuel cells, optimizing gas transport efficiency involves strategically regulating concentration gradients, diffusion coefficients, and medium structures, thereby enhancing overall system performance.

3 Charge transfer

The ability of an electrolyte solution to conduct electric current is based on the directional electromigration of charged solvated ions in the electrolyte solution under an electric field between the two electrodes. Ions with charge ze_0 undergo electromigration under the action of electric field strength (E) and are also subjected to frictional resistance (F) from the solution environment medium; the faster the ion movement velocity (v), the greater the frictional resistance it experiences. For a simple spherical ion with radius (r_1), the frictional force on the moving ion can be expressed by Stokes' equation: $F = 6\pi\eta r_1 v$, where η is the viscosity of the medium in which the ion is located. Therefore, after a process of curvilinear acceleration, the ion velocity will reach a limiting value v_{max}, At this point, the electric field force acting on the ion is equal to the frictional force.

$$ze_0|E| = 6\pi\eta r_1 v_{max}$$

The terminal velocity of solvated ions is

$$v_{max} = \frac{ze_0 |E|}{6\pi\eta r_1}$$

For given values of η and, each ion has a characteristic transport velocity related to its own charge and solvated ion radius, while its electromigration direction depends on the sign of the charge carried by the ion itself.

3.1 Ion transport carriers and directions

Ion transport carriers and directions are core factors determining charge balance, reaction rate, and energy efficiency. The types of ion carriers and their transport directions in different systems vary depending on electrolyte properties, reaction mechanisms, and application scenarios (see Table 2.1).

Ion types: OH^- is the conductive ion in AFCs, generated by cathode ORR and migrating through the electrolyte to the anode to participate in HOR.

(1) Transport direction:

As shown in Figure 2.1: OH^- moves from the cathode through the electrolyte to the anode, forming a closed current loop.

(2) Transmission path:

The transport path characteristics of alkaline exchange membrane fuel cells are determined by their solid-state anion exchange membranes and alkaline working environment. The core lies in the directional conduction of hydroxide ions (OH^-) and the collaborative transport at the gas-liquid-solid multiphase interface.

TABLE 2.1 Characteristics of transport pathways in alkaline anion exchange membrane fuel cells.

Pathway Components	Function	Challenge
AEM	Provide solid-state OH^- conduction channels	High alkaline resistance is required
Electrode membrane interface	Close contact reduces interface resistance	Catalysts need to adapt to high-pH environments
Three-phase reaction interface	The three-phase contact points of gas, solid, and ion determine the reaction efficiency	The electrode needs a porous structure to ensure gas diffusion and ion pathways

3.2 Electrode catalyst ion transport

3.2.1 Anode catalysts

The sluggish kinetics of the HOR remains one of the primary obstacles in the development of AEMFCs. Therefore, the development of novel, highly active materials for HOR is crucial. Platinum group metals (PGMs) and their alloys, including Pt, Ru, and Pd, have been extensively investigated as HOR catalysts. Additionally, non-PGM nickel-based catalysts have gained increasing attention in recent years.

Adabi et al. synthesized PtRu catalysts using a controlled surface tension method by adding acetone to water to reduce the solvent's surface tension (Adabi et al., 2023). The reduction in solvent surface tension minimizes capillary effects, generating smaller solvent droplets and consequently yielding ultra-small supported metal clusters.

Zeng et al. experimentally discovered that mesoporous structures can form internal hydrophobic surfaces and external hydrophilic surfaces within microporous frameworks. Relatively hydrophobic channels facilitate H_2 diffusion. Furthermore, due to spatial confinement effects, mesoporous structures prevent the agglomeration of Ru nanoparticles and reduce their exposure to air during synthesis, resulting in a higher proportion of metallic Ru nanoparticles (Zeng et al., 2020).

In 2008, Professor Zhuang's team first prepared a PGM-free AEMFC with a Cr-modified Ni (CND) anode catalyst. DFT calculations showed that the effects of H and O adsorption on the local density of states (LDOS) of the Ni(111) surface were different. H adsorption almost exclusively changed the bottom of the sp band, while O adsorption primarily altered the d band of the Ni(111) surface. The surface electronic properties of Ni can be adjusted by changes in CrO and LDOS. The density of states at the Fermi level of the d band of Ni(111) decreases, but the bottom of the sp band remains basically unaffected. This leads to a significant weakening of the Ni–O bond but has little effect on the Ni–H bond, which benefits from the peak temperature of Oad desorption on CND. Compared with the bare Ni catalyst, the peak temperature of CND exhibits a negative shift of 130°C. (S. Lu et al., no date).

3.2.2 Cathode catalysts

The development of oxygen-dependent fuel cells is often hindered by the high cost of cathode materials and the sluggish kinetics of the ORR. Therefore, exploring low-cost, highly active catalysts with accelerated mass transport is crucial (Chen et al., 2025). In practical applications of catalysts, improving the accessibility of catalytic sites and enhancing the transport of substances (i.e., ions, electrolytes, and O_2) is also essential (Ramli et al., 2025).

The overpotential of ORR in alkaline media is lower than that in acidic media, which is the most significant advantage of AEMFCs relative to PEMFCs.

Therefore, earth-abundant PGM-free materials can be used as ORR catalysts. Currently, carbon-supported nanoparticle materials, atomically dispersed materials, and carbon-based metal-free materials are mainly studied as platinum-free ORR catalysts in AEMFCs.

Wang et al. further studied the bimetallic Co and Mn electrocatalysts, in which the authors observed an unusual phenomenon (Y. Wang et al., 2019b). In the rotating disk electrode(RDE) test, the half-wave potential of the Mn–Co spinel (MCS) catalyst was 50 mV lower than that of the commercial Pt/C. However, in the AEMFC test at 100% RH, the MCS cathode produced a peak power density of 1.10 W cm^{-2}, which was comparable to that of the fuel cell using a Pt cathode. After detailed analysis, the mechanism revealed that, compared with the RDE test, H_2O was a minor component in the gas phase during the AEMFC test. The MCS catalyst has dual active sites (Mn and Co) to activate O_2, especially H_2O, which makes the MCS catalyst less sensitive to the reduced H_2O and proton transfer process than the Pt/C catalyst (F. Lu et al., no date). Based on the above mechanism and in-situ attenuated total reflection Fourier transform infrared spectroscopy data, a mechanism synergistic with the MCS catalyst was proposed. Specifically, the Mn site promotes the binding and cleavage of O_2 molecules to generate MnO_2. The Co site enriches the surface H_2O and activates it to generate Co-OH_2, thereby promoting the proton-coupled electron transfer process of ORR.

Adabi et al. adopted two strategies to enhance mass transport in the catalyst (Adabi et al., 2023). The first was using a silica template to expand the average pore size of the catalyst to the mesoporous range. The second was reducing the catalyst's hydrophilicity by increasing its graphitization level. The synthesized Fe/N/C catalyst (PMF11904) had a specific surface area of 550 m^2 g^{-1} and a pore size of 5–30 nm. In RRDE tests, the synthesized Fe/N/C exhibited a half-wave potential of 0.846 V vs. reversible hydrogen electrode (RHE), an electron transfer number of 3.967, and an H_2O_2 yield below 1% in 0.1 mol L^{-1} KOH, comparable to commercial 40 wt.% Pt/C.

Ordered mesoporous carbon catalysts are among the most promising ORR catalysts in alkaline media. Due to their ordered mesoporous structure, which provides interconnected channels and a high surface area for the diffusion of electroactive species and the protection of active sites, they exhibit high activity and long-term stability for ORR (To et al., 2017). The mesoporous N-doped hierarchical carbon (NHC) catalyst reported by To et al. was synthesized using a triblock copolymer as a template. After pyrolysis in an inert atmosphere and NH_3 activation, the catalyst was found to have a nitrogen content of 3.6 wt.%. In RDE tests, NHC showed a half-wave potential of 0.882 V vs. RHE in O_2-saturated 0.1 mol L^{-1} KOH. This activity can be attributed to its hierarchical structure and N-heteroatoms. In H_2/O_2 AEMFC tests, the cell with an NHC cathode achieved a peak power density of 0.23 W cm^{-2} at 70°C, comparable to that with a Pt cathode (To et al., 2017).

3.3 The influence of membranes

Anion exchange membranes serve as gas separators and solid electrolytes in AEMFCs, transporting anions. Recent breakthroughs in membrane design (L. Wang et al., 2019b) and electrode fabrication (Ul Hassan et al., 2020) have accelerated improvements in the achievable peak power density and lifespan of AFCs. For example, Huang et al. (Huang et al., 2019) used quaternized poly(norbornene) copolymers as anion exchange membranes (AEMs) and PtRu/C and Pt/C electrodes to achieve a peak power density of $3.4\ W \cdot cm^{-2}$. Ul Hassan and colleagues (Ul Hassan et al., 2020) increased the peak power density by 100% (from $1.6\ W \cdot cm^{-2}$ to $3.2\ W \cdot cm^{-2}$) and demonstrated 2000 hours of stable operation through electrode engineering. Chen et al. (N. Chen et al., 2021b) synthesized poly(diphenyl-terphenyl piperidinium) AEMs and AEIs containing aliphatic chains; in H2/O2 and H2/air, AEMFCs using PtRu/C anodes and Pt/C cathodes reached 2.58 W · cm−2 and 1.38 W · cm−2 at 80°C, respectively, with a specific power (PPD/catalyst loading) exceeding 8 W · mg-Pt−1, which is the highest among Pt-based AEMFCs to date.

One nonnegligible factor leading to the poor durability of AEMFC is the degradation of the AEM under high pH conditions (Park et al., 2024). AEMs typically consist of a polymer backbone and functional cationic groups, which serve as both ion conductors and electrode separators. Under high pH conditions, hydroxide ion erosion causes polyelectrolyte degradation, particularly affecting the cationic moieties and the backbone. Fortunately, AEMs with ether-free backbones, which are currently under intensive study, do not exhibit backbone degradation under alkaline conditions (Arges and Ramani, 2013). Therefore, in-depth research has been conducted to explore alkali-stable organic cations (Marino and Kreuer, 2015). Among them, N-containing organic cations, including alkyl quaternary ammonium (Lee et al., 2015) piperidine (Zhu et al., 2021) quinoline (Zeng et al., 2023) imidazolium (Hugar et al., 2015) and pyrazolium, are considered alkali-stable anion conductors.

The membrane thickness is also an important factor, as it directly affects the anion resistance (ohmic loss) and the diffusion rate of water between the anode and cathode. Reducing the membrane thickness usually improves cell performance, and thinner membranes further enhance this improvement. The reason is that reducing the membrane thickness can not only reduce ohmic loss but also balance the water in the anode and cathode, thereby reducing mass transfer loss at high current densities. It should be noted that the membrane thickness must also meet durability requirements. The influence of the electrode contact angle on the liquid water distribution also indicates that changing the wettability of the electrode on one side may have some impact on the liquid water distribution on the other side. In order to achieve the goals of membrane hydration, reducing cathode mass transfer loss (sufficient hydrogen in the anode CL), and reducing cathode mass transfer loss (sufficient oxygen and water in the cathode CL), it is necessary to carefully consider the membrane thickness and electrode wettability to balance the water in the anode and cathode (Jiao et al., 2014).

4 Water management

In an operating AEMFC, H_2 (or alternative fuels) and air (or O_2) are typically humidified before being supplied to the anode and cathode, respectively. Water is generated by the electrochemical reaction in the anode CL, consuming H_2. In an alkaline electrolyte, water molecules can usually combine with OH^-, so water can be dragged from the cathode to the anode through the movement of OH^- across the membrane. Additionally, if the anode contains more water, water may back-diffuse from the anode to the cathode, and vice versa (Huo et al., 2012). With changes in pressure, temperature, and water amount, phase transitions of water may also occur. Water management failures during fuel cell operation, including flooding and drying, hinder mass transport and accelerate membrane degradation. These issues seriously impair performance and significantly shorten the service life of the fuel cell (Huang et al., 2025).

Water management is an important issue for AFCs because it plays a vital role in the energy conversion process. Water management in AFCs is a key factor in ensuring their efficient operation and long lifespan, mainly involving water generation, distribution, discharge, and the balance of electrolyte concentration. As shown in Fig. 2.7, in AFCs, water is generated at the anode and consumed at the cathode. Water is also transported from the cathode to the anode through electroosmosis and back-diffuses under the driving force of the concentration gradient. Additionally, water is transported from the cathode to the anode by electroosmosis and back-diffuses to the cathode, driven by the concentration gradient. Due to these challenges, water management in AFCs is crucial to prevent electrodes from becoming too wet or too dry and to ensure that MEAs remain moist during long-term stable output (Lei et al., 2025).

In high current densities, mass transport limitations of water (consumed by reactions in the cathode and electroosmotic drag) become the primary performance loss, and the situation worsens if oxygen is not humidified. As mentioned above, water management in AFC is very complex and crucial because membrane hydration is required to maintain sufficient hydroxide conductivity, along with water consumption in the cathode and reverse electroosmotic drag (from the cathode to the anode). Unlike the research on water management in PEM fuel cells, studies on water management in AFCs are rarely found in the literature. Weinzierl and Krewer developed a mathematical model to analyze water management in AEM direct methanol fuel cells and showed that the demand for cathode water supply increases with the increase in current density. In addition, steady-state and transient 3D models of the negative electrode of hydrogen AAEM fuel cells were developed. Based on this, Jiao et al. developed a full-cell 3D multiphase model for hydrogen AAEM fuel cells. The results show that cathode humidification is more important than anode humidification, and the liquid water supply in the cathode flow channel has a positive impact on performance.

When the cell experiences flooding, the performance of AEMFC significantly declines due to mass transfer losses caused by excessive water accumulation in

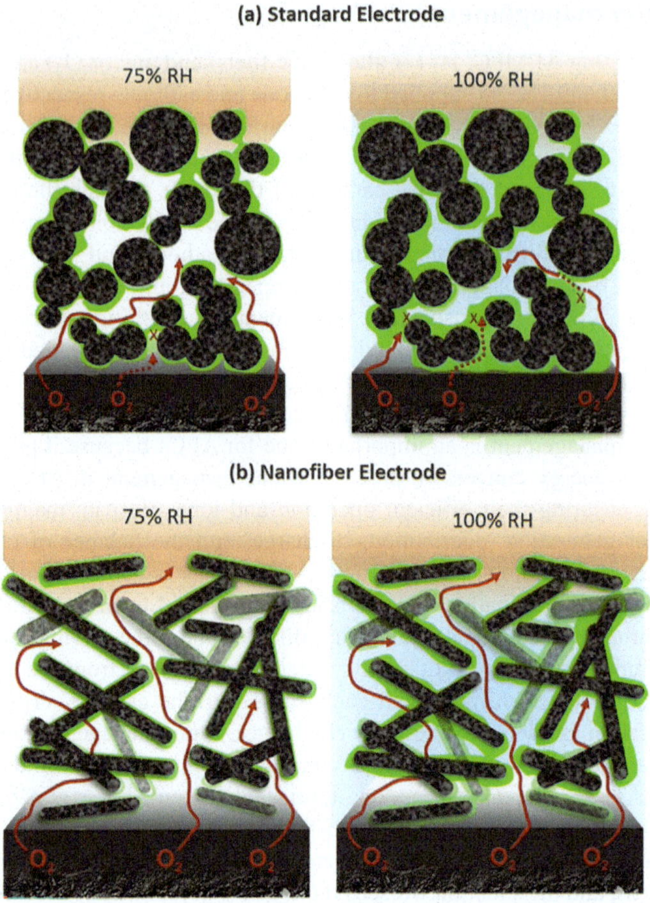

FIG. 2.7 Schematic representation of ionomer distribution and gas transport at high and low RH in (a) standard and (b) nanofiber electrode layers (Kabir et al., 2020). *(From Kabir, S., Medina, S., Wang, G., Bender, G., Pylypenko, S., Neyerlin, K.C., 2020. Improving the bulk gas transport of Fe–N–C platinum group metal-free nanofiber electrodes via electrospinning for fuel cell applications, Nano Energy 73, 104791. https://doi.org/10.1016/j.nanoen.2020.104791.)*

the anode, which hinders the normal diffusion of gases to the active sites. On the other hand, in the case of insufficient moisture, that is, when the cathode electrode is dry, the cell performance in the mass transfer region of the polarization curve also decreases due to the lack of reactant water at the cathode. The dynamic balance of water content can be achieved by adjusting the dew point and flow rate of the inlet gas, optimizing the electrode structure, and using suitable membranes to enhance the reverse diffusion of water. In order to diagnose the cell water management problems in a timely and effective manner, Huang et al. (Huang et al., 2025) adopted in-situ electrochemical impedance spectroscopy (EIS) measurement and distribution of relaxation times (DRT) for analysis to detect water management faults in the anode and cathode electrodes. By performing EIS measurements under different operating conditions, three different frequency ranges related to ion transport, charge transfer, and mass transport processes were revealed, and their contributions to voltage loss were clarified. On this basis, DRT analysis was further applied to explore the polarization impedance behavior and its changes under different water management fault conditions.

(1) Flooding: Excessive accumulation of liquid water in electrode pores or GDLs blocks gas diffusion pathways (see Table 2.2).

Mechanism: Oversaturation of liquid water beyond capillary limits obstructs reactant transport to catalytically active sites.

Consequence: Mass transfer restrictions cause concentration polarization, resulting in performance degradation or system failure.

(2) Dry-out:

Dry-out refers to the phenomenon where excessive water loss leads to increased electrolyte concentration, elevated viscosity, and localized dehydration, thereby impeding ion transport and causing performance degradation (see Table 2.3).

TABLE 2.2 Impact of flooding on alkaline fuel cell performance.

Impact dimension	Specific Manifestations
Gas transmission is obstructed	Liquid water clogs electrode pores, impairing H_2/O_2 transport to catalytic surfaces and triggering concentration polarization.
Active Site Coverage	Water film formation on the catalyst surface reduces the effective area of the triple-phase boundary (gas-liquid-solid).
Increased Ohmic Losses	Liquid water floods electrode pores, prolonging OH^- transport pathways and increasing internal resistance.
Accelerated material corrosion	Flooded zones create localized oxygen-deficient environments that exacerbate electrochemical corrosion of carbon supports or metallic bipolar plates.

TABLE 2.3 Impact of dry-out on alkaline fuel cell performance.

Impact Dimensions	Specific Manifestations
Ionic conductivity decline	Electrolyte dehydration increases viscosity, hinders OH^- migration, and results in a significant rise in Ohmic polarization
Electrode activity reduction	The catalyst surface fails to maintain the triple-phase boundary (gas-liquid-solid) under water deficiency, leading to suppressed reaction rates
Material degradation	KOH crystallization penetrates electrode pores or membrane structures; AEM dehydration induces embrittlement and cracking, causing permanent damage
Thermal runaway risk	Dry-out zones exhibit elevated resistance, where localized overheating accelerates material degradation, establishing a vicious cycle

5 Conclusions

This chapter systematically describes the mass transport behavior of AFCs. The mass transport process in AFCs is a core factor determining cell performance and durability, encompassing the diffusion of reactive gases to catalytic sites, migration of OH^- ions in the electrolyte, removal of product water, and maintenance of electrolyte distribution and concentration. Through relevant studies, the research progress on mass transport in AFCs has been reviewed, and the mass transfer processes in AFCs have been revealed at a fundamental theoretical level. The core challenges during AFC operation include hindered gas diffusion, ion transport in the catalyst layer and membrane, and water management difficulties. These can be addressed by optimizing the design of hierarchical porous electrode structures to promote transport and enhancing electrolyte management. Nevertheless, AFCs remain in the initial development stage, requiring more R&D efforts to achieve higher membrane conductivity and durability, as well as to develop superior ionomers. Additionally, the creation of efficient and stable nonprecious metal group (non-PGM) catalysts for HOR and ORR that can meet or exceed PGM-based performance benchmarks is essential. Although encouraging performances of transition metal and heteroatom-doped materials have been reported, substantial work remains, particularly regarding long-term stability (over 1000 hours). Overall, catalysts, ionomers, electrolytes, and other electrode components require simplified design strategies to avoid mismatches in electrochemical properties. Overcoming these mass transport bottlenecks is the core pathway to unlocking the potential of AFCs in specific application scenarios.

References

Adabi, H., Shakouri, A., Zitolo, A., Asset, T., Khan, A., Bohannon, J., Chattot, R., Williams, C., Jaouen, F., Regalbuto, J.R., Mustain, W.E., 2023. Multi-atom Pt and PtRu catalysts for high performance AEMFCs with ultra-low PGM content. Appl. Catal. B Environ. 09263373. 325. Available from: www.elsevier.com/inca/publications/store/5/2/3/0/6/6/index.htt. doi:10.1016/j.apcatb.2023.122375.

Adler, S.B., Lane, J.A., Steele, B.C.H., 1996. Electrode kinetics of porous mixed-conducting oxygen electrodes. J. Electrochem. Soc. 0013-4651. 143 (11), 3554–3564. doi:10.1149/1.1837252.

Arges, C.G., Ramani, V., 2013. Two-dimensional NMR spectroscopy reveals cation-triggered backbone degradation in polysulfone-based anion exchange membranes. Proc. Natl. Acad. Sci. U.S.A. 10916490. 110 (7), 2490–2495. Available from: http://www.pnas.org/content/110/7/2490.full.pdf+html. doi:10.1073/pnas.1217215110.

Bunyan, S.T., Dhahad, H.A., Khudhur, D.S., Yusaf, T., 2023. The effect of flow field design parameters on the performance of PEMFC: a review. Sustainability 2071-1050. 15 (13), 10389. doi:10.3390/su151310389.

Campos-Roldán, C.A., Calvillo, L., Granozzi, G., Alonso-Vante, N., 2020. Alkaline hydrogen electrode and oxygen reduction reaction on PtxNi nanoalloys. J. Electroanal. Chem. 15726657. 857, 113449. doi:10.1016/j.jelechem.2019.113449.

Chen, K., Liang, Y., Pan, D., Huang, J., Gao, J., Lu, Z., Liu, X., Chen, J., Zhang, H., Hu, X., Wen, Z., 2025. Mass transportation facilitated porous Fe/Co dual-site catalytic cathodes for ultrahigh-power-density Al–Air fuel cells. Adv. Energy Mater. 16146840. 15 (9). Available from: http://onlinelibrary.wiley.com/journal/10.1002/(ISSN)1614-6840. doi:10.1002/aenm.202404140.

Chen, N., Hu, C., Wang, H.H., Kim, S.P., Kim, H.M., Lee, W.H., Bae, J.Y., Park, J.H., Lee, Y.M., 2021b. Poly(alkyl-terphenyl piperidinium) ionomers and membranes with an outstanding alkaline-membrane fuel-cell performance of 2.58 W cm−2. Angew. Chem. Int. Ed. 15213773. 60 (14), 7710–7718. Available from: http://onlinelibrary.wiley.com/journal/10.1002/(ISSN)1521-3773. doi:10.1002/anie.202013395.

Hassan, N.U., Mandal, M., Huang, G., Firouzjaie, H.A., Kohl, P.A., Mustain, W.E., 2020. Achieving high-performance and 2000 h stability in anion exchange membrane fuel cells by manipulating ionomer properties and electrode optimization. Adv. Energy Mater. 16146840. 10 (40). Available from: http://onlinelibrary.wiley.com/journal/10.1002/(ISSN)1614-6840. doi:10.1002/aenm.202001986.

Huang, G., Mandal, M., Peng, X., Yang-Neyerlin, A.C., Pivovar, B.S., Mustain, W.E., Kohl, P.A., 2019. Composite poly(norbornene) anion conducting membranes for achieving durability, water management and high power (3.4 W/cm2) in hydrogen/oxygen alkaline fuel cells. J. Electrochem. Soc. 19457111. 166 (10), F637. Available from: http://jes.ecsdl.org/content/166/10/F637.full.pdf+html. doi:10.1149/2.1301910jes.

Huang, H., Zhang, Z., Xiao, C., Liu, J., Li, Z., Jiang, Y., Wei, L., Zhao, T., Ciucci, F., Zeng, L., 2025. Water management fault diagnosis by operando distribution of relaxation times analysis for anion exchange membrane fuel cells. Adv. Sci. 21983844. Available from: http://onlinelibrary.wiley.com/journal/10.1002/(ISSN)2198-3844. doi:10.1002/advs.202505304.

Hugar, K.M., Kostalik, H.A., Coates, G.W., 2015. Imidazolium cations with exceptional alkaline stability: a systematic study of structure-stability relationships. J. Am. Chem. Soc. 15205126. 137 (27), 8730–8737. Available from: http://pubs.acs.org/journal/jacsat. doi:10.1021/jacs.5b02879.

Huo, S., Deng, H., Chang, Y., Jiao, K., 2012. Water management in alkaline anion exchange membrane fuel cell anode. Int. J. Hydrogen Energy. 03603199. 37 (23), 18389–18402. doi:10.1016/j.ijhydene.2012.09.074.

Jiao, K., He, P., Du, Q., Yin, Y., 2014. Three-dimensional multiphase modeling of alkaline anion exchange membrane fuel cell. Int. J. Hydrogen Energy. 39. doi:10.1016/j.ijhydene.2014.01.180. 03603199 11 5981 5995.

Kabir, S., Medina, S., Wang, G., Bender, G., Svitlana Pylypenko, K.C.N., 2020. Improving the bulk gas transport of Fe-N-C platinum group metal-free nanofiber electrodes via electrospinning for fuel cell applications. Nano Energy 22112855. 73, 104791. doi:10.1016/j.nanoen.2020.104791.

Lee, W.H., Kim, Y.S., Bae, C., 2015. Robust hydroxide ion conducting poly(biphenyl alkylene)s for alkaline fuel cell membranes. ACS Macro Lett. 21611653. 4 (8), 814–818. Available from: http://pubs.acs.org/journal/amlccd. doi:10.1021/acsmacrolett.5b00375.

Lei, H., Yang, X., Chen, Z., Rawach, D., Du, L., Liang, Z., Li, D.S., Zhang, G., Tavares, A.C., Sun, S., 2025. Multiscale understanding of anion exchange membrane fuel cells: mechanisms, electrocatalysts, polymers, and cell management. Adv. Mater. 15214095. 37 (8). Available from: http://onlinelibrary.wiley.com/journal/10.1002/(ISSN)1521-4095. doi:10.1002/adma.202410106.

Li, F., Cai, S., Tu, Z., 2024. Flow channel geometry optimization and novel criterion for improved water and thermal management in anion-exchange membrane fuel cell. Int. J. Hydrogen Energy. 03603199. 87, 199–213. Available from: https://www.sciencedirect.com/science/journal/03603199. doi:10.1016/j.ijhydene.2024.09.012.

Lu, F., Zhang, Y., Liu, S., Lu, D.S., Liu, M., Zhang, Y., Liu, P., Wang, J.X., Adzic, R.R., Gang, O., 2017. Surface proton transfer promotes four-electron oxygen reduction on gold nanocrystal surfaces in alkaline solution. J. Am. Chem. Soc. 139 (21), 7310-7317. doi:10.1021/jacs.7b01735.

Ma, Z., Cano, Z.P., Yu, A., Chen, Z., Jiang, G., Fu, X., Yang, L., Wu, T., Bai, Z., Lu, J., 2020. Enhancing oxygen reduction activity of Pt-based electrocatalysts: from theoretical mechanisms to practical methods. Angew. Chem. Int. Ed. 15213773. 59 (42), 18334–18348. Available from: http://onlinelibrary.wiley.com/journal/10.1002/(ISSN)1521-3773. doi:10.1002/anie.202003654.

Marino, M.G., Kreuer, K.D., 2015. Alkaline stability of quaternary ammonium cations for alkaline fuel cell membranes and ionic liquids. ChemSusChem. 1864-5631. 8 (3), 513–523. doi:10.1002/cssc.201403022.

Mu, M., Liu, W., Wenjie, X., Yu, A., Shi, L., 2024. Numerical investigation of anisotropic gas diffusion layers with graded porosity and wettability in anion exchange membrane fuel cells. Int. J. Heat Mass Transf. 00179310. 226, 125493. doi:10.1016/j.ijheatmasstransfer.2024.125493.

Najmi, A.U.H., Anyanwu, I.S., Xie, X., Liu, Z., Jiao, K., 2021. Experimental investigation and optimization of proton exchange membrane fuel cell using different flow fields. Energy. 03605442. 217. Available from: https://www.journals.elsevier.com/energy. doi:10.1016/j.energy.2020.119313.

Park, E.J., Jannasch, P., Miyatake, K., Bae, C., Noonan, K., Fujimoto, C., Holdcroft, S., Varcoe, J.R., Henkensmeier, D., Guiver, M.D., Kim, Y.S., 2024. Aryl ether-free polymer electrolytes for electrochemical and energy devices. Chem. Soc. Rev. 14604744. 53 (11), 5704–5780. Available from: http://pubs.rsc.org/en/journals/journal/cs. doi:10.1039/d3cs00186e.

Peron, J., Shi, Z., Holdcroft, S., 2011. Hydrocarbon proton conducting polymers for fuel cell catalyst layers. Energy Environ. Sci. 1754-5692. 4 (5), 1575. doi:10.1039/c0ee00638f.

Ramli, Z.A.C., Pasupuleti, J., Osman, S.H., Zaiman, N.F.H.N., Ishak, N.A.I.M., Ainirazali, N., Wan Isahak, W.N.R., Kiong, S.T., Koh, S.P., Shaari, N., 2025. Transitioning from platinum: a comprehensive review of alternative cathodic catalysts in direct alcohol fuel cells. Process Saf. Environ. Prot. 09575820. 199. Available from: https://www.sciencedirect.com/science/journal/09575820. doi:10.1016/j.psep.2025.107240.

Rheinländer, P.J., Herranz, J., Durst, J., Gasteiger, H.A., 2014. Kinetics of the hydrogen oxidation/evolution reaction on polycrystalline platinum in alkaline electrolyte reaction order with respect to hydrogen pressure. J. Electrochem. Soc. 19457111. 161 (14), F1448. Available from: http://jes.ecsdl.org/content/by/year. doi:10.1149/2.0501414jes.

Strmcnik, D., Lopes, P.P., Genorio, B., Stamenkovic, V.R., Markovic, N.M., 2016. Design principles for hydrogen evolution reaction catalyst materials. Nano Energy. 22112855. 29, 29–36. Available from: http://www.journals.elsevier.com/nano-energy/. doi:10.1016/j.nanoen.2016.04.017.

To, J.W.F., Ng, J.W.D., Siahrostami, S., Koh, A.L., Lee, Y., Chen, Z., Fong, K.D., Chen, S., He, J., Bae, W.G., Wilcox, J., Jeong, H.Y., Kim, K., Studt, F., Nørskov, J.K., Jaramillo, T.F., Bao, Z., 2017. High-performance oxygen reduction and evolution carbon catalysis: from mechanistic studies to device integration. Nano Res. 19980000. 10 (4), 1163–1177. Available from: http://www.springer.com/materials/nanotechnology/journal/12274. doi:10.1007/s12274-016-1347-8.

Truong, V.M., Duong, N.B., Yang, H., 2021. Effect of gas diffusion layer thickness on the performance of anion exchange membrane fuel cells. Processes. 22279717. 9 (4). Available from: https://www.mdpi.com/2227-9717/9/4/718/pdf. doi:10.3390/pr9040718.

Wan, K., Yu, Z.P., Li, X.H., Liu, M.Y., Yang, G., Piao, J.H., Liang, Z.X., 2015. pH effect on electrochemistry of nitrogen-doped carbon catalyst for oxygen reduction reaction. ACS Catal. 21555435. 5 (7), 4325–4332. Available from: http://pubs.acs.org/page/accacs/about.html. doi:10.1021/acscatal.5b01089.

Wang, Y., Yang, Y., Jia, S., Wang, X., Lyu, K., Peng, Y., Zheng, H., Wei, X., Ren, H., Xiao, L., Wang, J., Muller, D.A., Abruña, H.D., Hwang, B.J., Lu, J., Zhuang, L., 2019b. Synergistic Mn-Co catalyst outperforms Pt on high-rate oxygen reduction for alkaline polymer electrolyte fuel cells. Nat. Commun. 20411723. 10 (1). Available from: http://www.nature.com/ncomms/index.html. doi:10.1038/s41467-019-09503-4.

Ya, Y., Xu, Y.S., Elbanna, A.M., Liu, Y., Sun, B., Cheng, X., 2025. Review of direct ammonia solid oxide fuel cells: low temperature cell structure and ammonia decomposition strategies. Renew. Sustain. Energy Rev. 18790690. 213. Available from: https://www.sciencedirect.com/science/journal/13640321. doi:10.1016/j.rser.2025.115350.

Zeng, L., Peng, H., Liu, W., Yin, J., Xiao, L., Juntao, L., Zhuang, L., 2020. Extraordinary activity of mesoporous carbon supported Ru toward the hydrogen oxidation reaction in alkaline media. J. Power Sources. 03787753. 461, 228147. doi:10.1016/j.jpowsour.2020.228147.

Zeng, M., He, X., Wen, J., Zhang, G., Zhang, H., Feng, H., Qian, Y., Li, M., 2023. N-methylquinuclidinium-based anion exchange membrane with ultrahigh alkaline stability. Adv. Mater. 15214095. 35 (51). Available from: http://onlinelibrary.wiley.com/journal/10.1002/(ISSN)1521-4095. doi:10.1002/adma.202306675.

Zhu, H., Sun, Z., Cao, H., Wang, B., Zhao, J., Pan, J., Xu, G., Jin, Z., Yan, F., 2021. Highly conductive and dimensionally stable anion exchange membranes based on poly(dimethoxybenzene-co-methyl 4-formylbenzoate) ionomers. Macromolecules. 15205835. 54 (12), 5557–5566. Available from: http://pubs.acs.org/journal/mamobx. doi:10.1021/acs.macromol.1c00704.

Chapter 3

Catalysts for hydrogen oxidation in alkaline media: Mechanisms, classifications, designs and challenges

Wenzhi Li, Yun Liu, Xiaohong Zou, and Liang An
Department of Mechanical Engineering, The Hong Kong Polytechnic University, Hung Hom, Hong Kong SAR, China

1 Introduction

Hydrogen has emerged as a promising alternative to fossil fuels due to its high energy density, sustainability, and zero carbon emissions, offering a viable solution to global energy and environmental challenges (Odenweller et al., 2022; Turner, 2004; Seh et al., 2017). The efficient utilization of hydrogen relies heavily on advanced conversion technologies, among which alkaline fuel cells have gained significant attention for their high energy conversion efficiency and compatibility with nonprecious metal catalysts compared to proton exchange membrane fuel cells (Wang et al., 2013; Xiao et al., 2021). However, the sluggish kinetics of the hydrogen oxidation reaction (HOR) in alkaline media about 2–3 orders of magnitude slower than in acidic environments, necessitating high loadings of noble metal catalysts and increasing system costs (Sheng et al., 2010). This limitation underscores the urgent need to develop highly active and cost-effective alkaline HOR catalysts, along with elucidating the underlying reaction mechanisms to guide their design. The kinetics of HOR in alkaline media are modulated by the adsorption/desorption of key intermediates (e.g., H_{ad}, OH_{ad}) and the associated energy barriers (Li et al., 2017; Sheng et al., 2013; Su et al., 2021). Two primary theories have been proposed to explain alkaline HOR activity: the hydrogen-binding energy (HBE) theory, which correlates activity with the strength of hydrogen adsorption, and the

Alkaline Fuel Cells. https://doi.org/10.1016/B978-0-443-36638-3.00002-4

bifunctional theory, which emphasizes the coadsorption of hydroxyl species to facilitate hydrogen desorption (Wu et al., 2025; Li et al., 2023). However, the mechanistic role of interfacial water and the synergistic effects of multiple intermediates remain debated, highlighting the complexity of alkaline HOR and the need for further mechanistic exploration (Li et al., 2017; Lamoureux et al., 2019).

Current alkaline HOR catalysts can be classified into noble metal-based materials (e.g., Pt (Alia et al., 2013), Pd (Bonnefont et al., 2013), Ir (Qin et al., 2024), Ru (John et al., 2015)) and nonnoble metal-based alternatives (e.g., Ni-based catalysts (Duan et al., 2020)). Noble metal-based catalysts, particularly Pt, exhibit exceptional activity but face challenges related to high cost, CO poisoning, and oxidative degradation (Mu et al., 2023; Ren et al., 2024). To address these issues, strategies such as size control (Ohyama et al., 2013a, 2013b), alloying (Li and Sun, 2019; Lu et al., 2020), and heterostructure engineering (Li et al., 2021b; Zhang et al., 2023) have been employed to optimize their electronic structures and reduce noble metal usage. Meanwhile, nonnoble metal-based catalysts, especially Ni-based catalysts, have shown potential due to their Pt-like electronic properties, but their strong hydrogen adsorption and poor stability under alkaline conditions limit their performance (Wang et al., 2019; Sheng et al., 2014). Recent efforts have focused on enhancing their intrinsic activity through doping, heterostructure engineering, and the introduction of oxophilic sites to modulate intermediate adsorption (Jiang et al., 2019; Huang et al., 2023a).

This chapter provides a comprehensive overview of alkaline HOR catalysts, covering fundamental mechanisms, material classification, and rational design strategies. We first discuss the prevailing HOR theories, followed by a systematic analysis of catalyst optimization approaches, including geometry engineering, composition engineering, substrate engineering, oxyphilic site engineering, and heterostructure engineering. Finally, we outline the remaining challenges, including ongoing mechanistic controversies, reliance on precious metals, as well as the insufficient activity and stability of nonnoble metal-based catalysts.

2 Hydrogen oxidation mechanisms

The HOR typically proceeds via two primary pathways: the Tafel-Volmer and Heyrovsky-Volmer mechanisms, both of which involve a two-step process (Yao et al., 2022). The first step is the formation of the adsorbed hydrogen intermediate (H_{ad}), referred to as the Tafel or Heyrovsky step, followed by the second step, the Volmer step, which involves water formation. The distinction between these pathways hinges on whether electron transfer occurs during the formation of the H_{ad} intermediate. In the Tafel pathway, molecular hydrogen adsorbs onto active sites of the catalyst and dissociates into chemically adsorbed H_{ad} without electron transfer. Conversely, in the Heyrovsky pathway, hydrogen dissociation occurs with the assistance of adjacent OH^-, forming H_{ads} and water molecules, coupled with an electron transfer from hydrogen to the catalyst. In the subsequent Volmer step, common to both pathways, the H_{ad} intermediate reacts with OH^- to produce water molecules and release electrons. Currently, two theoretical frameworks dominate the understanding of HOR mechanisms: the HBE theory, which emphasizes the strength of the H_{ad}-catalyst interaction, and the bifunctional theory, which highlights the synergistic role of active sites and hydroxyl species in facilitating the reaction (Mu et al., 2023). These theories provide critical insights into the factors governing HOR efficiency and catalyst design.

$$Tafel: H_2 + 2^* \rightarrow 2H_{ad}$$

$$Heyrovsky: H_2 + OH^- + ^* \rightarrow H_{ad} + H_2O + e^-$$

$$Volmer: OH^- + H_{ad} \rightarrow H_2O + e^- + ^*$$

2.1 Hydrogen binding energy theory

HBE theory serves as a key descriptor in hydrogen electrocatalysis, as the strength of hydrogen-metal bonding directly governs the HOR activity (Zheng et al., 2016a). High HBE promotes the Tafel and Heyrovsky reactions (hydrogen atom adsorption), while low HBE facilitates the Volmer reaction (hydrogen atom desorption). For instance, a low HBE enhances the Volmer reaction but hinders the Tafel/Heyrovsky reactions (Zhao et al., 2021c). Therefore, an optimal HBE enables catalysts to achieve superior HOR performance (Sheng et al., 2013; Parsons, 1958). This relationship can be visualized through a volcano plot (Fig. 3.1A) (Sheng et al., 2013), where the logarithm of the HOR exchange current density correlates closely with the standard free energy of HBE, peaking near HBE = 0. Cheng et al. (Cheng et al., 2018) employs Quantum Mechanics Molecular Dynamics (QMMD) simulations with explicit solvent and potential effects to demonstrate that increasing pH strengthens HBE on Pt(100) by ~ 0.13 eV, due to more negative potentials repelling interfacial water molecules, directly explaining the experimentally observed pH-dependent HOR/HER kinetics (Fig. 3.1B). As pH increases, the HBE is regarded as the sole descriptor for the HOR. Moreover, it is emphasized that solvated OH^-, rather than adsorbed OH (OH_{ad}), influences the HOR activity (Yang et al., 2019). PtNi (Lu and Zhuang, 2017) or PtRu (Wang et al., 2014) can achieve enhanced HOR activity compared to Pt, which is attributed to the optimized HBE.

2.2 Bi-functional theory

In contrast to the HBE theory, which primarily addresses the interaction between H_{ad} and the catalyst surface, the bifunctional theory focuses on the

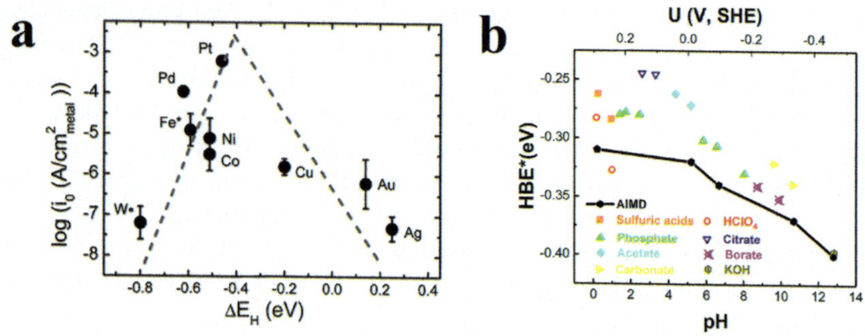

FIG. 3.1 (A) Exchange current densities versus calculated HBE for various monometallic surfaces. Reprinted with permission from (Sheng et al., 2013). Copyright 2013, Royal Society of Chemistry. (B) Comparison of experimental HBE data with QMMD predictions. Reprinted with permission from (Cheng et al., 2018). Copyright 2018, American Chemical Society.

catalytic role of adsorbed hydroxyl species (OH_{ad}) in the HOR. It elucidates the mechanistic role of the catalyst in the Volmer step by proposing that OH_{ad} species actively participate in the reaction pathway. It has been widely demonstrated that oxyphilic sites on the catalyst surface promote the adsorption of OH_{ad}, which subsequently interacts with adjacent H_{ad} species to accelerate the HOR kinetics (Li et al., 2017; Cong et al., 2019; Alesker et al., 2016). To verify the role of OH_{ad} in HOR, Markovic and coworkers investigated the HOR occurring on Au catalysts (Fig. 3.2) (Strmcnik et al., 2013). Typically, Au exhibits poor affinity for both Had and OH_{ad}. However, HOR can proceed on the Au surface when the potential exceeds 0.6 V (vs. RHE). Through cyclic voltammetry (CV) measurements, Strmcnik et al. (2020) observed that OH_{ad} adsorption could indeed occur on the Au surface. An OH_{ad} layer forms when the potential rises above 0.6 V. The findings of Markovic demonstrate that HOR on Au surfaces can be controlled by OH_{ad} adsorption. To further validate the bifunctional mechanism, they modulated OH_{ad} coverage on undercoordinated Pt sites through CO poisoning (to decrease OH_{ad}) and Pt ad-island deposition (to increase OH_{ad}). The CO-poisoned catalyst showed suppressed HOR activity, whereas the Pt-ad-island modified surface demonstrated enhanced HOR performance. Moreover, they proposed that enhancing OH^- adsorption could significantly improve HOR activity in alkaline media. They modified Pt with $Ni(OH)_2$ to facilitate the adsorption of OH^- onto the catalyst and observed that the resulting $Pt/Ni(OH)_2$ catalyst exhibited markedly higher HOR activity than pure Pt. This can be attributed to the rapid reaction between OH_{ad} and H_{ad} on the catalyst surface, forming water and thereby accelerating the reaction kinetics. Moreover, they found Ru exhibited stronger oxophilic ability than Pt, and when alloyed with Pt, could supply plentiful OH_{ad} near Pt active sites, enhancing alkaline HOR.

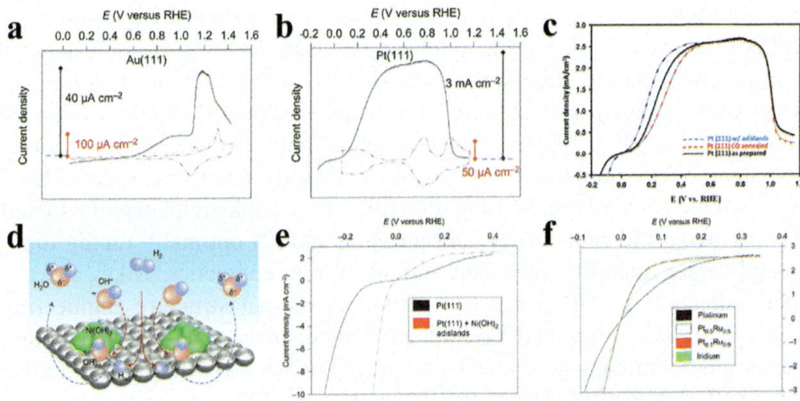

FIG. 3.2 (A) HOR polarization and CV curves for Au(111) in 0.1 M KOH electrolyte. (B) HOR polarization and CV curves for Pt(111) in 0.1 M KOH electrolyte. (C) HOR polarization curves for modified Pt(111) surfaces. (D) HOR mechanism on $Ni(OH)_2$-modified Pt(111) surfaces. (E) HER/HOR polarization plots of pristine and $Ni(OH)_2$-decorated Pt(111). (F) HER/HOR polarization plots of Pt, $Pt_{0.5}Ru_{0.5}$, and Ir. Reprinted with permission from (Strmcnik et al., 2013). Copyright 2013, Nature.

3 Materials for hydrogen oxidation

The HOR occurs at the anode in alkaline fuel cells; however, its inherently sluggish kinetics under alkaline conditions necessitate the development of highly efficient electrocatalysts. Current catalysts for HOR can be categorized into noble metal-based and nonnoble metal-based materials. Noble metal catalysts include Pt, Pd, Ir, and Ru, demonstrating higher activity for HOR, attributed to their optimal HBE and bifunctional mechanisms. Notably, Pt remains the benchmark catalyst due to its superior performance, whereas Pd exhibits enhanced tolerance to CO-poisoning, Ir displays higher intrinsic activity, and Ru presents a more cost-effective alternative. As for nonnoble metal catalysts, particularly Ni-based alloys (e.g., Ni–Cu, CoNiMo) and Ni-free systems (e.g., Co_2N, W_2C), they have garnered significant interest owing to their tunable electronic structures and competitive activity for HOR. This section systematically classifies electrocatalysts for HOR and critically reviews recent advancements in their development.

3.1 Noble metals

3.1.1 Pt-based catalysts

Pt-based catalysts remain the benchmark for the HOR in alkaline media due to their superior activity and stability (Su et al., 2025a). A significant breakthrough in atomic-level catalyst design was achieved by Wang et al. (2022c), who developed triphenylphosphine-stabilized Pt_6 nanoclusters for alkaline HOR. These structurally precise nanoclusters demonstrated exceptional mass activity and stability compared to both Pt single atoms and nanoparticles (NPs), while exhibiting remarkable CO tolerance, a critical advantage over commercial Pt/C catalysts. Density functional theory (DFT) calculations revealed the dual functionality of the triphenylphosphine ligand: (1) stabilizing the nanocluster configuration and (2) optimally tuning the Pt d-band center, thereby optimizing the adsorption energetics of key reaction intermediates (H_{ad}, OH_{ad}, and CO). Building on nanostructure engineering, Alia et al. (2013) employed a partial galvanic displacement approach to synthesize platinum-coated copper nanowires (Pt/CuNWs) with a well-defined core-shell architecture. The lattice mismatch-induced compressive strain in the Pt shell resulted in significantly enhanced HOR activity, with the catalysts exhibiting 3.5-fold higher area-specific and 1.9-fold greater mass-specific exchange current densities compared to conventional Pt/C. Further advancing the understanding of metal-support interactions, Yang et al. (2023a) systematically investigated atomic-scale Pt catalysts supported on tungsten nitride (Pt/WN) and tungsten oxide ($Pt/W_{18}O_{49}$). Their study revealed that moderate electron transfer in Pt/WN yielded 5.7-fold higher intrinsic HOR activity than the strongly interacting $Pt/W_{18}O_{49}$ catalyst. DFT analysis demonstrated that the metal-support electronic interaction precisely modulates the adsorption energies of reaction intermediates, establishing a fundamental principle for designing high-performance supported catalysts through controlled interfacial electronic effects. These studies collectively demonstrate that the rational design of Pt-based catalysts at atomic and nanoscale levels, coupled with precise control of electronic structure and support interactions, can significantly enhance HOR performance in alkaline media while overcoming traditional limitations of Pt catalysts.

3.1.2 Pd-based catalysts

Palladium (Pd) has emerged as a highly promising alternative to Pt for the HOR, offering comparable structural and electronic properties while demonstrating superior CO tolerance and greater natural abundance (Zhang et al., 2022b; Wang et al., 2022a). These inherent advantages have motivated extensive research into optimizing Pd-based catalysts. The fundamental relationship between nanoparticle size and catalytic activity was systematically investigated by Zheng et al. (Zheng et al., 2016b) through a comprehensive study of Pd NPs ranging from 3 to 42 nm. Their work revealed a critical size-activity correlation, where HOR performance improved with increasing particle size up to an optimal diameter of 19 nm, beyond which bulk-like behavior was observed. This phenomenon was attributed to the increasing proportion of surface sites exhibiting weakened HBE, a key descriptor for HOR activity. Comparative studies in both acidic and alkaline media demonstrated identical size-dependent trends, though the superior performance in acidic conditions was conclusively linked to the intrinsically lower HBE of Pd through detailed underpotential hydrogen deposition (H_{upd}) analysis and activation energy measurements. Further enhancement of HOR activity was achieved through electronic structure modification, as demonstrated by Kim et al. (2022) in their development of a boron-doped Pd nanocatalyst. Their innovative wet-chemistry synthesis approach revealed significant boron incorporation during catalyst formation. Subsequent ab initio calculations confirmed that this boron doping effectively optimized the adsorption energetics of both H_{ad} and OH_{ad} intermediates, addressing a critical challenge in HOR catalysis. The mechanistic aspects of Pd-based catalysts were elucidated by Bonnefont et al. (2013) through a combination of rotating disk electrode experiments and detailed kinetic modeling. By systematically varying Pd coverage on Au/C nanoparticle supports, they established that the Heyrovsky-Volmer mechanism predominates across all coverage levels. Notably, their work revealed an inverse relationship between Pd coverage and Tafel step contribution, providing valuable insights for the rational design of bimetallic HOR catalysts. These studies collectively demonstrate that through careful structural and electronic engineering, Pd-based catalysts can overcome traditional limitations while maintaining the advantages of greater CO tolerance and natural abundance. The development of such optimized Pd catalysts represents a significant step toward practical alternatives to Pt in fuel cell applications.

3.1.3 *Ir-based catalysts*

Among precious metal catalysts, iridium (Ir) demonstrates superior HOR activity, ranking second only to platinum in the established activity series: Pt > Ir > Rh > Pd (Durst et al., 2014). Despite its exceptional catalytic performance, the development of Ir-based catalysts faces significant challenges due to its higher cost compared to Pt. Current research on Ir-based catalysts primarily focuses on enhancing their HOR activity and reducing Ir loading through strategies such as alloying, doping, or modulating surface HBE and hydroxyl binding energy (OHBE) (Mu et al., 2023). The fundamental relationship between nanoparticle size and catalytic efficiency was systematically investigated by Zheng et al. (2015) through the fabrication of size-controlled Ir/C NPs (3–12 nm). Their comprehensive CV analysis revealed a critical structure-activity correlation: larger Ir particles, despite their lower electrochemical surface area, contained a higher proportion of active sites with optimal HBE (−0.33 eV). Remarkably, these high-efficiency sites, constituting only 15–30% of the total surface sites, were found to account for the majority of the observed catalytic activity, highlighting the importance of targeted surface engineering in catalyst design. Building on these insights, Qin et al. (2024) developed an innovative catalyst system comprising compressively-stressed Ir nanowires coupled with carbon dots (Ir NWs/CDs). This design simultaneously addressed two key challenges through: (1) strain engineering to optimize HBE via compressive stress and (2) interface modification to enhance charge transfer. The resulting catalyst exhibited exceptional bifunctional performance, demonstrating 1.5-fold higher HER mass activity than commercial Pt/C while maintaining 2-fold better CO tolerance than PtRu/C. Further advancing the field, Xu et al. (2024) engineered ultrathin Ir–Sb nanowires protected by SbO_x (Ir–Sb NWs/SbOx) as a high-performance bifunctional catalyst. This sophisticated architecture achieved remarkable catalytic performance, showing 3.5-fold and 4.8-fold higher HOR activity compared to Ir NWs and Pt/C, respectively. Moreover, the catalyst maintained excellent long-term stability in CO-containing environments, addressing a critical limitation of many precious metal catalysts. These studies demonstrate that through careful structural design and electronic modulation, Ir-based catalysts can achieve exceptional HOR performance while mitigating the challenges posed by material costs.

3.1.4 Ru-based catalysts

Ruthenium (Ru) has emerged as the most promising Pt substitute among precious metals for HOR, owing to its abundant reserves and relatively low cost (Su et al., 2025b). Recent studies have demonstrated that, through precise structural control, alloying strategies, and surface doping, the catalytic activity of Ru-based materials can be significantly enhanced, even surpassing that of commercial Pt/C catalysts. For instance, Lee et al. (2021) successfully synthesized uniform Ru NPs on carbon supports using fluidized bed reactor-atomic layer deposition with the 10-cycle Ru NPs (~2 nm) exhibiting a unique amorphous-crystalline structure and roughened surface. This particular nanostructure demonstrated superior HOR activity and mass activity compared to commercial Ru/C, primarily attributed to the weakened metal-hydrogen interactions resulting from optimal NP size and crystallinity. Further mechanistic studies revealed that the reduced HBE played a crucial role in enhancing the catalytic performance. In another significant development, John et al. (2015) designed carbon-supported Ru_xPt_y alloy catalysts, where the $Ru_{0.2}Pt_{0.8}$ composition exhibited exceptional HOR performance, showing a threefold higher exchange current density (1.42 versus 0.49 mA cm^{-2}) and significantly reduced Tafel slope (30 versus 125 mVdec^{-1}) compared to pure Pt. This remarkable enhancement was ascribed to the synergistic bifunctional and electronic effects arising from Ru–Pt alloying, which dramatically accelerated the rate-limiting hydrogen adsorption step. Moreover, Cheng et al. (2024) achieved a breakthrough by developing a single-atom Ce-doped Ru catalyst, where the oxophilic Ce atoms preferentially adsorbed OH^- species, thereby freeing adjacent Ru sites for H_{ad} adsorption and facilitating the formation of crucial H_3O^+ intermediates, as confirmed by in-situ Raman spectroscopy. This Ce–Ru synergistic effect resulted in unprecedented catalytic activity (27–59 times that of Pt/C) at a low overpotential of 30 mV, along with significantly improved stability and CO tolerance. These collective findings underscore the tremendous potential of Ru-based catalysts and provide valuable insights for the rational design of high-performance, cost-effective alternatives to Pt-based catalysts.

3.2 Nonnoble metals

3.2.1 Nickel-based catalysts

Nickel-based materials have emerged as promising low-cost electrocatalysts for the alkaline HOR, yet their widespread application is hindered by two critical limitations: excessively strong HBE, which restricts catalytic activity, and insufficient stability under operational conditions. To address these challenges, researchers have extensively explored alloying and doping strategies to optimize the electronic structure and surface properties of nickel-based catalysts (Davydova et al., 2018; Roy et al., 2018). A significant advancement was demonstrated by Wang et al. (2019), who developed Ni–Cu alloy films through combinatorial magnetron cosputtering, revealing that Cu incorporation not only enhances intrinsic activity but also improves oxidation resistance. Their systematic investigation established a volcano-type relationship between Cu content and catalytic performance, with the optimal composition (40 at.% Cu) exhibiting a fourfold increase in exchange current density compared to pure Ni. Building upon this alloy design concept, Sheng et al. (2014) achieved a further breakthrough by developing a CoNiMo ternary catalyst via pulse-electrodeposition, which demonstrated superior HOR activity among nonprecious metals—about 20 times higher than that of pure Ni. Through comprehensive characterization combining electrochemical measurements, DFT calculations, and H_2 temperature-programmed desorption analysis, they elucidated that the multimetallic synergy in CoNiMo optimally modulates HBE to approach Pt-like levels while retaining cost advantages. Remarkable progress was reported by Duan et al. (2020) with their development of phase-engineered tetragonal $MoNi_4$ nanoalloys, which set a new record by achieving an exchange current density (3.41 mA/cm^2) that surpasses commercial Pt/C by 40%. This exceptional performance stems from a bifunctional mechanism where Ni sites optimize hydrogen adsorption and Mo sites facilitate hydroxyl binding, thereby synergistically accelerating the rate-limiting Volmer step. These successive developments collectively demonstrate that rational alloy design and phase control can effectively overcome the inherent limitations of nickel-based catalysts, offering a viable pathway to develop high-performance, cost-effective alternatives to precious metal catalysts for alkaline HOR applications.

3.2.2 Nickel-free catalysts

Nonnoble metal electrocatalysts for the HOR that do not rely on Ni as active sites remain relatively underexplored, with most reported systems adopting heterogeneous or interfacial structures to enhance catalytic performance (Zhao et al., 2021b). A notable advancement in this field was achieved by Song et al. (2019), who developed a bifunctional Co_2N/Co interfacial catalyst through controlled nitridation of cobalt NPs. This catalyst exhibited HER/HOR performance comparable to Pt-based benchmarks, attributed to the synergistic effect of the Co_2N/Co interfaces, which simultaneously optimize hydrogen adsorption energy and facilitate water dissociation, a critical step in the alkaline HOR mechanism. Importantly, this catalyst also demonstrated exceptional CO tolerance, a key requirement for practical fuel cell applications. Building on the concept of hybrid catalysts, Huang et al. (2020) designed a Co_2P/CoP composite derived from cobalt carbonate hydroxide precursors, which leverages the complementary properties of its constituent phases: CoP provides optimal hydrogen adsorption characteristics, while Co_2P ensures high electrical conductivity. This dual-phase catalyst delivered outstanding performance, requiring only 61 mV overpotential to achieve 10 mA cm^{-2} for the HER (with a low Tafel slope of 55.1 mV dec^{-1}) and exhibiting a comparable performance to Pt/C with exchange current density (0.82 vs. 1.14 mA cm^{-2}) for the HOR. Further expanding the scope of non-Ni-based HOR catalysts, Xiong et al. (2019) introduced a novel class of tungsten carbide (W_2C) electrocatalysts through a facile one-step synthesis of N,P-doped carbon-encapsulated W_2C NPs (W_2C@N,P–C). This catalyst demonstrated remarkable versatility, achieving high exchange current densities in acidic (1.03 mA cm^{-2}), alkaline (0.91 mA cm^{-2}), and neutral (0.84 mA cm^{-2}) media at 0.05 V vs. RHE, along with exceptional stability (10,000 cycles) and CO tolerance. The superior performance was attributed to the synergistic effects of N,P codoping, which enhances the reducibility of W_2C and creates abundant active sites, while the carbon encapsulation improves durability. Collectively, these studies highlight the potential of interfacial engineering, heteroatom doping, and nanostructure design in developing high-performance, nonprecious HOR catalysts that circumvent the limitations of Ni-based catalysts.

4 Design strategies

The development of high-performance catalysts for the HOR in alkaline media is crucial for advancing alkaline fuel cell technologies. This section discusses five key catalyst design strategies: geometry engineering, composition engineering, substrate engineering, oxyphilic site engineering, and heterostructure engineering. These approaches systematically optimize catalyst properties by precisely controlling nanomaterial morphology, electronic structure, support interactions, and active site configuration. Through rational manipulation of particle size, crystallographic phase, support materials, and interfacial characteristics, researchers have achieved enhanced catalytic activity, improved stability, and reduced costs. These advancements have led to HOR catalysts that outperform conventional platinum-based catalysts while enabling the design of efficient noble-metal-free alternatives.

4.1 Geometry engineering

Nanomaterials exhibit distinct physical and chemical properties, such as ultrahigh catalytic activity and expanded specific surface area, making them ideal candidates for enhancing HOR performance. Researchers optimize these properties by carefully designing the size and morphology of nanomaterials. For instance, Ohyama et al. (2013a) engineered carbon-supported Ru nanoparticles (Ru/C) with sizes ranging from 2 to 7 nm, achieving peak HOR activity at around 3 nm (Fig. 3.3A). This size, surpassing commercial Pt/C (~2 nm), was attributed to an amorphous-to-nanocrystalline transition, which resulted in

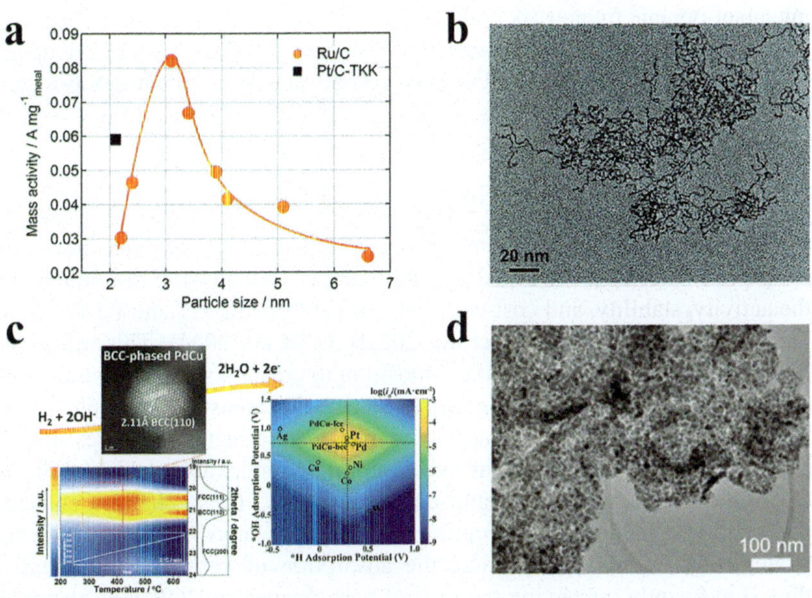

FIG. 3.3 (A) Mass activity of the Ru/C against the mean particle size. Reprinted with permission from (Ohyama et al., 2013a). Copyright 2013, American Chemical Society. (B) TEM images of Ir ONA. Reprinted with permission from (Yang et al., 2017). Copyright 2017, Royal Society of Chemistry. (C) BCC-Phased PdCu Alloy. Reprinted with permission from (Qiu et al., 2018). Copyright 2018, American Chemical Society. (D) TEM images of Ni–H_2-2%. Reprinted with permission from (Ni et al., 2020). Copyright 2020, Wiley.

roughened surfaces rich in coordinatively unsaturated Ru atoms. These features enhanced H and OH adsorption, thereby improving HOR kinetics. Similarly, Yang et al. (2017) synthesized ultrafine worm-like Ir-oriented nanocrystalline assemblies (ONAs), optimizing the electrocatalytic performance for HOR in alkaline media (Fig. 3.3B). The worm-like structure, with its extended low-index crystalline planes, facilitated rapid mass and charge transfer, significantly boosting catalytic efficiency. In another approach, Maiti et al. (2020) designed N, Ru co-doped pellet drum bundle-like antimony sulfide (Sb_2S_3) through a reflux method, leveraging its unique morphology to enhance HOR performance in 1 M KOH. The pellet drum-like structure, combined with geometric lattice defects and surface oxygen vacancies, created abundant active sites, resulting in higher HOR activity. Qiu et al. (2018) synthesized BCC-phased PdCu alloy nanoparticles using a wet-chemistry method followed by critical thermal treatment, achieving up to 20-fold higher HOR mass and specific activities compared to FCC-phased PdCu (Fig. 3.3C). Their findings, supported by in situ HE-XRD and DFT, revealed that the BCC phase optimized H and OH binding strengths, akin to Pt, underscoring crystal phase engineering as a crucial strategy for developing efficient anode catalysts for alkaline fuel cells. Ni et al. (2020) developed a Ni-based catalyst, Ni–H_2-2%, through pyrolysis of a Ni–containing metal-organic framework under a N_2/H_2 atmosphere (Fig. 3.3D). The strain engineering in Ni–H_2-2% nanoparticles optimized HBE and maximized active sites, resulting in the highest HOR mass activity among platinum-group-metal-free catalysts.

4.2 Composition engineering

Composition engineering has emerged as a important strategy for improving the performance of electrocatalysts, particularly in the context of enhancing the activity, stability, and cost-efficiency of catalysts like Pt and Pd by incorporating other metals (Wang et al., 2022b; Li et al., 2024). This technique involves modifying the catalyst composition to create novel compounds with unique crystal phases, which requires substantial adjustments to electronic interactions within the material. Compared to pure metal structures, multi-component catalysts exhibit increased atomic complexity, causing shifts in electronic occupancy and d-band center positions. These changes enhance the creation of additional adsorption sites on the catalyst surface, which are strategically tailored to optimize the adsorption of various intermediates, thus significantly improving the overall performance in HOR. For example, Gao et al. (2022) developed a nickel–molybdenum–niobium metallic catalyst ($Ni_{52}Mo_{13}Nb_{35}$). Through composition engineering, they achieved an intrinsic exchange current density of 0.35 mA cm^{-2}, surpassing that of Pt disk catalysts (0.30 mA cm^{-2}), and demonstrated excellent stability up to 0.8 V versus the reversible hydrogen electrode. This catalyst, when used as an anode, enabled power densities of 390 mW cm^{-2} in hydrogen fuel cells, with minimal degradation after 50 h of operation. Similarly, Men et al. (2023) designed a high-entropy alloy (HEA) catalyst, PdNiRuIrRh, that achieved a remarkable mass activity of 3.25 mA μg^{-1} for alkaline HOR, an 8-fold improvement over commercial Pt/C (Fig. 3.4A). Furthermore, Su et al. (2023) developed an ordered Ru_3Sn_7/C intermetallic compound, leveraging composition engineering to improve HOR activity, stability, and CO-tolerance compared to disordered RuSn alloys (Fig. 3.4B). The strong electron interactions of this intermetallic structure were found to enhance hydroxyl-binding energy, resulting in superior alkaline HOR performance and a reduced kinetic gap between acidic and alkaline conditions. In another study, Li et al., (2023) engineered a phase-structured Ni-based electrocatalyst with a hexagonal-close-packed (hcp)/face-centered-cubic (fcc) interface, leading to a significant enhancement of alkaline HOR performance in fuel cells (Fig. 3.4C). By optimizing the electronic structure and d-band center at the hcp/fcc-Ni–C interface, they

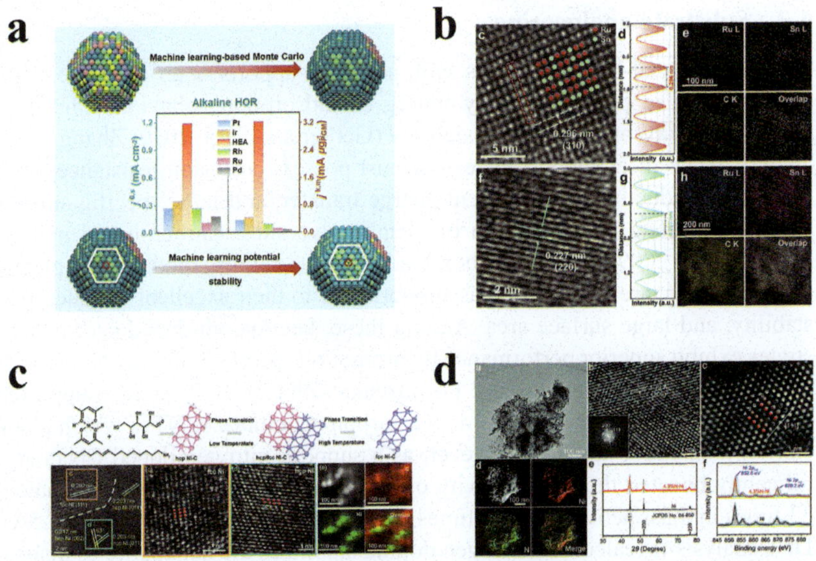

FIG. 3.4 (A) Atomic structure of a high-entropy alloy (HEA-PdNiRuIrRh) regulated by machine learning. Reprinted with permission from (Men et al., 2023). Copyright 2023, Wiley. (B) HAADF-STEM-EDX elemental mappings for Ru_3Sn_7/C and RuSn/C. Reprinted with permission from (Su et al., 2023). Copyright 2023, Wiley. (C) HRTEM-EDX mapping of hcp/fcc-Ni-C. Reprinted with permission from (Li et al., 2023). Copyright 2022, American Chemical Society. (D) HAADF-STEM-EDX image, XRD and XPS results of 4.3% N–Ni nanosheets. Reprinted with permission from (Zhao et al., 2022). Copyright 2022, Royal Society of Chemistry.

achieved ideal binding energies for H and OH, as well as a reduction in water formation energy, resulting in a 6-fold increase in mass activity compared to conventional fcc-Ni catalysts. Zhao et al. (2022) adopted a novel strategy by incorporating nitrogen into Ni nanosheets, using composition engineering to finely tune orbital hybridization and strain fields. This induced strong d-sp orbital interactions and reduced tensile strain, optimizing hydrogen and hydroxyl adsorption, which led to a 53-fold increase in mass activity (77.13 $A\ g^{-1}$) compared to pure Ni nanosheets, surpassing many nonnoble metal catalysts for HOR (Fig. 3.4D). Huang et al. (2023b) introduced a defect-derived composition engineering approach by creating phosphorus vacancies (P-vacancies) on Rh_2P electrocatalysts. This approach exposed subsurface Rh atoms as dedicated hydrogen adsorption sites and leveraged peripheral P-sites for OH adsorption. Their density functional theory calculations confirmed that P-vacancies enhanced H_{ad} and OH_{ad} interactions, accelerating HOR kinetics. The resulting defective Rh_2P nanowires exhibited mass activities that were 5.7 and 14.3 times greater than perfect Rh_2P nanocubes and commercial Pt/C, respectively.

4.3 Substrate engineering

Decorating metal-based catalysts with suitable supports is an effective strategy to optimize performance by tuning the adsorption behavior, enhancing catalytic activity, and improving stability (Gerber and Serp, 2020; Zhang et al., 2022a, 2021). Ideal support materials must possess corrosion resistance, high electrical conductivity for efficient charge transfer, and large specific surface areas to facilitate the dispersion of metal active sites. Carbon-based materials (Antolini, 2009), such as carbon black, porous carbon, modified graphene, and nanotubes, are widely used as supports due to their excellent conductivity, stability, and large surface area. Among these, heteroatom-doped carbon substrates exhibit superior performance by promoting specific bonding interactions and enhancing anchoring effects. For instance, Zhuang et al. (2016) supported nickel nanoparticles on nitrogen-doped carbon nanotubes (CNTs), which alone exhibit poor HOR activity. However, as a support, nitrogen-doped CNTs significantly boosted the HOR activity of nickel nanoparticles by approximately 33 times in mass activity and 21 times in exchange current density (Fig. 3.5A). DFT analysis revealed that nitrogen doping stabilizes the nanoparticles against reconstruction and tunes the adsorption sites by modulating the d-orbitals of nickel. Similarly, Li et al. (2021a) developed MoO_2-supported Ir catalysts with varied interfacial bonds (Ir–O, Ir–Mo, and mixed Ir–Mo/O) to optimize HOR activity (Fig. 3.5B). The MoO_2 support plays a critical role in regulating the electronic configurations of the catalysts. The Ir/O–MoO_2 interface, with localized electron distribution via Ir–O bonds, downshifts the Ir d-band center, weakening H and OH adsorption, thereby enhancing HOR activity (1.96 mA/cm^2). In contrast, Ir/Mo–MoO_2, with delocalized electrons via Ir–Mo bonds, strengthens

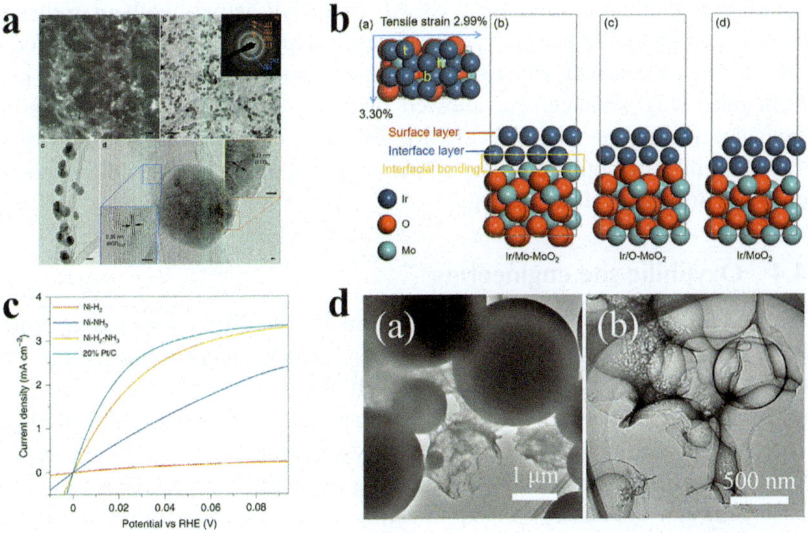

FIG. 3.5 (A) SEM, TEM and HRTEM images of nickel particles and CNT. Reprinted with permission from (Zhuang et al., 2016). Copyright 2016, Nature. (B) Top view and side view of geometric structures of Ir/Mo–MoO_2 with interfacial Ir–Mo bonds. Reprinted with permission from (Li et al., 2021a). Copyright 2021, American Chemical Society. (C) HOR polarization curves of Ni–H_2-NH_3, Ni-NH_3, Ni–H_2 and commercial Pt/C (20 wt%). Reprinted with permission from (Ni et al., 2022). Copyright 2022, Nature. (D) TEM images of SiO_2@TN-CNS-0.3 and TN-CNS-0.3. Reprinted with permission from (Jiang et al., 2019). Copyright 2019, Elsevier.

adsorption, leading to reduced activity (1.47 mA/cm^2). The mixed Ir–Mo/O bonds result in irregular adsorption and further reduced activity (1.10 mA/cm^2), highlighting the significant role of support in tuning catalytic performance. Furthermore, Jiang et al. (2019) employed a dual-templating method to create N-doped carbon nanosheets with varying nitrogen doping levels as supports for nickel nanoparticles in HOR catalysis (Fig. 3.5D). The nitrogen-doped carbon support critically influenced HOR activity by modulating the HBE on the nickel nanoparticles. A volcano-shaped relationship between HOR activity and nitrogen content was observed, with the catalyst at 8.7% N achieving optimal activity, comparable to some platinum-group catalysts. Additionally, Yang et al., (2023b) supported Ru onto defective-carbon-supported catalysts (Ru@C), achieving exceptional activity and CO tolerance for alkaline HOR. The degree of defects in the carbon support significantly tuned the Ru–O–C interfacial chemistry, redistributing the d-electrons of Ru, which downshifted its d-band center, weakened hydrogen adsorption, and suppressed CO binding. As a result, Ru@C exhibited a 6.6-fold increase in mass activity (@25 mV) compared to conventional Ru/C, outperforming leading catalysts while tolerating 20,000 ppm CO. Ni et al. (2022) developed a nickel-based catalyst (Fig. 3.5C), embedding nickel nanoparticles in a nitrogen-doped carbon support, achieving record-high platinum group metal (PGM)-free HOR activity with an exchange current density of 70 μA cm^{-2}. The nitrogen-doped carbon support played a critical role by enabling electronic interactions that enhanced the catalytic performance. This advancement allowed platinum group metal-free hydroxide exchange membrane fuel cells (HEMFCs) to reach a peak power density of 488 mW cm^{-2}, demonstrating the pivotal role of support in advancing efficient, cost-effective fuel cell technology.

4.4 Oxyphilic site engineering

In the HOR, forming OH_{ad} requires additional active sites compared to hydrogen evolution, making the adsorption of oxygen-containing species critical (Pan et al., 2019; Cui et al., 2021). Designing HOR catalysts often focuses on oxyphilic site engineering to enhance this capability. Some researchers advocate constructing dual active sites to tune catalyst hydrophilicity and boost surface OHad participation in HOR. By engineering oxyphilic sites, catalysts can better balance H and OH adsorption, significantly improving HOR performance. Cui et al. (2021) developed high-efficiency electrocatalysts for HOR by leveraging oxyphilic site engineering. Using density functional theory (DFT), they demonstrated that partial substitution of Ru (Rh) clusters with oxyphilic metal (OM) species (e.g., 0.13 wt% Co, 0.23 wt% Mn, 2.4 wt% Cr, or 1.3 wt% Fe) disrupts the long-range order of surface atoms, triggering active atom reconstruction at buried interfaces. This reduces oxidation intermediates, optimizing HBE and forming OH^- adsorption promoters, enhancing HOR, HER, and oxygen evolution reaction (OER) activities in alkaline media. Guided by these findings, Mu et al. synthesized ultralow-loaded RuRh-OM catalysts via a mixed-solvent strategy. These catalysts exhibited superior HOR activity in 0.1 M KOH, outperforming pristine RuRh and commercial Pt by ~2 times, and surpassing most reported HOR catalysts.

4.5 Heterostructure engineering

The presence of heterojunctions between different metal compounds or between metal compounds and pure metals facilitates charge separation and transfer, thereby enhancing the efficiency of catalytic processes (Li et al., 2021b; Feng et al., 2022). The interface between different components in a heterojunction can create unique electronic environments that are absent in individual components, optimizing the adsorption energies of various reactants and intermediates while significantly improving catalytic activity and selectivity (Dou et al., 2019; Zhao et al., 2021a). Ru/RuO_2 in-plane heterostructures are developed by Zhang et al. (2023) as efficient electrocatalysts for the hydrogen oxidation reaction (HOR) in alkaline media (Fig. 3.6A). The Ru–RuO_2 interfaces, with Ru and RuO_2 as optimal H and OH adsorption sites, respectively, enable fast HOR electrocatalysis at the Ru(100)/RuO_2(200) interface. Electron-deficient Ru domains, due to strong RuO_2 interaction, ensuring durability. The heterostructure exhibits a high exchange current density of 8.86 mA cm^{-2} and robust stability, attributed to the multifunctional Ru–RuO_2 interfaces, highlighting their critical role in enhancing alkaline HOR performance. Chen et al. (Yang et al., 2024) designed a Ni/MoO_xH_y heterostructure catalyst to enhance the alkaline HOR by leveraging the hydrogen spillover effect (Fig. 3.6B). This facilitates Had migration from Ni to the interface, where it reacts with OHad on MoO_xH_y, improving catalytic efficiency. The heterostructure engineering optimizes intermediate transfer pathways, preserves the metallic phase of Ni, and achieves high activity. The Ni/$Ni_{0.2}Mo_{0.8}N$/rGO heterostructure developed by Huang et al. (2023a) significantly enhances the HOR activity in alkaline conditions in anion exchange membrane fuel cells. The heterostructure achieves a low overpotential of 100 mV at 2.9 mA cm^{-2}, high kinetics (exchange current density of 2.74 mA cm^{-2}), and exceptional durability (over 40 h or 10,000 cycles). The Ni–$Ni_{0.2}Mo_{0.8}N$ interface optimizes hydrogen and hydroxyl binding energies through tuned d-band electronic structures, as revealed by characterizations and DFT simulations, driving superior HOR performance. Ni_3B/Ni heterostructures fabricated by Yang et al. (2020) exhibit exceptional alkaline HOR performance, with mass activity ~ 10 times higher than individual Ni_3B or Ni, ranking among top platinum-group-metal-free electrocatalysts. Electron transfer from Ni_3B to Ni at the heterostructure interface interregulates d-band centers, optimizing

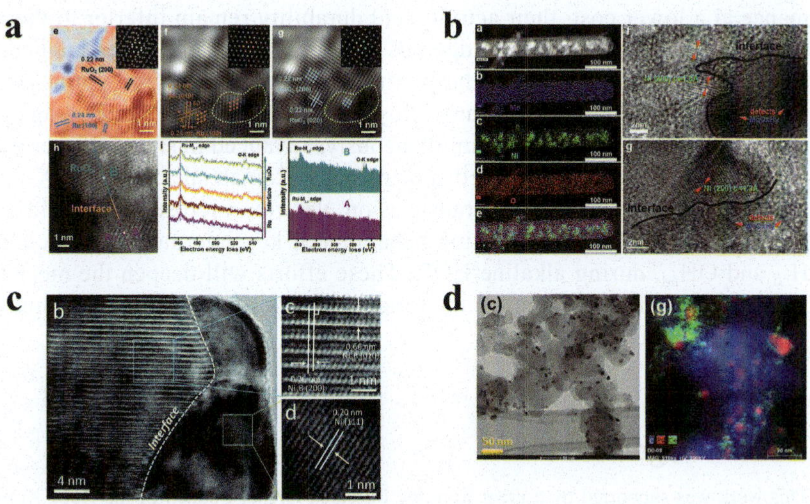

FIG. 3.6 (A) HRTEM, IFFT and HAADF-STEM images for Ru/RuO_2. Reprinted with permission from (Zhang et al., 2023). Copyright 2022, Wiley. (B) HAADF-STEM image of the Ni/MoO_xH_y heterostructure. Reprinted with permission from (Yang et al., 2024). Copyright 2024, American Chemical Society. (C) HRTEM images of NiB-300. Reprinted with permission from (Yang et al. 2020). Copyright 2020, Royal Society of Chemistry. (D) HRTEM and corresponding overlay STEM/EDS maps for 0.38 CeO_x-Pd/C. Reprinted with permission from (Singh et al., 2020). Copyright 2020, Wiley.

hydrogen and hydroxyl binding energies, as confirmed by experimental results and theoretical calculations (Fig. 3.6C). CeO_x-Pd/C heterostructure catalysts made by Singh et al. (2020) by selectively depositing CeO_x on Pd nanoparticles, enhance alkaline HOR for anion-exchange membrane fuel cells (Fig. 3.6D). The intimate CeO_x-Pd interface, confirmed by advanced characterizations, optimizes oxophilic properties and binding energies, boosting HOR activity. The catalyst with a $Ce_{0.38}$/Pd ratio achieves a record HOR specific exchange current (51.5 mA mg^{-1} Pd) and peak power density (1,169 mW cm^{-2} $mgPd^{-1}$). The CeO_x-Pd interface drives superior electrocatalytic performance, highlighting its critical role in enhancing HOR kinetics.

5 Challenges and perspectives

Despite the development of numerous advanced catalysts exhibiting high HOR activity and stability, their performance still falls short of practical application requirements. A critical issue lies in the incomplete understanding of the HOR mechanism, with ongoing debates between the HBE theory and the bi-functional theory. Notably, modulating catalyst oxophilicity alone cannot independently alter HBE, making it difficult to experimentally validate the role of HBE theory in HOR activity, thereby necessitating innovative approaches to both catalyst design and mechanistic investigations. While nonnoble metal catalysts have shown promising HOR performance at a lower cost, their activity and durability remain inferior to those of noble metal catalysts. To bridge this gap, strategic material design methodologies, including doping, heterostructure engineering for optimizing HBE, nanoscale size control, and alloying to reduce noble metal usage—must be rigorously explored. Simultaneously, advanced in situ characterization techniques, such as infrared spectroscopy, Raman spectroscopy, X-ray absorption spectroscopy, and transmission electron microscopy, should be employed to probe the adsorption behaviors of key intermediates, such as H_{ad} and OH_{ad}, during alkaline HOR. These efforts will deepen the mechanistic understanding and guide the rational design of next-generation high-performance HOR catalysts.

Acknowledgement

This work was supported by a grant from the Research Grants Council of the Hong Kong Special Administrative Region, China (Project No. 15308024); a grant from the Research Institute for Smart Energy (CDB2) at The Hong Kong Polytechnic University; a grant from the Research Institute for Advanced Manufacturing at The Hong Kong Polytechnic University (CDJQ); and a grant from the Research Center for Carbon-Strategic Catalysis (RC-CSC) at The Hong Kong Polytechnic University (CE2X).

References

Alesker, M., Page, M., Shviro, M., Paska, Y., Gershinsky, G., Dekel, D.R., Zitoun, D., 2016. Palladium/nickel bifunctional electrocatalyst for hydrogen oxidation reaction in alkaline membrane fuel cell. J. Power Sources 304, 332–339. doi:https://doi.org/10.1016/j.jpowsour.2015.11.026.

Alia, S.M., Pivovar, B.S., Yan, Y., 2013. Platinum-coated copper nanowires with high activity for hydrogen oxidation reaction in base. ACS Publ. doi:https://doi.org/10.1021/ja405598a.

Antolini, E., 2009. Carbon supports for low-temperature fuel cell catalysts. Appl. Catal. B: Environ. 88, 1–24. doi:https://doi.org/10.1016/j.apcatb.2008.09.030.

Bonnefont, A., Simonov, A.N., Pronkin, S.N., Gerasimov, E.Y., Pyrjaev, P.A., Parmon, V.N., Savinova, E.R., 2013. Hydrogen electrooxidation on PdAu supported nanoparticles: an experimental RDE and kinetic modeling study. Catal. Today 202, 70–78. doi:https://doi.org/10.1016/j.cattod.2012.03.076.

Cheng, T., Wang, L., Merinov, B.V., Goddard, W.A.I., 2018. Explanation of dramatic ph-dependence of hydrogen binding on noble metal electrode: greatly weakened water adsorption at high ph. J. Am. Chem. Soc. 140, 7787–7790. doi:https://doi.org/10.1021/jacs.8b04006.

Cheng, Z., Yang, Y., Zheng, F., Chen, S., Wang, P., Wang, P., Huang, H., Wang, C., Wang, D., Chen, Q., 2024. Fast kinetics of hydrogen oxidation reaction on single-atom Ce-alloyed Ru in alkaline electrolytes. ACS Catal. 14, 3845–3854. doi:https://doi.org/10.1021/acscatal.3c05954.

Cong, Y., McCrum, I.T., Gao, X., Lv, Y., Miao, S., Shao, Z., Yi, B., Yu, H., Janik, M.J., Song, Y., 2019. Uniform Pd0.33Ir0.67 nanoparticles supported on nitrogen-doped carbon with remarkable activity toward the alkaline hydrogen oxidation reaction. J. Mater. Chem. A 7, 3161–3169. doi:https://doi.org/10.1039/C8TA11019K.

Cui, Y., Xu, Z., Chen, D., Li, T., Yang, H., Mu, X., Gu, X., Zhou, H., Liu, S., Mu, S., 2021. Trace oxophilic metal induced surface reconstruction at buried RuRh cluster interfaces possesses extremely fast hydrogen redox kinetics. Nano Energy 90, 106579. doi:https://doi.org/10.1016/j.nanoen.2021.106579.

Davydova, E.S., Zaffran, J., Dhaka, K., Toroker, M.C., Dekel, D.R., 2018. Hydrogen oxidation on Ni-based electrocatalysts: the effect of metal doping. Catalysts 8, 454. doi:https://doi.org/10.3390/catal8100454.

Dou, S., Wang, X., Wang, S., 2019. Rational design of transition metal-based materials for highly efficient electrocatalysis. Small Methods 3, 1800211. doi:https://doi.org/10.1002/smtd.201800211.

Duan, Y., Yu, Z.-Y., Yang, L., Zheng, L.-R., Zhang, C.-T., Yang, X.-T., Gao, F.-Y., Zhang, X.-L., Yu, X., Liu, R., Ding, -H.-H., Gu, C., Zheng, X.-S., Shi, L., Jiang, J., Zhu, J.-F., Gao, M.-R., Yu, S.-H., 2020. Bimetallic nickel-molybdenum/tungsten nanoalloys for high-efficiency hydrogen oxidation catalysis in alkaline electrolytes. Nat. Commun. 11, 4789. doi:https://doi.org/10.1038/s41467-020-18585-4.

Durst, J., Simon, C., Hasché, F., Gasteiger, H.A., 2014. Hydrogen oxidation and evolution reaction kinetics on carbon supported Pt, Ir, Rh, and Pd electrocatalysts in acidic media. J. Electrochem. Soc 162, F190. doi:https://doi.org/10.1149/2.0981501jes.

Feng, Y., Guan, Y., Zhou, E., Zhang, X., Wang, Y., 2022. Nanoscale double-heterojunctional electrocatalyst for hydrogen evolution. Adv. Sci. 9, 2201339. doi:https://doi.org/10.1002/advs.202201339.

Gao, F.-Y., Liu, S.-N., Ge, J.-C., Zhang, X.-L., Zhu, L., Zheng, Y.-R., Duan, Y., Qin, S., Dong, W., Yu, X., Bao, R.-C., Yang, -P.-P., Niu, -Z.-Z., Ding, Z.-G., Liu, W., Lan, S., Gao, M.-R., Yan, Y., Yu, S.-H., 2022. Nickel–molybdenum–niobium metallic glass for efficient hydrogen oxidation in hydroxide exchange membrane fuel cells. Nat. Catal. 5, 993–1005. doi:https://doi.org/10.1038/s41929-022-00862-8.

Gerber, I.C., Serp, P., 2020. A theory/experience description of support effects in carbon-supported catalysts. Chem. Rev. 120, 1250–1349. doi:https://doi.org/10.1021/acs.chemrev.9b00209.

Huang, C., Feng, M., Peng, Y., Zhang, B., Huang, J., Yue, X., Huang, S., 2023a. In situ construction of Ni/Ni0.2Mo0.8N heterostructure to enhance the alkaline hydrogen oxidation reaction by balancing the binding of intermediates. Adv. Funct. Mater. 33, 2300593. doi:https://doi.org/10.1002/adfm.202300593.

Huang, G., Liang, W., Wu, Y., Li, J., Jin, Y.Q., Zeng, H., Zhang, H., Xie, F., Chen, J., Wang, N., Jin, Y., Meng, H., 2020. Co2P/CoP hybrid as a reversible electrocatalyst for hydrogen oxidation/evolution reactions in alkaline medium. J. Catal. 390, 23–29. doi:https://doi.org/10.1016/j.jcat.2020.07.023.

Huang, H., Liu, K., Yang, F., Cai, J., Wang, S., Chen, W., Wang, Q., Fu, L., Xie, Z., Xie, S., 2023b. Breaking surface atomic monogeneity of Rh2P nanocatalysts by defect-derived phosphorus vacancies for efficient alkaline hydrogen oxidation. Angew. Chem. Int. Ed. 62, e202315752. doi:https://doi.org/10.1002/anie.202315752.

Jiang, S., Cheng, Q., Zou, L., Zou, Z., Li, Y., Zhang, Q., Gao, Y., Yang, H., 2019. Ni nanoparticles supported on carbon nanosheets with tunable N doping content for hydrogen oxidation reaction. Chem. Phys. Lett. 728, 19–24. doi:https://doi.org/10.1016/j.cplett.2019.04.072.

John, S.S., Atkinson, R.W.I., Unocic, R.R., Zawodzinski, T.A.J., Papandrew, A.B., 2015. Ruthenium-alloy electrocatalysts with tunable hydrogen oxidation kinetics in alkaline electrolyte. J. Phys. Chem. C 119, 13481–13487. doi:https://doi.org/10.1021/acs.jpcc.5b03284.

Kim, S.-H., Yoo, S.-H., Shin, S., El-Zoka, A.A., Kasian, O., Lim, J., Jeong, J., Scheu, C., Neugebauer, J., Lee, H., Todorova, M., Gault, B., 2022. Controlled doping of electrocatalysts through engineering impurities. Adv. Mater. 34, 2203030. doi:https://doi.org/10.1002/adma.202203030.

Lamoureux, P.S., Singh, A.R., Chan, K., 2019. pH effects on hydrogen evolution and oxidation over Pt(111): insights from first-principles. ACS Catal. 9, 6194–6201. doi:https://doi.org/10.1021/acscatal.9b00268.

Lee, W.-J., Bera, S., Woo, H.-J., An, J.-W., Bae, J.-S., Oh, I.-K., Kwon, S.-H., 2021. Controllable size and crystallinity of Ru nanoparticles on a carbon support synthesized by fluidized bed reactor-atomic layer deposition for enhanced hydrogen oxidation activity. J. Mater. Chem. A 9, 17223–17230. doi:https://doi.org/10.1039/D1TA03678E.

Li, J., Ghoshal, S., Bates, M.K., Miller, T.E., Davies, V., Stavitski, E., Attenkofer, K., Mukerjee, S., Ma, Z.-F., Jia, Q., 2017. Experimental proof of the bifunctional mechanism for the hydrogen oxidation in alkaline media. Angew. Chem. Int. Ed. 56, 15594–15598. doi:https://doi.org/10.1002/anie.201708484.

Li, J., Sun, S., 2019. Intermetallic nanoparticles: synthetic control and their enhanced electrocatalysis. Acc. Chem. Res. 52, 2015–2025. doi:https://doi.org/10.1021/acs.accounts.9b00172.

Li, M., Xie, Z., Zheng, X., Li, L., Li, J., Ding, W., Wei, Z., 2021a. Revealing the regulation mechanism of Ir–MoO2 interfacial chemical bonding for improving hydrogen oxidation reaction. ACS Catal. 11, 14932–14940. doi:https://doi.org/10.1021/acscatal.1c04440.

Li, X., Han, X., Yang, Z., Wang, S., Yang, Y., Wang, J., Chen, J., Chen, Z., Jin, H., 2024. Lattice-distorted Pt wrinkled nanoparticles for highly effective hydrogen electrocatalysis. Nano Res. 17, 3819–3826. doi:https://doi.org/10.1007/s12274-023-6328-0.

Li, Y., Li, Z., Cong, H., Yang, C., Cheng, G., Luo, W., 2022. Boosting hydrogen oxidation performance of phase-engineered Ni electrocatalyst under alkaline media. ACS Sustain. Chem. Eng. 10, 3682–3689. doi:https://doi.org/10.1021/acssuschemeng.2c00037.

Li, Z., An, L., Song, M., Zhao, T., Zhang, J., Zhang, C., Li, Z., Wang, D., 2023. Tuning the hydrogen and hydroxyl adsorption on Ru nanoparticles for hydrogen electrode reactions via size controlling. Chin. Chem. Lett. 34, 107622. doi:https://doi.org/10.1016/j.cclet.2022.06.045.

Li, Z., Hu, M., Wang, P., Liu, J., Yao, J., Li, C., 2021b. Heterojunction catalyst in electrocatalytic water splitting. Coord. Chem. Rev. 439, 213953. doi:https://doi.org/10.1016/j.ccr.2021.213953.

Lu, L., Peng, L., Li, L., Li, J., Huang, X., Wei, Z., 2020. Improved hydrogen oxidation reaction under alkaline conditions by Au–Pt alloy nanoparticles. J. Energy Chem. 40, 52–56. doi:https://doi.org/10.1016/j.jechem.2019.02.008.

Lu, S., Zhuang, Z., 2017. Investigating the influences of the adsorbed species on catalytic activity for hydrogen oxidation reaction in alkaline electrolyte. J. Am. Chem. Soc. 139, 5156–5163. doi:https://doi.org/10.1021/jacs.7b00765..

Maiti, A., Srivastava, S.K., Ru, N., 2020. Codoped pellet drum bundle-like Sb2S3: an efficient hydrogen evolution reaction and hydrogen oxidation reaction electrocatalyst in alkaline medium. ACS Appl. Mater. Interfaces 12, 7057–7070. doi:https://doi.org/10.1021/acsami.9b17368.

Men, Y., Wu, D., Hu, Y., Li, L., Li, P., Jia, S., Wang, J., Cheng, G., Chen, S., Luo, W., 2023. Understanding alkaline hydrogen oxidation reaction on PdNiRuIrRh high-entropy-alloy by machine learning potential. Angew. Chem. Int. Ed. 62, e202217976. doi:https://doi.org/10.1002/anie.202217976.

Mu, X., Liu, S., Chen, L., Mu, S., 2023. Alkaline hydrogen oxidation reaction catalysts: insight into catalytic mechanisms, classification, activity regulation and challenges. Small Struct. 4, 2200281. doi:https://doi.org/10.1002/sstr.202200281.

Ni, W., Wang, T., Héroguel, F., Krammer, A., Lee, S., Yao, L., Schüler, A., Luterbacher, J.S., Yan, Y., Hu, X., 2022. An efficient nickel hydrogen oxidation catalyst for hydroxide exchange membrane fuel cells. Nat. Mater 21, 804–810. doi:https://doi.org/10.1038/s41563-022-01221-5.

Ni, W., Wang, T., Schouwink, P.A., Chuang, Y.-C., Chen, H.M., Hu, X., 2020. Efficient hydrogen oxidation catalyzed by strain-engineered nickel nanoparticles. Angew. Chem. 132, 10889–10893. doi:https://doi.org/10.1002/ange.201916314.

Ni3B/Ni heterostructure, 2020. Inter-regulated d-band centers of the Ni3B/Ni heterostructure for boosting hydrogen electrooxidation in alkaline media. Chem. Sci. 11, 12118–12123. doi:https://doi.org/10.1039/d0sc03917a

Odenweller, A., Ueckerdt, F., Nemet, G.F., Jensterle, M., Luderer, G., 2022. Probabilistic feasibility space of scaling up green hydrogen supply. Nat. Energy 7, 854–865. doi:https://doi.org/10.1038/s41560-022-01097-4.

Ohyama, J., Sato, T., Satsuma, A., 2013b. High performance of Ru nanoparticles supported on carbon for anode electrocatalyst of alkaline anion exchange membrane fuel cell. J. Power Sources 225, 311–315. doi:https://doi.org/10.1016/j.jpowsour.2012.10.051.

Ohyama, J., Sato, T., Yamamoto, Y., Arai, S., Satsuma, A., 2013a. Size specifically high activity of Ru nanoparticles for hydrogen oxidation reaction in alkaline electrolyte. J. Am. Chem. Soc. 135, 8016–8021. doi:https://doi.org/10.1021/ja4021638.

Pan, Y., Hu, G., Lu, J., Xiao, L., Zhuang, L., 2019. Ni(OH)2-Ni/C for hydrogen oxidation reaction in alkaline media. J. Energy Chem. 29, 111–115. doi:https://doi.org/10.1016/j.jechem.2018.02.011.

Parsons, R., 1958. The rate of electrolytic hydrogen evolution and the heat of adsorption of hydrogen. Trans. Faraday Soc. 54, 1053–1063. doi:https://doi.org/10.1039/TF9585401053.

Qin, K., Zhu, W., Wang, M., Wu, J., Ma, M., Chen, J., Liao, F., Kang, Z., Shao, M., 2024. Carbon dots tailoring the interfacial proton and charge transfer of iridium nanowires with stress strain for boosting bifunctional hydrogen catalysis. J. Colloid Interface Sci. 653, 434–442. doi:https://doi.org/10.1016/j.jcis.2023.09.088.

Qiu, Y., Xin, L., Li, Y., McCrum, I.T., Guo, F., Ma, T., Ren, Y., Liu, Q., Zhou, L., Gu, S., Janik, M.J., Li, W., 2018. BCC-phased PdCu alloy as a highly active electrocatalyst for hydrogen oxidation in alkaline electrolytes. J. Am. Chem. Soc. 140, 16580–16588. doi:https://doi.org/10.1021/jacs.8b08356.

Ren, J.-T., Chen, L., Wang, H.-Y., Feng, Y., Yuan, Z.-Y., 2024. Hydrogen oxidation electrocatalysts for anion-exchange membrane fuel cells: activity descriptors, stability regulation, and perspectives. Energy Environ. Sci. 17, 3960–4009. doi:https://doi.org/10.1039/D3EE04251K.

Roy, A., Talarposhti, M.R., Normile, S.J., Zenyuk, I.V., Andrade, V.D., Artyushkova, K., Serov, A., Atanassov, P., 2018. Nickel–copper supported on a carbon black hydrogen oxidation catalyst integrated into an anion-exchange membrane fuel cell. Sustain. Energy Fuels 2, 2268–2275. doi:https://doi.org/10.1039/C8SE00261D.

Seh, Z.W., Kibsgaard, J., Dickens, C.F., Chorkendorff, I., Nørskov, J.K., Jaramillo, T.F., 2017. Combining theory and experiment in electrocatalysis: insights into materials design. Science 355, eaad4998. doi:https://doi.org/10.1126/science.aad4998.

Sheng, W., Bivens, A.P., Myint, M., Zhuang, Z., Forest, R.V., Fang, Q., Chen, J.G., Yan, Y., 2014. Non-precious metal electrocatalysts with high activity for hydrogen oxidation reaction in alkaline electrolytes. Energy Environ. Sci. 7, 1719–1724. doi:https://doi.org/10.1039/C3EE43899F.

Sheng, W., Gasteiger, H.A., Shao-Horn, Y., 2010. Hydrogen oxidation and evolution reaction kinetics on platinum: acid vs alkaline electrolytes. J. Electrochem. Soc 157, B1529. doi:https://doi.org/10.1149/1.3483106.

Sheng, W., Myint, M., Chen, J.G., Yan, Y., 2013. Correlating the hydrogen evolution reaction activity in alkaline electrolytes with the hydrogen binding energy on monometallic surfaces. Energy Environ. Sci. 6, 1509–1512. doi:https://doi.org/10.1039/C3EE00045A.

Sheng, W., Zhuang, Z., Gao, M., Zheng, J., Chen, J.G., Yan, Y., 2015. Correlating hydrogen oxidation and evolution activity on platinum at different pH with measured hydrogen binding energy. Nat. Commun. 6, 5848. doi:https://doi.org/10.1038/ncomms6848.

Singh, R.K., Davydova, E.S., Douglin, J., Godoy, A.O., Tan, H., Bellini, M., Allen, B.J., Jankovic, J., Miller, H.A., Alba-Rubio, A.C., Dekel, D.R., 2020. Synthesis of CeOx-decorated Pd/C catalysts by controlled surface reactions for hydrogen oxidation in anion exchange membrane fuel cells. Adv. Funct. Mater. 30, 2002087. doi:https://doi.org/10.1002/adfm.202002087.

Song, F., Li, W., Yang, J., Han, G., Yan, T., Liu, X., Rao, Y., Liao, P., Cao, Z., Sun, Y., 2019. Interfacial sites between cobalt nitride and cobalt act as bifunctional catalysts for hydrogen electrochemistry. ACS Energy Lett. doi:https://doi.org/10.1021/acsenergylett.9b00738.

Strmcnik, D., Uchimura, M., Wang, C., Subbaraman, R., Danilovic, N., van der Vliet, D., Paulikas, A.P., Stamenkovic, V.R., Markovic, N.M., 2013. Improving the hydrogen oxidation reaction rate by promotion of hydroxyl adsorption. Nat. Chem. 5, 300–306. doi:https://doi.org/10.1038/nchem.1574.

Su, L., Cui, C., Zhou, S., Wu, H., Zhang, S., Pang, H., 2025b. Alkaline hydrogen oxidation reaction on ruthenium-based catalysts: from mechanism insights to catalyst advances. Nano Energy 140, 111051. doi:https://doi.org/10.1016/j.nanoen.2025.111051.

Su, L., Fan, X., Jin, Y., Cong, H., Luo, W., 2023. Hydroxyl-binding energy-induced kinetic gap narrowing between acidic and alkaline hydrogen oxidation reaction on intermetallic RU3SN7 catalyst. Small 19, 2207603. doi:https://doi.org/10.1002/smll.202207603.

Su, L., Gong, D., Yao, N., Li, Y., Li, Z., Luo, W., 2021. Modification of the intermediate binding energies on Ni/Ni3N heterostructure for enhanced alkaline hydrogen oxidation reaction. Adv. Funct. Mater. 31, 2106156. doi:https://doi.org/10.1002/adfm.202106156.

Su, L., Wu, H., Zhang, S., Cui, C., Zhou, S., Pang, H., 2025a. Insight into intermediate behaviors and design strategies of platinum group metal-based alkaline hydrogen oxidation catalysts. Adv. Mater. 37, 2414628. doi:https://doi.org/10.1002/adma.202414628.

Turner, J.A., 2004. Sustainable hydrogen production. Science 305, 972–974. doi:https://doi.org/10.1126/science.1103197.

Wang, G., Li, W., Huang, B., Xiao, L., Lu, J., Zhuang, L., 2019. Exploring the composition–activity relation of Ni–Cu binary alloy electrocatalysts for hydrogen oxidation reaction in alkaline media. ACS Appl. Energy Mater. 2, 3160–3165. doi:https://doi.org/10.1021/acsaem.8b02206.

Wang, M., Li, L., Wang, M., Huang, X., 2022a. Recent progress in palladium-nonmetal nanostructure development for fuel cell applications. NPG Asia Mater. 14, 1–8. doi:https://doi.org/10.1038/s41427-022-00423-2.

Wang, Q., Liu, J., Li, T., Zhang, T., Arbiol, J., Yan, S., Wang, Y., Li, H., Cabot, A., 2022b. Pd2Ga nanorods as highly active bifunctional catalysts for electrosynthesis of acetic acid coupled with hydrogen production. Chem. Eng. J. 446, 136878. doi:https://doi.org/10.1016/j.cej.2022.136878.

Wang, X., Zhao, L., Li, X., Liu, Y., Wang, Y., Yao, Q., Xie, J., Xue, Q., Yan, Z., Yuan, X., Xing, W., 2022c. Atomic-precision Pt6 nanoclusters for enhanced hydrogen electro-oxidation. Nat. Commun. 13, 1596. doi:https://doi.org/10.1038/s41467-022-29276-7.

Wang, Y.-J., Qiao, J., Baker, R., Zhang, J., 2013. Alkaline polymer electrolyte membranes for fuel cell applications. Chem. Soc. Rev. 42, 5768–5787. doi:https://doi.org/10.1039/C3CS60053J.

Wang, Y., Wang, G., Li, G., Huang, B., Pan, J., Liu, Q., Han, J., Xiao, L., Lu, J., Zhuang, L., 2014. Pt–Ru catalyzed hydrogen oxidation in alkaline media: oxophilic effect or electronic effect? Energy Environ. Sci. 8, 177–181. doi:https://doi.org/10.1039/C4EE02564D.

Wu, J., Kang, Z., Zhang, X., 2025. Recent progress in alkaline electrocatalytic hydrogen oxidation reaction: mechanisms, catalysts design and perspectives. Coord. Chem. Rev. 528, 216427. doi:https://doi.org/10.1016/j.ccr.2024.216427.

Xiao, F., Wang, Y.-C., Wu, Z.-P., Chen, G., Yang, F., Zhu, S., Siddharth, K., Kong, Z., Lu, A., Li, J.-C., Zhong, C.-J., Zhou, Z.-Y., Shao, M., 2021. Recent advances in electrocatalysts for proton exchange membrane fuel cells and alkaline membrane fuel cells. Adv. Mater. 33, 2006292. doi:https://doi.org/10.1002/adma.202006292.

Xiong, B., Zhao, W., Chen, L., Shi, J., 2019. One-step synthesis of W2C@N,P-C nanocatalysts for efficient hydrogen electrooxidation across the whole pH range. Adv. Funct. Mater. 29, 1902505. doi:https://doi.org/10.1002/adfm.201902505.

Xu, B., Huang, X., Liu, S., Hu, Z., Kao, C.-W., Chan, T.-S., Geng, H., Zhang, Y., Huang, X., 2024. Antimony oxides-protected ultrathin Ir-Sb nanowires as bifunctional hydrogen electrocatalysts. Nano Res. 17, 1042–1049. doi:https://doi.org/10.1007/s12274-023-5996-0.

Yang, F., Fu, L., Cheng, G., Chen, S., Luo, W., 2017. Ir-oriented nanocrystalline assemblies with high activity for hydrogen oxidation/evolution reactions in an alkaline electrolyte. J. Mater. Chem. A 5, 22959–22963. doi:https://doi.org/10.1039/C7TA07635E.

Yang, F., Han, P., Yao, N., Cheng, G., Chen, S., Luo, W., 2020. Inter-regulated d-band centers of the Ni3B/Ni heterostructure for boosting hydrogen electrooxidation in alkaline media. Chem. Sci. 11, 12118–12123. doi:https://doi.org/10.1039/D0SC03917A.

Yang, X., Nash, J., Oliveira, N., Yan, Y., Xu, B., 2019. Understanding the pH dependence of underpotential deposited hydrogen on platinum. Angew. Chem. 131, 17882–17887. doi:https://doi.org/10.1002/ange.201909697.

Yang, Y., Zhao, L., Sun, Y., Lin, Y., Yang, L., Mao, K., Li, C., Xu, K., 2023a. Tuning electron transfer in atomic-scale Pt-supported catalysts for the alkaline hydrogen oxidation reaction. Inorg. Chem 62, 5032–5039. doi:https://doi.org/10.1021/acs.inorgchem.3c00293.

Yang, Y., Zheng, W., Wang, P., Cheng, Z., Wang, P., Yang, J., Wang, C., Chen, J., Qu, Y., Wang, D., Chen, Q., 2024. Tailoring the hydrogen spillover effect in ni-based heterostructure catalysts for boosting the alkaline hydrogen oxidation reaction. ACS Nano 18, 24458–24468. doi:https://doi.org/10.1021/acsnano.4c07738.

Yang, Z., Lai, W., He, B., Wang, J., Yu, F., Liu, Q., Liu, M., Zhang, S., Ding, W., Lin, Z., Huang, H., 2023b. Tailoring interfacial chemistry of defective carbon-supported Ru catalyst toward efficient and CO-tolerant alkaline hydrogen oxidation reaction. Adv. Energy Mater. 13, 2300881. doi:https://doi.org/10.1002/aenm.202300881.

Yao, Z.-C., Tang, T., Jiang, Z., Wang, L., Hu, J.-S., Wan, L.-J., 2022. Electrocatalytic hydrogen oxidation in alkaline media: from mechanistic insights to catalyst design. ACS Nano 16, 5153–5183. doi:https://doi.org/10.1021/acsnano.2c00641.

Zhang, J., Ma, J., Choksi, T.S., Zhou, D., Han, S., Liao, Y.-F., Yang, H.B., Liu, D., Zeng, Z., Liu, W., Sun, X., Zhang, T., Liu, B., 2022a. Strong metal–support interaction boosts activity, selectivity, and stability in electrosynthesis of H2O2. J. Am. Chem. Soc. 144, 2255–2263. doi:https://doi.org/10.1021/jacs.1c12157.

Zhang, J., Qu, X., Shen, L., Li, G., Zhang, T., Zheng, J., Ji, L., Yan, W., Han, Y., Cheng, X., Jiang, Y., Sun, S., 2021. Engineering the near-surface of PtRu3 nanoparticles to improve hydrogen oxidation activity in alkaline electrolyte. Small 17, 2006698. doi:https://doi.org/10.1002/smll.202006698.

Zhang, Y., Liu, X., Liu, T., Ma, X., Feng, Y., Xu, B., Cai, W., Li, Y., Su, D., Shao, Q., Huang, X., 2022b. Rhombohedral Pd–Sb nanoplates with Pd-terminated surface: an efficient bifunctional fuel-cell catalyst. Adv. Mater. 34, 2202333. doi:https://doi.org/10.1002/adma.202202333.

Zhao, C.-X., Wang, H.-F., Li, B.-Q., Zhang, Q., 2021a. Multianion transition metal compounds: synthesis, regulation, and electrocatalytic applications, ACC. Mater. Res. 2, 1082–1092. doi:https://doi.org/10.1021/accountsmr.1c00136.

Zhao, G., Chen, J., Sun, W., Pan, H., 2021b. Non-platinum group metal electrocatalysts toward efficient hydrogen oxidation reaction. Adv. Funct. Mater. 31, 2010633. doi:https://doi.org/10.1002/adfm.202010633.

Zhao, R., Yue, X., Li, Q., Fu, G., Lee, J.-M., Huang, S., 2021c. Recent advances in electrocatalysts for alkaline hydrogen oxidation reaction. Small 17, 2100391. doi:https://doi.org/10.1002/smll.202100391.

Zhao, X., Li, X., An, L., Iputera, K., Zhu, J., Gao, P., Liu, R.-S., Peng, Z., Yang, J., Wang, D., 2022. Nitrogen-inserted nickel nanosheets with controlled orbital hybridization and strain fields for boosted hydrogen oxidation in alkaline electrolytes. Energy Environ. Sci. 15, 1234–1242. doi:https://doi.org/10.1039/D1EE03482K.

Zheng, J., Sheng, W., Zhuang, Z., Xu, B., Yan, Y., 2016a. Universal dependence of hydrogen oxidation and evolution reaction activity of platinum-group metals on pH and hydrogen binding energy. Sci. Adv. 2, e1501602. doi:https://doi.org/10.1126/sciadv.1501602.

Zheng, J., Zhou, S., Gu, S., Xu, B., Yan, Y., 2016b. Size-dependent hydrogen oxidation and evolution activities on supported palladium nanoparticles in acid and base. J. Electrochem. Soc 163, F499. doi:https://doi.org/10.1149/2.0661606jes.

Zheng, J., Zhuang, Z., Xu, B., Yan, Y., 2015. Correlating hydrogen oxidation/evolution reaction activity with the minority weak hydrogen-binding sites on Ir/C catalysts. ACS Catal. 5, 4449–4455. doi:https://doi.org/10.1021/acscatal.5b00247.

Zhang, X., Xia, L., Zhao, G., Zhang, B., Chen, Y., Chen, J., Gao, M., Jiang, Y., Liu, Y., Pan, H., Sun, W., 2023. Fast and durable alkaline hydrogen oxidation reaction at the electron-deficient ruthenium–ruthenium oxide interface. Adv. Mater. 35, 2208821. doi:https://doi.org/10.1002/adma.202208821.

Zhuang, Z., Giles, S.A., Zheng, J., Jenness, G.R., Caratzoulas, S., Vlachos, D.G., Yan, Y., 2016. Nickel supported on nitrogen-doped carbon nanotubes as hydrogen oxidation reaction catalyst in alkaline electrolyte. Nat. Commun. 7, 10141. doi:https://doi.org/10.1038/ncomms10141.

Chapter 4

Anion exchange membranes for alkaline fuel cells: Synthesis, properties and perspectives

Anastasiia Konovalova[a], David Aili[a], and Dirk Henkensmeier[b,c,d]
[a]*Department of Energy Conversion and Storage, Technical University of Denmark, Kongens Lyngby, Denmark*
[b]*KIST School, Energy and Environment Technology, University of Science and Technology, Seoul, South Korea*
[c]*Center for Hydrogen and Fuel Cells, Korea Institute of Science and Technology, Seoul, South Korea*
[d]*Graduate School of Energy and Environment, Korea University, Seoul, South Korea*

1 Introduction to anion-exchange membranes

In the 1930s, Francis Bacon began exploring fuel cell technology and, in 1939, presented a working alkaline fuel cell (AFC), which employed asbestos cloth soaked with a 27% potassium hydroxide (KOH) solution to separate the anode and cathode. A major challenge was that the KOH solution needed to be retained in the separator to maintain conductivity and prevent the mixing of gases. Another challenge was that carbon dioxide from the air was absorbed by KOH electrolytes, precipitating as potassium bicarbonate in the pores of the electrodes and separator. After further development, which included the transition from porous diaphragms to ion-exchange membranes, proton exchange membrane fuel cells were used as an auxiliary power source in the Gemini space flights (Bacon, 1969). The development of ion-exchange membranes was a game changer, not only for fuel cells. In 1962, Walther Kangro published a paper on flow batteries, in which the separator between the anode and cathode was a plexiglass disc with 60 drilled holes per cm^2 (Kangro and Pieper, 1962). This design led to significant electrolyte crossover. The first flow batteries using a membrane were patented in 1976 (Thaller, 1976), and many consider this the birth of modern flow batteries. So, what is the difference between diaphragms and ion-exchange membranes? While diaphragms have an open porous structure and conduct potassium cations and hydroxide ions by absorbing KOH electrolyte into the pores, anion-exchange membranes (AEM) are nonporous

Alkaline Fuel Cells. https://doi.org/10.1016/B978-0-443-36638-3.00005-X

materials with the potential to selectively conduct hydroxide ions. However, at high salt concentrations, this selectivity breaks down.

Chemically, AEMs are made of polymers that are functionalized with cationic groups. To maintain charge neutrality, each cationic group possesses an anionic counter ion. While the cationic groups are immobilized on the polymer, the anions become mobile when the ionic groups dissociate upon hydration. They are only loosely associated with the cationic groups and transport charge across the membrane when an electric field is applied. The most common type of cationic group is the quaternary ammonium group, such as the tetraalkyl/aryl ammonium group. Other examples include sulfonium or phosphonium, but ammonium-based groups can be synthesized more easily and often exhibit higher alkaline stability.

The cationic groups are usually attached to the polymer backbone or a side chain (Fig. 4.1A), but they can also be part of the polymer backbone itself (Fig. 4.1B). The latter class of polymers is called ionenes. Although it might be assumed that the position of the charge is not particularly important, it has a significant impact on the membranes' morphology. Let's first consider a polymer that has a hydrophobic backbone and side chains with hydrophilic cationic groups. Especially if the polymer backbone is flexible (i.e., mainly C–C single bonds), the hydrophilic and hydrophobic segments of the polymer separate from each other and agglomerate into hydrophilic and hydrophobic domains. In contact with humidity or liquid water, these hydrophilic regions absorb water molecules and swell into 3–9 nm large domains (Luo et al., 2020; Wang et al., 2020; Dang and Jannasch, 2016). Although the phase separation is less pronounced in most AEMs (lower intensity in SAXS), the dimensions are similar to the phase-separated morphology observed in perfluorosulfonic acid membranes (3–6 nm, (Luo et al., 2020) and sulfonated hydrocarbon membranes (2–3 nm, (Jutemar et al., 2011). However, note that there is significant structural variation between different membranes, so varying morphologies are to be expected. If the density of hydrophobic groups is high enough, these domains connect with each other and form a percolating water-filled network through which the mobile counterions can move freely in all directions. Ionenes, which have the positive

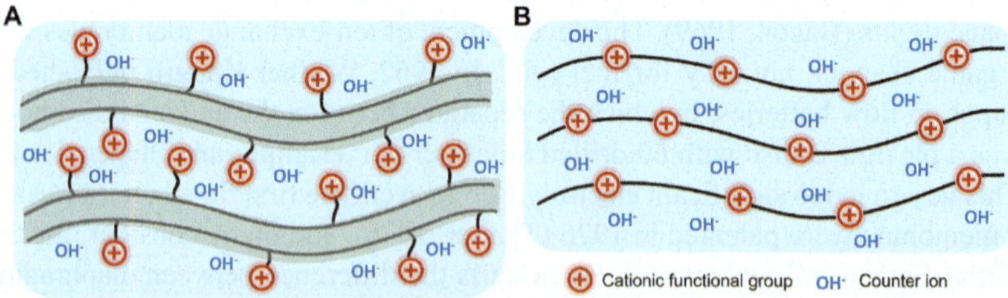

FIG. 4.1 Structure of cationic functional group containing anion exchange polymers. (A) side-change functionalized polymers with pendant cationic groups and (B) ionenes which have the cationic groups within the backbone.

charge on the backbone, lack distinct hydrophilic and hydrophobic segments and therefore struggle to form clearly separated hydrophilic and hydrophobic domains (Fig. 4.1B). Water absorbed by the hydrophilic groups fills the polymer free volume and may also increase the free volume by pushing neighboring polymer chains away from each other. However, the average distance between the polymer chains is expected to remain in the range of ≈ 0.5–2 nm (Schibli et al., 2018; Mara Ikhsan et al., 2022).

1.1 Function of the membrane and its requirements

In the AFC, the membrane electronically separates the anode and cathode but offers an ion-conductive pathway for hydroxide ions. It also prevents the anode and cathode gases from mixing. However, although AEMs are "dense" membranes, polymers have a finite permeability for oxygen and especially hydrogen, which is the smallest of all molecules. In fact, the polymer free volume and the hydrophilic domains offer potential pathways for the diffusion of gas molecules dissolved in the aqueous phase within the membrane. AEMs can be considered nanofiltration membranes, which means that they can transport particles or molecules with a diameter $<$ 2 nm ("Terminology for membranes and membrane processes (IUPAC Recommendation 1996)," Koros et al., 1996). This diffusive crossover of hydrogen and oxygen is so slow that it normally poses no safety risk because the crossed-over gases are diluted by the applied gas stream and oxidized by the platinum-based catalysts, which are often used as they are the most reactive ones, before explosive gas mixtures can build up. However, gas crossover is a main reason why the open cell voltage of AFCs is lower than the thermodynamic value of 1.23 V.

In addition to the explained functions, AEMs need to fulfill several other requirements. They must be thermally and mechanically stable at the fuel cell operation temperature of up to 80°C and withstand the chemical environment, which is highly alkaline, oxidative, and humid (for example, aliphatic polyesters would hydrolyze in the fuel cell). The water uptake needs to be in an optimal range: Since water uptake swells the hydrophilic domains to a point where they form a percolating network, conductivity generally increases with the water uptake. At some point, however, the hydrophilic regions become well swollen and fully connected, and further water uptake starts to have an adverse effect because it reduces the concentration of hydroxide ions and thus decreases the conductivity. A related point is the dimensional stability: Fuel cells produce water, and changes in the current density, temperature, humidification level of fuel and oxidant, shut-down, and start-up of cells alter the local water content. This swells and shrinks the membranes, and local differences in water uptake can result in serious stresses. One example is membrane fracture at the edge of the active area (Krishnan et al., 2014; Krishnan et al., 2016). During operation, the active membrane area between the electrodes is humidified, and over time, water diffuses into the inactive area of the membrane, which helps to seal the cell. When cell operation is stopped and the cell is flushed with dry nitrogen to avoid the buildup of explosive hydrogen/oxygen mixtures, the membrane area between the electrodes rapidly shrinks, whereas the outer areas remain wet. Since water plasticizes AEMs, the wet outer area is softer than the shrinking dry inner area, and the edge area becomes perforated and breaks. Therefore, the AEMs should exhibit small dimensional changes between dry and wet states. Furthermore, membranes need to demonstrate excellent mechanical properties (i.e., high tensile strength and high elongation at break) to allow robust handling and counter the stresses during humidity changes.

1.2 Alkaline stability

While the AFC demonstrated by Francis Bacon in 1939 used a 27 wt% KOH electrolyte at 100°C, the system used in NASA's Apollo program even operated with a 70–85 wt% KOH electrolyte at 204°C. Both are highly alkaline conditions (Bacon, 1969). Since modern AFCs substitute the porous diaphragms with "dense" AEM, they don't operate with KOH-based liquid electrolytes but are only fed with humidified gas streams. To differentiate the two systems, many people refer to modern AFCs as Alkaline AEM Fuel Cells (AAEMFC) (Varcoe and Slade, 2005). The absence of KOH electrolyte alleviates material challenges for other cell components like tubing, sealings, pumps, etc., but the mobile hydroxide ions in the AEM still provide a locally high pH inside the polymer, which is a major challenge for the long-term durability of AEMs. For the commercially available FAA-3-50 membrane, the hydroxide ion concentration within the aqueous phase corresponds to around 0.7 mol/L under fully hydrated conditions (Aili et al., 2023). At lower degrees of hydration, the hydroxide ion concentration increases.

The first generation of AEMs developed for AAEMFC applications was based on quaternary ammonium-functionalized polystyrene, which is a rather brittle material but can be stabilized by radiation grafting, a process in which a polymer film made of low-density polyethylene, poly(ethylene-tetrafluoroethylene), fluorinated poly(ethylene propylene), poly(tetrafluoroethylene) (PTFE), etc., is irradiated by gamma rays or an electron beam to introduce radicals. When vinylbenzyl chloride diffuses into the polymer film, the radical sites serve as the starting point for growing poly(vinylbenzyl chloride) side chains (Varcoe and Slade, 2005). Finally, the benzyl chloride groups are reacted with an amine to quaternary ammonium groups. Since the grafting process is kinetically controlled, it is strongly affected by the process parameters, as well as the concentration and nature of radical scavengers present in commercial vinyl monomers. A more versatile and well-controllable approach is the use of poly(arylene ether sulfone)s (Yan and Hickner, 2010). They provide several sites and reaction pathways for functionalization and yield strong materials with high organosolubility and, therefore, beneficial processability. Unfortunately, poly(arylene ether sulfones) have ether groups next to a sulfone group, and the electron-withdrawing nature of the sulfone linkages makes the arylene ether linkages susceptible to nucleophilic attack by hydroxide ions (Arges and Ramani, 2013). This leads to rapid chain scission and severe backbone degradation under technologically relevant conditions, as illustrated in Fig. 4.2 (top). The focus has therefore shifted to backbone chemistries devoid of polarizable bonds in the backbone, with heteroatom-free structures based on all-carbon architectures (Park et al., 2024).

The mechanisms for the main degradation modes of the cationic groups are shown in Fig. 4.2. For the functional cationic groups, a main reaction between quaternary ammonium groups and hydroxide ions is the nucleophilic substitution reaction, in which hydroxide ions react with an *N*-methylene or *N*-methyl

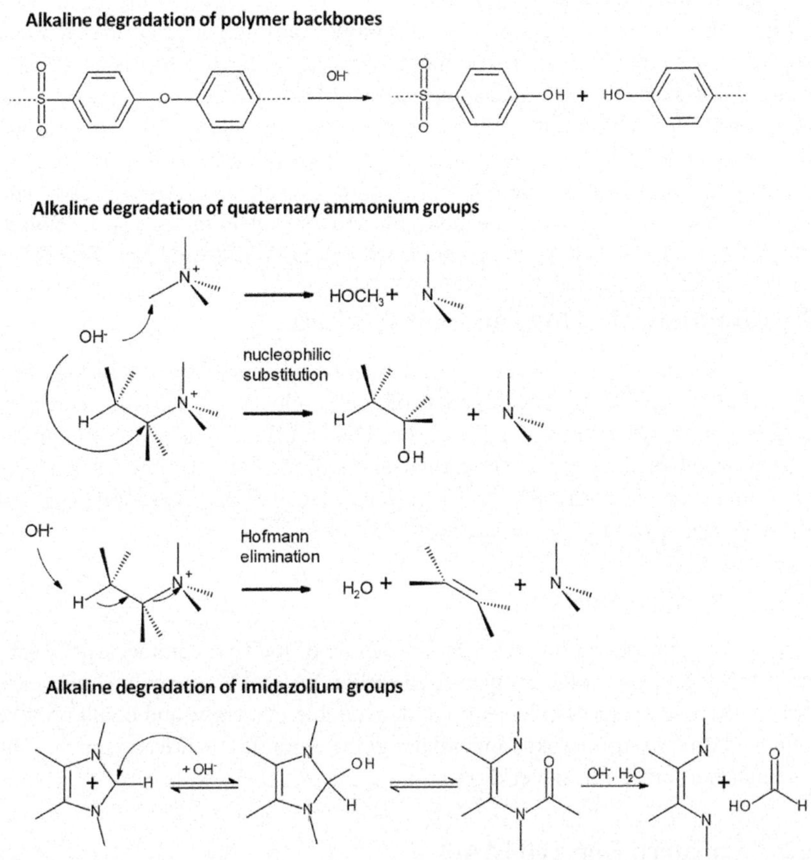

FIG. 4.2 Alkaline degradation of AEMs by nucleophilic substitution reaction, Hofmann elimination, hydrolysis of imidazolium ions, and hydrolysis of aromatic ether groups. CC-BY 4.0, American Society of Mechanical Engineers. *(From Henkensmeier et al., 2021. Overview: State-of-the Art Commercial Membranes for Anion Exchange Membrane Water Electrolysis. J. Electrochem. Energy Conv. Storage, 18(2). https://doi.org/10.1115/1.4047963)*

group, resulting in the formation of an alcohol and an amine. The direct consequence is the loss of mobile hydroxide ions and reduced ion exchange capacity (IEC). This also means that the polarity of the membrane decreases, and therefore the water content decreases. The result is that the conductivity of AEM decreases over time. If the quaternary ammonium group contains N-CH_2-CH_2- groups, Hofmann elimination is also possible, with the same consequence on conductivity as described above. For imidazolium ions, in addition to the substitution reaction on the methyl groups, a ring-opening reaction was found, initiated by the attack of hydroxide ions on the C2 position.

The low alkaline stability of quaternary ammonium groups has been a major challenge, and systematic investigations into the stability of different functional groups have revealed that quaternary ammonium groups tethered to a C4 to C6 alkyl chain have higher stability than ammonium groups tethered to ethyl groups, due to steric hindrance and rotational barriers (Marino and Kreuer, 2015). Hofmann elimination can be effectively mitigated by using cations devoid *of β*-hydrogens (Henkensmeier et al., 2021). Another strategy is the use of heterocyclic ammonium ions. The bulky structures slow down substitution reactions, and the fixed conformation of the rings prevents Hofmann elimination, which requires an antiperiplanar orientation of the involved C–H bond and the C–N bond. Examples include piperidinium ions, quaternized 1,4-diazabicyclo[2.2.2]octane (DABCO), and similar structures (Marino and Kreuer, 2015).

2 Commercial AEMs and their synthesis

There is a wide commercial market for membranes used in water purification technologies, while the development of ion-conducting materials, especially AEMs, for energy-related technologies has taken off relatively recently. In this part of the chapter, we will cover known anion-conducting membranes that have been investigated in AEMFC or AEM Water Electrolysis (AEMWE) applications, and their chemistry.

2.1 Tokuyama A201

The A201 membrane from Tokuyama was one of the first commercially available AEMs and was extensively used in earlier AAEMFC research. However, it was never widely accessible, as it is not available currently, and could only be obtained directly from Tokuyama after signing a nondisclosure agreement. The chemical structure was never disclosed.

2.2 Fumatech Fumasep FAA-3

Similar to A201, there is no literature that reliably discloses the chemistry of Fumatech's Fumasep FAA-3. However, a recent work by Giovanelli et al. reports the analysis of an FAA-3 membrane by solid-state NMR, and a structure is suggested (Fig. 4.3) (Giovanelli et al., 2024). The most striking feature is the arylene

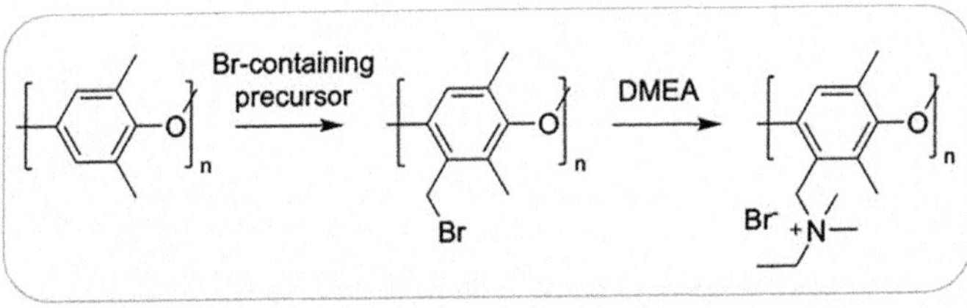

FIG. 4.3 Potential structure of Fumatech's Fumasep FAA-3, as suggested by Giovanelli et al., and a possible synthesis pathway by bromomethylation of PPO followed by amination with DMEA.

ether group in the backbone, which is one of the known breaking points of AEM and indicates that the membrane will not have long durability in highly alkaline systems. It was also reported (Khalid et al., 2022) that FAA-3 membranes have relatively large swelling in water, which is disadvantageous for use in AAEMFC, where humidity fluctuates. However, FAA-3 membranes are widely available at a reasonable cost and have been used by many researchers, so data for comparison with other work can easily be found. Synthetically, bromomethylation might be easily possible by the reaction of poly(phenyleneoxide) with NaBr and sulfuric acid to produce HBr, and paraformaldehyde (Kubiczek and Neugebauer, 1950).

2.3 Aemion

Aemion membranes were developed in the group of Steven Holdcroft (Wright and Holdcroft, 2014; Wright et al., 2016) and are produced by the Canadian company Ionomr Innovations. The original developments suggest that the membranes are made of polybenzimidazole, which is methylated to change the uncharged imidazole groups into dimethylbenzimidazolium hydroxide units. While initial work showed that methylated *meta*-polybenzimidazole rapidly degrades due to the attack of hydroxide ions on the C2 position - the carbon atom between the two nitrogen atoms (Henkensmeier et al., 2011) - Holdcroft and coworkers demonstrated that the introduction of methyl substituents in the ortho position to the phenyl-imidazole link effectively blocks access of hydroxide ions to the C2 position through steric hindrance (Thomas et al., 2012). In another study, the separation of the two positively charged imidazolium units was shown to enhance alkaline stability by reducing the positive charge density (Henkensmeier et al., 2012). The combination of spatial separation of the imidazolium ions and steric protection resulted in Aemion (Fig. 4.4), the first AEM with significantly improved alkaline stability. It can be validly stated that this was a breakthrough for AEM technology.

2.4 Aemion+

AEMION + is a next generation of the AEMION membrane that is currently available on the market, offering significantly improved alkaline stability (Fan

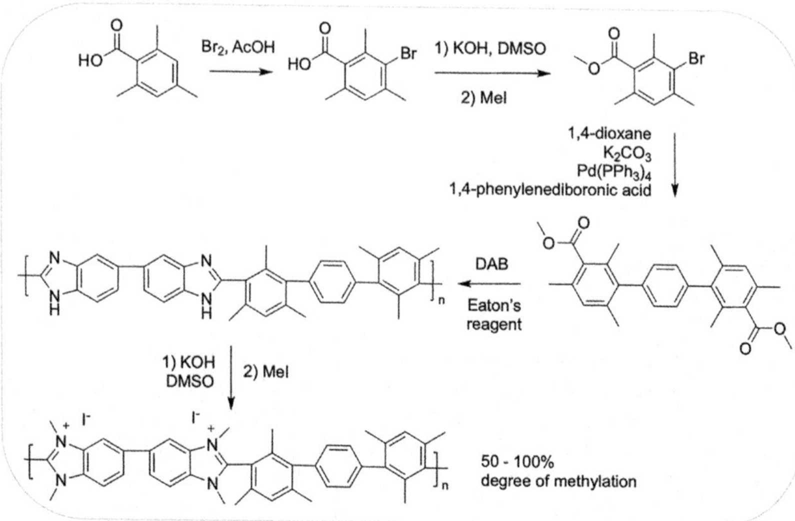

FIG. 4.4 Synthesis of HMT-PDMBI ("Aemion").

et al., 2019). Various grades and reinforced options are available. Experiments with model compounds revealed that the half-lives of methylated 2-phenyl benzimidazole, sterically protected methylated 2-(2,4,6-trimethylphenyl) benzimidazole, and sterically protected methylated 2-(2,4,6-trimethylphenyl)-4,5-diphenylimidazole in 3 M $NaOD/D_2O/CD_3OD$ at 80°C are < 0.1 h, 436 h, and 7790 h, respectively (Fan et al., 2019). Note that the high stability of diphenylimidazolium is also utilized in Sustainion membranes, which are functionalized with 2,3,4,5-trimethylimidazolium units (Kaczur et al., 2018). The producer of Aemion + explicitly recommends the use of Aemion + for AAEMFC, and 8900 h of stable operation in a water electrolyzer has been proven (Moreno-González et al., 2023). A potentially weak point is the multistep synthesis pathway (Fig. 4.5), which might increase the cost of production. However, technically, Aemion + appears to be one of the most promising membranes for use in AAEMFC.

2.5 Sustainion

Sustainion membranes are commercialized by Dioxide Materials and are available in many different grades. Although 12,000-hour operation in an electrolyzer has been reported (Motealleh et al., 2021) none of the available Sustainion membrane types is recommended for use in fuel cells. The reason is their chemical composition. Basically, Sustainion membranes are synthesized by copolymerization of styrene and 4-vinylbenzyl chloride, followed by functionalization with tetramethylimidazolium groups (Fig. 4.6) (Kaczur et al., 2018). Unfortunately, polystyrenic membranes have a tendency to be brittle,

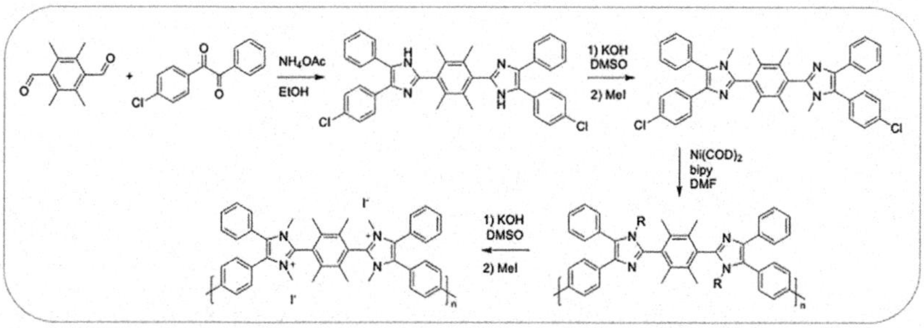

FIG. 4.5 Synthesis of Aemion+.

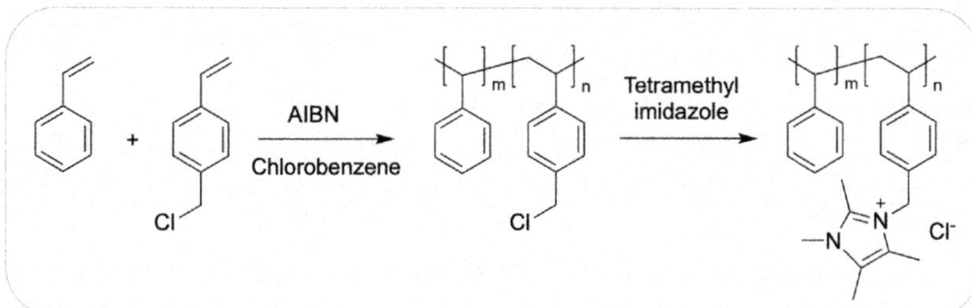

FIG. 4.6 Synthesis of Sustainion.

and Sustainion easily cracks in the dry state, which makes them unattractive for use in fuel cells. In the electrolyzer, membranes are plasticized by water, and pristine membranes contain a water-soluble plasticizer.

2.6 Orion polymer TM1

Orion Polymer's TM1 membrane was developed by Chulsung Bae and his group (Lee et al., 2017) and was reported to be the most stable membrane in a study by NREL, which compared over 50 different membranes (Meek et al., 2019). The TM1 membrane (mTPN in Lee et al., 2017, Fig. 4.7) was designed to be alkaline-stable by avoiding heteroatoms like ether groups in the polymer backbone and tethering the quaternary ammonium group via a pentyl chain to the backbone, which potentially allows for some degree of phase separation and suppresses Hofmann elimination, which would be strong for ethyl chains (Marino and Kreuer, 2015). While TM1 is not fluorine-free, and the synthesis requires a large excess of fluorinated superacids and cosolvents, it is expected that a ban on fluorinated substances in the European Union and the US would exclude materials that have only a CF_3 group, presumably because these materials will not have the high persistence in the environment as fully fluorinated materials. However, the cost of the monomer 7-bromo-1,1,1-trifluoro-2-heptanone appears to be high and might influence the membrane cost.

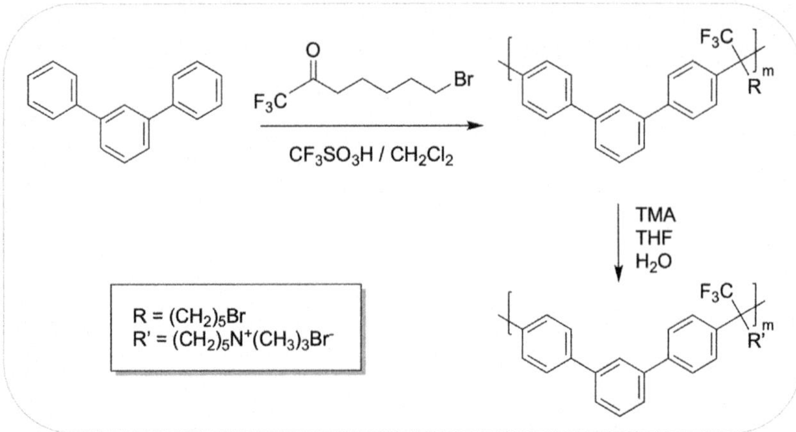

FIG. 4.7 Synthesis of Orion Polymer's TM1 (mTPN).

2.7 PiperION

Building further upon the pioneering work on poly(arylene piperidinium) in the group of Patric Jannasch (Olsson et al., 2018), PiperION was initially developed by Yushan Yan and his group at the University of Delaware and is now commercialized by Versogen (Wang et al., 2019). To increase the mechanical stability of thin membranes suitable for use in fuel cells, ePTFE-reinforced 15 μm thick PiperION membranes are also available and are specifically recommended for use in AAEMFC. Similar to TM1, the isolated trifluoromethyl groups are currently not expected to affect the commercialization of PiperION when per- and polyfluoroalkyl substances are banned, because CF_3 groups appear to be exempted. Synthetically, PiperION is made in a superacid-catalyzed polyhydroxyalkylation using a large excess of trifluoromethanesulfonic acid (TFSA) and trifluoroacetic acid as cosolvent, similar to TM1 (Fig. 4.8). Although the monomer trifluoroacetophenone is not very cheap, it is only used in a small ratio with piperidinone (ca. 15:85). Other monomers are reasonably cheap, and the cost for the superacid catalyst TFSA seems to be the main cost driver and imposes work environment risks and environmental concerns. Water uptake at room temperature is $<50\%$, and swelling is in the range of 7%, which is promising for use in fuel cells.

2.8 Hollex

Hollex membrane is commercialized by TailorMem. It is a triblock copolymer of styrene, ethylene/butylene, and styrene (SEBS), which is chloromethylated and functionalized by DABCO (Fig. 4.9) (Hnát et al., 2017). Although polystyrene is a rather brittle polymer, the flexible ethylene/butadiene block completely changes the properties and gives SEBS a ductile elastomeric behavior.

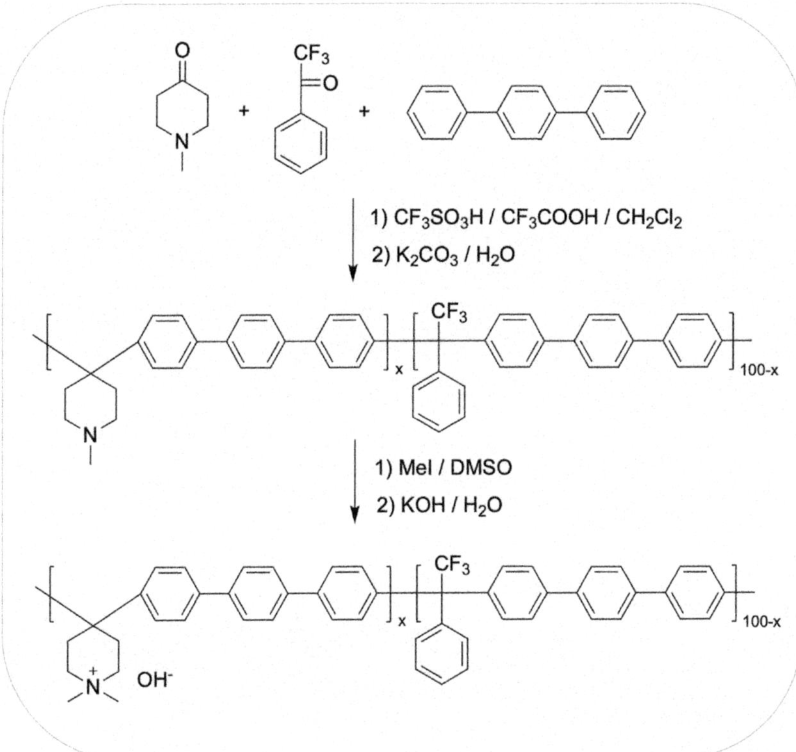

FIG. 4.8 Structure and synthesis of Piperion.

Because the DABCO group has a rigid cage in which C–N and β-C–H bonds are not in an antiperiplanar orientation, Hofmann elimination is suppressed, and the all-carbon backbone does not offer points for hydrolytic chain scission under alkaline conditions. Furthermore, the starting materials needed to prepare Hollex membranes are widely available and relatively cheap, so the upscaling of membrane production is not expected to be limited by the starting materials. A potentially weak point for use in fuel cells is that SEBS-based polymers often swell strongly. However, it appears that quaternary ammonium-based SEBS absorbs less water than sulfonated SEBS (Žitka et al., 2019; Wang et al., 2016). Data for the use of Hollex membranes in fuel cells are not yet available in the scientific literature.

2.9 Duraion

Duraion is an AEM developed by Evonik Industries and is apparently ready to be produced roll-to-roll, at least on a demonstration scale (Evonik, 2025). However, only limited information is available on Duraion, and its structure has not been disclosed yet.

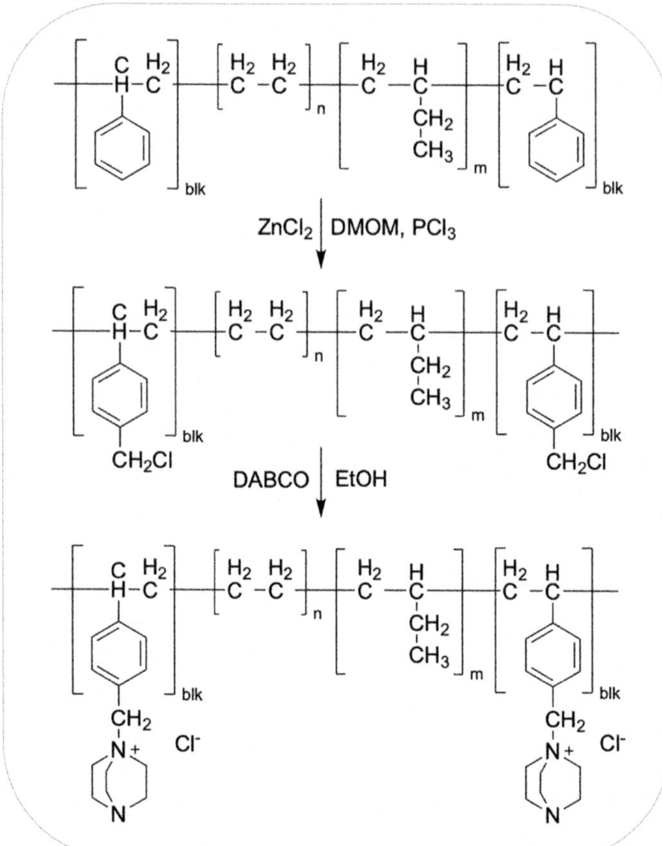

FIG. 4.9 Synthesis of Hollex membranes (DABCO-functionalized SEBS).

2.10 Branion

Branion was initially developed in the group of Xile Hu at EPFL (Wu et al., 2022) and is commercialized by NovaMEA. Structurally, it is very similar to PiperION, but with the difference that no fluorinated monomers are used. Furthermore, the polymer structure is not linear but branched by copolymerizing piperidinone and terphenyl with 1,3,5-triphenylbenzene (Fig. 4.10). Consequently, the IEC is in the range of 2.8, but <2.4 in the case of PiperION, and the swelling ratio at 40°C is $<20\%$. Still, 500 hours of stable operation in the fuel cell have been reported (Wu et al., 2022). Interestingly, even though Branion is said to be based on the polymer shown in (Wu et al., 2022) (Zheng et al., 2023), the producer's specification gives a swelling ratio of $<10\%$ at 80°C. It appears that the exact membrane formulation or fabrication process is not disclosed yet, which underscores how small adjustments can have a large impact.

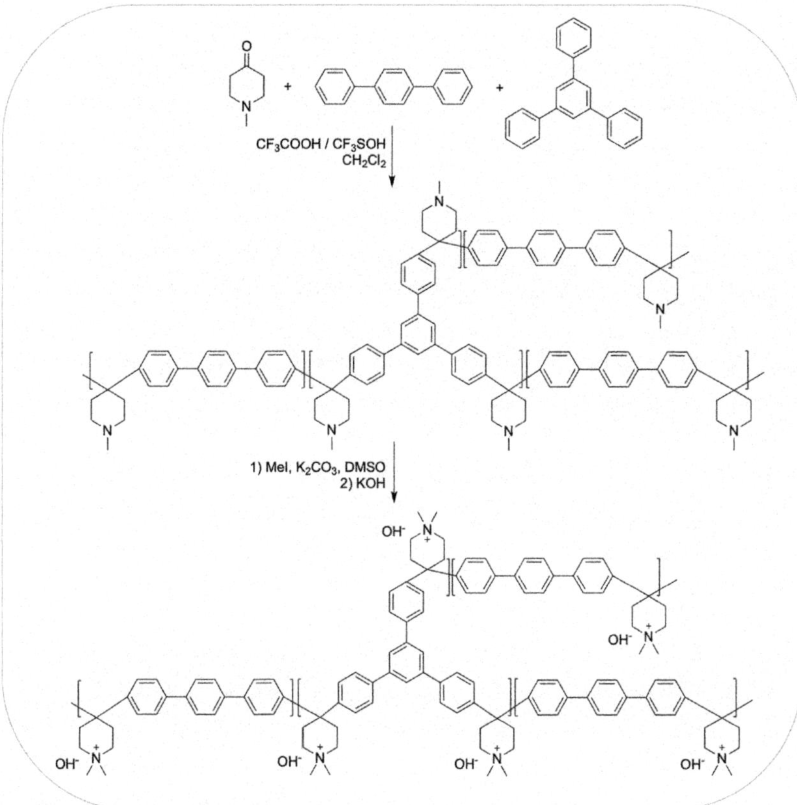

FIG. 4.10 Synthesis of Branion from NovaMEA.

3 Properties of AEMs

The investigated properties of the materials listed above are summarized in Table 4.1. It is important to note that a standardized protocol for AEM characterization does not exist, and the conditions under which measurements were performed are specified next to the values. Each highlighted intrinsic property is briefly described below.

By definition, IEC describes the amount of charged functional groups per gram of dry membrane and is measured in meq (mmol) g^{-1}. IEC can be calculated based on NMR data of a functionalized polymer, and measured experimentally via acid-base or Mohr titrations (OH^- or Cl^- detection) or by potentiometry with ion-selective electrodes. Ionic conductivity of the membrane is typically calculated based on the specific sample dimensions and the resistance measured in a 2-electrode or 4-electrode fixture connected to the potentiostat, and experessed in mS cm^{-1}. It elucidates the ability of the polymer to conduct a specific ion (Wang et al., 2022; Fisher and Kunin, 1955). In the case of AEMs, it could

TABLE 4.1 Comparison of commercial membrane properties; data is from specification sheets or literature.

Membrane	Producer	Type	Thickness (dry, μm)	IEC (mmol g^{-1})	*Conductivity* (mS cm^{-1})	Swelling (*dimensional* change)	References
A201	Tokuyama (Japan)	–	28	1.7	OH: 42 (23°C 90% RH) OH: 93 (80°C in water) HCO_3: 7.2 (80°C 98% RH)	4 % (in-plane, n/a) 50°C 90% RH, Cl: 7% (in-plane) 80°C 90% RH, Cl: 9% (in-plane)	(Yanagi and Fukuta, 2008; Duan et al., 2013; Ouadah et al., 2017)
FAA-3-50	Fumatech (Germany)		45-55	1.6–2.1	Cl: 3–8 (25°C in 0.5 M NaCl)	Br: 2% (in-plane, n/a)	(Fumatech, 2025)
Aemion	Ionomr Innovations (Canada)	AF1-HNN8-50	50	2.1–2.5	OH: 102 (50°C 90% RH)	8% (in-plane, n/a) 25% (through-plane, n/a)	(Fortin et al., 2020; Ionomr, 2025a)
Aemion+	Ionomr Innovations (Canada)	AH2-INN8-50-	50	2.3–2.6	Cl: 18 (70°C 95% RH)	25°C, Cl: 57%	(Wei et al., 2022; Ionomr, 2025b)

Sustainion	Dioxide Materials (USA)	X-37	50	1-1.4	OH: 140 (80°C in 1 M KOH)	25°C, Cl: 15% (in-plane)(Rodriguez et al., 2023)	(Motealleh et al., 2021)
TM1	Orion Polymers (USA)		35-40	2.2	OH: 144 (80°C)	80°C, OH: 23%	(Orion Polymer, 2025; Lee et al., 2017)
PiperION	Versogen (USA)		20, 40, 60, 80	2.35	OH: 150 (80°C)	95°C, OH: 11%	(Versogen, 2025; Wang et al., 2019)
Hollex	Tailormem (Czech Republic)		50-200	>0.94	OH: 75 (30°C)	20–50% range (through-plane, n/a)	(TailorMem, 2025; Hnát et al., 2017)
Duraion	Evonik (Germany)		60	–	OH: 108	–	(Evonik, 2025)
Branion	NovaMEA (Switzerland)		30, 50, 80	2.7–2.9	OH: >80 (40°C) OH: >150 (80°C)	80°C,OH: <33%	(NovaMea, 2025; Wu et al., 2022)

be halogen-ions, hydroxide, carbonate or bicarbonate ions. The swelling ratio of the membrane indicates the expansion of the material under the exposure to water or electrolyte and could be characterized volumetrically, in-plane or through-plane (Díaz and Kamcev, 2021).

4 Specific technical challenges of AEMs for alkaline AEM fuel cells

4.1 Effect of CO_2

For quite some time, it was considered challenging to measure the conductivity of hydroxide-exchanged membranes because the absorption of CO_2 from air changes all hydroxide ions into (bi)carbonates within just a few minutes (Yanagi and Fukuta, 2008). Some researchers measured through-plane conductivity in KOH solutions, but the water uptake of membranes changes in contact with solutions that have a high ionic strength. The absorbed electrolyte also contributes to the conductivity, which alters both conductivity and selectivity. These problems were solved by Ziv and Dekel, who developed a method to determine the "true hydroxide conductivity" (Ziv and Dekel, 2018). In this method (see Fig. 4.11), a membrane strip in an undefined mixed hydroxide/(bi)carbonate form is assembled into an in-plane conductivity cell, in which the membrane is contacted with four electrodes. The cell is assembled into a temperature and humidity-controlled nitrogen atmosphere (for example, in the Scribner MTS740 membrane test station) or immersed in nitrogen-purged water (Khalid et al., 2022) and a small current is applied over the outer electrodes. At the cathode, water is reduced into hydrogen and hydroxide ions, which then move toward the anode. The hydroxide and (bi) carbonate ions within the membrane migrate toward the anode to form CO_2 and oxygen, which are carried away by, e.g., the applied nitrogen stream. Over time, all (bi)carbonates are electrochemically purged from the membrane, and a 100% pure hydroxide form AEM is obtained. Since hydroxide ions absorb more water into the membrane than carbonates and have a higher molar conductivity, the conductivity increases until all (bi)carbonate ions are removed from the membrane.

The strong influence of CO_2 on membrane conductivity means that fuel cells, which have CO_2 in the air or oxygen feed, experience reduced cell performance. This was initially a problem, but with the development of the field, the performance of AAEMFCs has constantly improved, and the achievable current density at 400 mV increased from approximately 300 mA cm^{-2} in 2007 to 3000 mA cm^{-2} in 2017 (Dekel, 2018). Mechanistically, it appears that CO_2 is mainly absorbed from the air stream into the alkaline membrane. Therefore, while the absorption rate of CO_2 is not significantly affected by the current, the amount of produced hydroxide ions directly scales with the current. In other words, a tenfold increase in current density should decrease the (bi)carbonate concentration in the membrane by a factor of 10 (ignoring other effects like

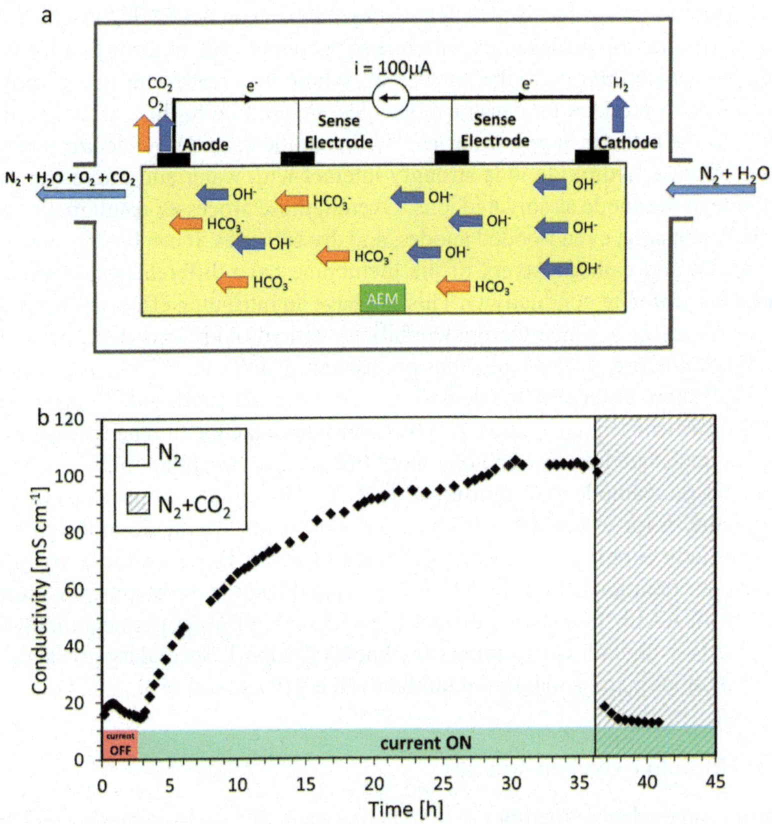

FIG. 4.11 The Ziv-Dekel method to measure the "true hydroxide conductivity". (A) Electrochemical purging of AEMs to obtain perfectly (bi)carbonate-free AEMs; (B) development of the conductivity during purging. Reproduced with permission from Elsevier. *(From Ziv, N., Dekel, D.R., 2018. A practical method for measuring the true hydroxide conductivity of anion exchange membranes. Electrochem. Commun. 88, 109–113. https://doi.org/10.1016/j.elecom.2018.01.021)*

flow rate), and the effect of CO_2 on state-of-the-art fuel cells operating at high current density appears to be negligible.

4.2 Water management and alkaline stability

The Ziv-Dekel method for testing the true hydroxide conductivity was also applied to monitor the alkaline stability of AEMs for fuel cells (Haj-Bsoul et al., 2022). In detail, AEM samples are installed into the Scribner MTS740 test station, converted into the hydroxide form, and then kept at a certain temperature and humidity for prolonged periods. Since the hydroxide ions are less solvated under dry conditions, their nucleophilicity increases, and degradation accelerates, as proven by decreasing conductivity over time.

It is important to recall the chemical reactions in an AAEMFC. At the cathode, two water molecules react with one oxygen molecule to form four hydroxide ions, which migrate to the anode side, where they react with two hydrogen molecules to produce four water molecules. In addition to the consumption of water at the cathode, some water is also lost at the cathode by electro-osmotic drag, because hydroxide ions strongly interact with water and move from the cathode to the anode as solvated ions. Overall, these processes result in wet, and in the worst case, even flooded anodes, and dry cathodes. It can be expected that the anode and cathode layers of the membrane have different water contents and, thus, different conductivity. This can have an intriguing effect on the stability of AAEMFCs. Since the nucleophilicity of hydroxide ions depends on the lambda value (i.e., water molecules per hydroxide ion), the AEM's dry cathode side will show faster alkaline degradation than the wet anode side (Yassin et al., 2024). As a consequence, there can be cases where the membrane appears to be almost unchanged by fuel cell operation but, in fact, has a thin, highly resistant layer on the cathode side. Fortunately, if the membrane is thin and has high permeability for water, back-diffusion of water from the anode to the cathode can partially compensate for the movement of water. Based on these and some other considerations, Yassin et al. recommended that the development of future AEMs for AAEMFCs should primarily focus on high water permeability ($\geq 6 \times 10^{-10}$ m^2 s^{-1} and a low membrane thickness (≤ 10 μm). Secondary goals are an IEC ≥ 3 mmol/g and a maximum lambda value ≤ 20 (Yassin et al., 2024).

4.3 Reactive oxygen species

For proton exchange membrane (PEM) fuel cells, reactive oxygen species like hydroxyl radicals were identified as the main source of chemical degradation. Especially at high fuel cell voltages, such as under open circuit conditions, the incomplete reaction between hydrogen and crossed-over oxygen results in peroxide, which further degrades into hydroxyl radicals that then react with the membrane polymer. A facile ex-situ test for assessing the chemical stability of PEMs is the Fenton test, in which membranes are immersed in a diluted peroxide solution that may also contain iron salts in ppm concentrations as a catalyst to break up peroxides into radicals. One drawback of this test is that the ex-situ stability of polymers against peroxide may not accurately represent the in-situ stability, where the amount of crossed-over oxygen is different. For example, Nafion has approximately 10 times higher oxygen permeability than many sulfonated aromatic polymers.

For AEMs, it must be considered that hydrogen peroxide rapidly decomposes under alkaline conditions, that halogenides react with peroxide to form halogens, and sulfates to peroxosulfates. Therefore, hydroxide, halogenide and sulfate-form AEMs may give misleading results. Furthermore, one of the main reactive species in the standard Fenton test, the hydroxyl radical, does not exist under alkaline conditions. It is believed that the reactive oxygen species in AFCs are superoxide anion radicals, and Wierzbicki et al. proved their presence by analyzing the radicals formed in an AAEMFC with electron paramagnetic resonance spectroscopy (Wierzbicki et al., 2020). Recently, T. Nemeth et al. reported a new method to test the stability of AEM in alkaline solutions. AEM samples are immersed in an alkaline solution that contains dispersed TiO_2 particles and is illuminated by UV, producing reactive oxygen species in situ (Solyom et al., 2025).

An intriguing observation was made by Parrondo et al., who noticed that an alkaline stability test gave different results for nitrogen and oxygen-purged solutions. They hypothesized that the faster degradation of AEMs in oxygen-saturated solutions is related to the formation of superoxides when oxygen reacts with ylides, which can form when hydroxide ions deprotonate the α-carbon atom of quaternary ammonium ions (Fig. 4.12). Peng et al. reported that antioxidants, as additives for AEMs, can increase their lifetime in alkaline stability tests (Peng et al., 2021).

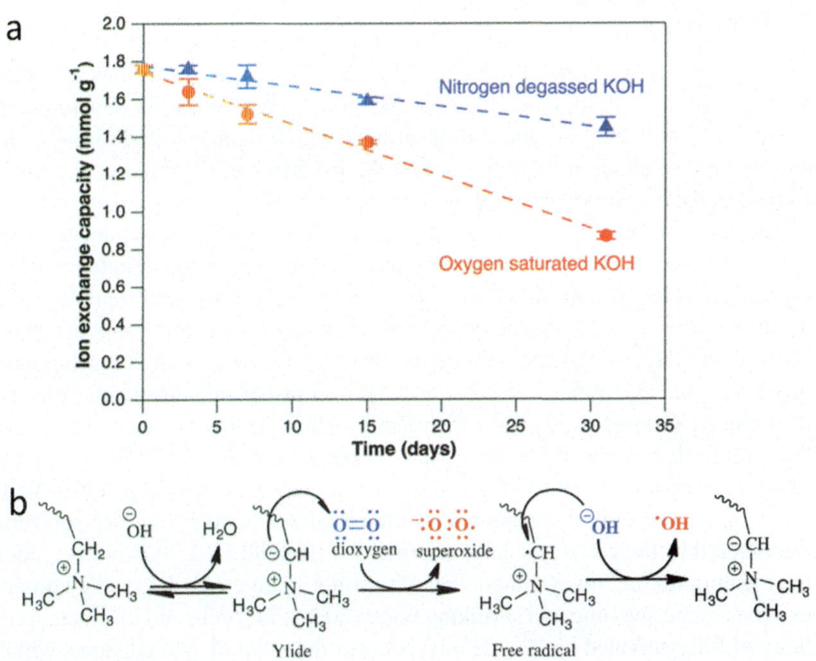

FIG. 4.12 Effect of reactive oxygen species on AEM. (A) Loss of IEC when a trimethylammonium-functionalized polyphenylene oxide (TMA-PPO) type membrane was stored at 60°C either in nitrogen-degassed or oxygen-saturated 1 M KOH; (B) proposed mechanism for quaternary ammonium-catalyzed formation of superoxides and hydroxyl radicals from oxygen. Reproduced with permission from Royal Society of Chemistry. *(From Parrondo, J., Wang, Z., Jung, M., Suk, J., Ramani, V., 2016. Reactive oxygen species accelerate degradation of anion exchange membranes based on polyphenylene oxide in alkaline environments. Phys. Chem. Chem. Phys., 18 (29), 19705–19712. https://doi.org/10.1039/c6cp01978a)*

4.4 Other fuels

Fuel cells with AEM can also be operated with liquid fuels like methanol. The main challenge is that the hydroxide-containing membranes absorb CO_2 as carbonates, which are constantly produced at the anode, resulting in reduced conductivity. However, if the current density is high enough, the carbonate ions will be replaced by hydroxide ions produced at the cathode. It is not clear which current density would be needed to achieve carbonate-free membranes and whether this is achievable with alkaline direct methanol fuel cells (alkaline DMFC). A potential advantage for alkaline DMFC is that the solvated charge-balancing hydroxide ions are transported from the cathode to the anode, whereas methanol is fed to the anode and diffuses to the cathode. The expectation is that the migration of hydroxide ions can reduce the unwanted methanol crossover, because methanol can interact with hydroxide ions either indirectly by becoming part of their solvation shell or can be deprotonated to methanolate ions, which would move in the applied electric field back to the anode (Yu and Scott, 2004).

5 Perspective

For many years, the low alkaline stability of AEMs was a significant hurdle for their use in real applications. A breakthrough came with the development of sterically protected polybenzimidazolium, which demonstrated to the community that the alkaline stability issue could be addressed (Thomas et al., 2012) and led to the development of Aemion membranes. Work by Marino and Kreuer (Marino and Kreuer, 2015) showed that the conformation of cyclic amines can effectively hinder Hofmann elimination, which led to the development of piperidinium-based and DABCO-based membranes. They also demonstrated that ammonium groups connected to an alkyl chain of 3–6 CH_2 groups are more stable than CH_2CH_2 tethers, a guideline that was followed in the development of Orion TM1 membranes (Lee et al., 2017). Further advancements resulted in Aemion+ (Fan et al., 2019) which again exhibited higher alkaline stability than Aemion. A water electrolysis system operating with an AEMION + membrane was reported to run for over 8900 hours (Moreno-González et al., 2023) and Enapter, the only company that commercializes AEM water electrolyzers, estimates that their systems have a projected lifetime of 35,000 hours. Since the membranes are always in contact with water in an electrolyzer, they do not experience the swelling and shrinking observed in a fuel cell, and the nucleophilicity of fully solvated hydroxide ions is lower than that of "naked" ions, which may be found in the dry regions of a fuel cell.

An alternative to AEM is ion-solvating membranes, which are inert polymer matrices filled with absorbed KOH solution (Aili et al., 2013; Dayan et al., 2023; Xing and Savadogo, 2000). While these materials are very promising for use in water electrolyzers, it seems to be difficult to use them in fuel cells, where they are not in contact with KOH solution and are expected to lose the absorbed KOH solution over time. Therefore, the use of AEMs in AFCs seems to be unavoidable. To achieve long durability, the membrane water permeability and thickness, as well as the fuel cell operating conditions, should be optimized to avoid local dehydration.

Future research may focus on chemistries that bind water to avoid "naked" hydroxide ions. In addition, more knowledge about degradation by reactive oxygen species under alkaline conditions is needed. This subject has been less investigated because the main target was to improve alkaline stability. PEM uses cerium salts or cerium oxide nanoparticles to destroy reactive oxygen species. A similar approach could also be used to increase the chemical durability of AEM.

References

Aili, D., Hansen, M.K., Renzaho, R.F., Li, Q., Christensen, E., Jensen, J.O., Bjerrum, N.J., 2013. Heterogeneous anion conducting membranes based on linear and crosslinked KOH doped polybenzimidazole for alkaline water electrolysis. J. Membr. Sci. 18733123. 447, 424–432. doi:10.1016/j.memsci.2013.07.054.

Aili, D., Kraglund, M.R., Rajappan, S.C., Serhiichuk, D., Xia, Y., Deimede, V., Kallitsis, J., Bae, C., Jannasch, P., Henkensmeier, D., Jensen, J.O., 2023. Electrode separators for the next-generation alkaline water electrolyzers. ACS Energy Lett. 23808195. 8 (4), 1900–1910. doi:10.1021/acsenergylett.3c00185.

Arges, C.G., Ramani, V., 2013. Two-dimensional NMR spectroscopy reveals cation-triggered backbone degradation in polysulfone-based anion exchange membranes. Proc. Natl. Acad. Sci. U. S. A. 10916490. 110 (7), 2490–2495. doi:10.1073/pnas.1217215110.

Bacon, F.T., 1969. Fuel cells, past, present and future. Electrochim. Acta. 00134686. 14 (7), 569–585. doi:10.1016/0013-4686(69)87042-8.

Dang, H.S., Jannasch, P., 2016. Alkali-stable and highly anion conducting poly(phenylene oxide)s carrying quaternary piperidinium cations. J. Mater. Chem. A. 20507496. 4 (30), 11924–11938. doi:10.1039/c6ta01905f.

Dayan, A., Trisno, M.L.A., Yang, C., Kraglund, M.R., Almind, M.R., Hjelm, J., Jensen, J.O., Aili, D., Park, H.S., Henkensmeier, D., 2023. Quaternary ammonium-free membranes for water electrolysis with 1 M KOH. Adv. Energy Mater. 16146840. 13 (46). doi:10.1002/aenm.202302966.

Dekel, D.R., 2018. Review of cell performance in anion exchange membrane fuel cells. J. Power Sources. 03787753. 375, 158–169. doi:10.1016/j.jpowsour.2017.07.117.

Díaz, J.C., Kamcev, J., 2021. Ionic conductivity of ion-exchange membranes: measurement techniques and salt concentration dependence. J. Membr. Sci. 03767388. 618, 118718. doi:10.1016/j.memsci.2020.118718.

Duan, Q., Shanhai, G., Wang, C.-Y., 2013. Water uptake, ionic conductivity and swelling properties of anion-exchange membrane. J. Power Sources. 03787753. 243, 773–778. doi:10.1016/j.jpowsour.2013.06.095.

Evonik 2025 Datasheet https://products.evonik.com/assets/s_/us/Amicure_IC_221_us_TDS_EN_EN_TDS_PV_52042325_en_US_US.pdf

Fan, J., Willdorf-Cohen, S., Schibli, E.M., Paula, Z., Li, W., Skalski, T.J.G., Sergeenko, A.T., Hohenadel, A., Frisken, B.J., Magliocca, E., Mustain, W.E., Diesendruck, C.E., Dekel, D.R., Holdcroft, S., 2019. Poly(bis-arylimidazoliums) possessing high hydroxide ion exchange capacity and high alkaline stability. Nat. Commun. 20411723. 10 (1). doi:10.1038/s41467-019-10292-z.

Fisher, S., Kunin, R., 1955. Routine exchange capacity determinations of ion exchange resins. Anal. Chem. 0003-2700. 27 (7), 1191–1194. doi:10.1021/ac60103a052.

Fortin, P., Khoza, T., Cao, X., Martinsen, S.Y., Barnett, A.O., Holdcroft, S., 2020. High-performance alkaline water electrolysis using Aemion™ anion exchange membranes. J. Power Sources 03787753. 451, 227814. doi:10.1016/j.jpowsour.2020.227814.

Fumatech, D. (2025), Available from: https://www.fuelcellstore.com/spec-sheets/fumasep-faa-3-50-technical-specifications.pdf.

Giovanelli, A., Pozio, A., Pucci, A., Geppi, M., Martini, F., 2024. Fumasep FAA-3-PK-130: exploiting multinuclear solid-state NMR to shed light on undisclosed structural properties. Polymer 00323861. 311, 127536. doi:10.1016/j.polymer.2024.127536.

Haj-Bsoul, S., Varcoe, J.R., Dekel, D.R., 2022. Measuring the alkaline stability of anion-exchange membranes. J. Electroanal. Chem. 15726657. 908, 116112. doi:10.1016/j.jelechem.2022.116112.

Henkensmeier, D., Cho, H.-R., Kim, H.-J., Kirchner, C.N., Leppin, J., Dyck, A., Jang, J.H., Cho, E., Nam, S.-W., Lim, T.-H., 2012. Polybenzimidazolium hydroxides – structure, stability and degradation. Polym. Degrad. Stab. 01413910. 97 (3), 264–272. doi:10.1016/j.polymdegradstab.2011.12.024.

Henkensmeier, D., Kim, H., Lee, H., Lee, D.H., Oh, I., Hong, S., Nam, S., Lim, T., 2011. Polybenzimidazolium-based solid electrolytes. Macromol. Mater. Eng. 1438-7492. 296 (10), 899–908. doi:10.1002/mame.201100100.

Henkensmeier, D., Najibah, M., Harms, C., Žitka, J., Hnát, J., Bouzek, K., 2021. Overview: state-of-the art commercial membranes for anion exchange membrane water electrolysis. J. Electrochem. Energy Convers. Storage. 23816910. 18 (2). doi:10.1115/1.4047963.

Hnát, J., Plevová, M., Žitka, J., Paidar, M., Bouzek, K., 2017. Anion-selective materials with 1,4-diazabicyclo[2.2.2]octane functional groups for advanced alkaline water electrolysis. Electrochim. Acta. 00134686. 248, 547–555. doi:10.1016/j.electacta.2017.07.165.

Ionomr 2025a Datasheet https://ionomr.com/wp-content/uploads/2018/12/Ionomr-Hydrogen-Info-Sheet.pdf

Ionomr 2025b Datasheet 2. https://ionomr.com/wp-content/uploads/2021/04/FM-6028-A-Properties-of-Aemion-Water-Electrolysis-Membranes-copy.pdf

Jutemar, E.P., Takamuku, S., Jannasch, P., 2011. Sulfonated poly(arylene ether sulfone) ionomers containing di- and tetrasulfonated arylene sulfone segments. Polym. Chem. 1759-9954. 2 (1), 181–191. doi:10.1039/c0py00290a.

Kaczur, J.J., Yang, H., Liu, Z., Sajjad, S.D., Masel, R.I., 2018. Carbon dioxide and water electrolysis using new alkaline stable anion membranes. Front. Chem. 22962646. 6. doi:10.3389/fchem.2018.00263.

Kangro, W., Pieper, H., 1962. Zur frage der speicherung von elektrischer energie in flüssigkeiten. Electrochim. Acta. 00134686. 7 (4), 435–448. doi:10.1016/0013-4686(62)80032-2.

Khalid, H., Najibah, M., Park, H., Bae, C., Henkensmeier, D., 2022. Properties of anion exchange membranes with a focus on water electrolysis. Membranes. 2077-0375. 12 (10), 989. doi:10.3390/membranes12100989.

Koros, W. J., Ma, Y. H., Shimidzu, T., 1996. Terminology for membranes and membrane processes (IUPAC Recommendation 1996). J. Membr. Sci. 03767388. 120 (2), 149–159. doi:10.1016/0376-7388(96)82861-4.

Krishnan, N.N., Henkensmeier, D., Jang, J.H., Hink, S., Kim, H.-J., Nam, S.-W., Lim, T.-H., 2014. Locally confined membrane modification of sulfonated membranes for fuel cell application. J. Membr. Sci. 03767388. 454, 174–183. doi:10.1016/j.memsci.2013.12.020.

Krishnan, N.N., Henkensmeier, D., Park, Y.H., Jang, J.H., Kwon, T., Koo, C.M., Kim, H.J., Han, J., Nam, S.W., 2016. Blue membranes: sulfonated copper(II) phthalocyanine tetrasulfonic acid based composite membranes for DMFC and low relative humidity PEMFC. J. Membr. Sci. 18733123. 502, 1–10. doi:10.1016/j.memsci.2015.12.035.

Kubiczek, G., Neugebauer, L., 1950. Experimenteller beitrag zur "brommethylierung" aromatischer verbindungen. Monatshefte Für Chemie. 14344475. 81 (6), 917–920. doi:10.1007/BF00899333.

Lee, W.H., Park, E.J., Han, J., Shin, D.W., Kim, Y.S., Bae, C., 2017. Poly(terphenylene) anion exchange membranes: the effect of backbone structure on morphology and membrane property. ACS Macro Lett. 21611653. 6 (5), 566–570. doi:10.1021/acsmacrolett.7b00148.

Luo, X., Rojas-Carbonell, S., Yan, Y., Kusoglu, A., 2020. Structure-transport relationships of poly(aryl piperidinium) anion-exchange membranes: eeffect of anions and hydration. J. Membr. Sci. 03767388. 598, 117680. doi:10.1016/j.memsci.2019.117680.

Mara Ikhsan, M., Abbas, S., Do, X.H., Choi, S.Y., Azizi, K., Hjuler, H.A., Jang, J.H., Ha, H.Y., Henkensmeier, D., 2022. Polybenzimidazole membranes for vanadium redox flow batteries: effect of sulfuric acid doping conditions. Chem. Eng. J. 13858947. 435. doi:10.1016/j.cej.2022.134902.

Marino, M.G., Kreuer, K.D., 2015. Alkaline stability of quaternary ammonium cations for alkaline fuel cell membranes and ionic liquids. ChemSusChem. 1864-5631. 8 (3), 513–523. doi:10.1002/cssc.201403022.

Meek, K.M., Antunes, C.M., Strasser, D., Owczarczyk, Z.R., Neyerlin, A., Pivovar, B.S., 2019. High-throughput anion exchange membrane characterization at NREL. ECS Trans. 1938-6737. 92 (8), 723–731. doi:10.1149/09208.0723ecst.

Moreno-González, M., Mardle, P., Zhu, S., Gholamkhass, B., Jones, S., Chen, N., Britton, B., Holdcroft, S., 2023. One year operation of an anion exchange membrane water electrolyzer utilizing Aemion+® membrane: minimal degradation, low H_2 crossover and high efficiency. J. Power Sources Adv. 26662485. 19, 100109. doi:10.1016/j.powera.2023.100109.

Motealleh, B., Liu, Z., Masel, R.I., Sculley, J.P., Richard Ni, Z., Meroueh, L., 2021. Next-generation anion exchange membrane water electrolyzers operating for commercially relevant lifetimes. Int. J. Hydrogen Energy. 03603199. 46 (5), 3379–3386. doi:10.1016/j.ijhydene.2020.10.244.

NovaMea 2025 Datasheet. https://www.novamea.swiss/services-details

Olsson, J.S., Pham, T.H., Jannasch, P., 2018. Poly(arylene piperidinium) hydroxide ion exchange membranes: synthesis, alkaline stability, and conductivity. Adv. Funct. Mater. 1616-301X. 28 (2), doi:10.1002/adfm.201702758.

Orion Polymer 2025 Datasheet https://orionpolymer.com/technical-data/

Ouadah, A., Xu, H., Luo, T., Gao, S., Zhang, Z., Li, Z., Zhu, C., 2017. Synthesis of novel copolymers based on p-methylstyrene, N,N-butylvinylimidazolium and polybenzimidazole as highly conductive anion exchange membranes for fuel cell application. RSC Adv. 2046-2069. 7 (75), 47806–47817. doi:10.1039/C7RA06394F.

Park, E.J., Jannasch, P., Miyatake, K., Bae, C., Noonan, K., Fujimoto, C., Holdcroft, S., Varcoe, J.R., Henkensmeier, D., Guiver, M.D., Kim, Y.S., 2024. Aryl ether-free polymer electrolytes for electrochemical and energy devices. Chem. Soc. Rev. 14604744. 53 (11), 5704–5780. doi:10.1039/d3cs00186e.

Peng, X., Yang, Y., Ye, N., Xu, S., Zhang, D., Wan, R., He, R., 2021. Synergy effects of hindered phenol and diphosphite antioxidants on promoting alkali resistance of quaternary ammonium functionalized poly(4-vinylbenzyl chloride-styrene) anion exchange membranes. Electrochim. Acta 00134686. 380, 138249. doi:10.1016/j.electacta.2021.138249.

Rodriguez, C.A.G., Óladottir Joensen, B., Moss, A.B., Larrazábal, G.O., Whelligan, D.K., Seger, B., Varcoe, J.R., Willson, T.R., 2023. Influence of headgroups in ethylene-tetrafluoroethylene-based radiation-grafted anion exchange membranes for CO_2 electrolysis. ACS Sustain. Chem. Eng. 2168-0485. 11 (4), 1508–1517. doi:10.1021/acssuschemeng.2c06205.

Schibli, E.M., Wright, A.G., Holdcroft, S., Frisken, B.J., 2018. Morphology of anion-conducting ionenes investigated by X-ray scattering and simulation. J. Phys. Chem. B. 15205207. 122 (5), 1730–1737. doi:10.1021/acs.jpcb.7b10177.

Solyom, P., Nauser, T. and Nemeth T., 2025. Photochemical methods to study the radical-induced degradation of anion-exchange membranes. Membranes 15 (10), 305. doi:10.3390/membranes15100305.

TailorMem, D. (2025), Available from: https://tailormem.com/.

Thaller, L.H., US patent 3,996,064, Assignee: NASA. (1976).

Thomas, O.D., Soo, K.J.W.Y., Peckham, T.J., Kulkarni, M.P., Holdcroft, S., 2012. A stable hydroxide-conducting polymer. J. Am. Chem. Soc. 15205126. 134 (26), 10753–10756. doi:10.1021/ja303067t.

Varcoe, J.R., Slade, R.C.T., 2005. Prospects for alkaline anion-exchange membranes in low temperature fuel cells. Fuel Cells. 1615-6846. 5 (2), 187–200. doi:10.1002/fuce.200400045.

Versogen 2025 Datasheet. https://www.fuelcellstore.com/versogen-piperion-aem-self-supporting-80um-73800002

Wang, J., Zhao, Y., Setzler, B.P., Rojas-Carbonell, S., Ben Yehuda, C., Amel, A., Page, M., Wang, L., Hu, K., Shi, L., Gottesfeld, S., Xu, B., Yan, Y., 2019. Poly(aryl piperidinium) membranes and ionomers for hydroxide exchange membrane fuel cells. Nat. Energy. 20587546. 4 (5), 392–398. doi:10.1038/s41560-019-0372-8.

Wang, L., Rojas-Carbonell, S., Hu, K., Setzler, B.P., Motz, A.R., Ueckermann, M.E., Yan, Y., 2022. Standard operating protocol for ion-exchange capacity of anion exchange membranes. Front. Energy Res. 2296-598X. 10. doi:10.3389/fenrg.2022.887893.

Wang, Q., Lu, Y., Li, N., 2016. Preparation, characterization and performance of sulfonated poly(styrene-ethylene/butylene-styrene) block copolymer membranes for water desalination by pervaporation. Desalination. 00119164. 390, 33–46. doi:10.1016/j.desal.2016.04.005.

Wang, Y., Zhang, D., Liang, X., Shehzad, M.A., Xiao, X., Zhu, Y., Ge, X., Zhang, J., Ge, Z., Wu, L., Xu, T., 2020. Improving fuel cell performance of an anion exchange membrane by terminal pending bis-cations on a flexible side chain. J. Membr. Sci. 18733123. 595. doi:10.1016/j.memsci.2019.117483.

Wei, Q., Cao, X., Veh, P., Konovalova, A., Mardle, P., Overton, P., Cassegrain, S., Vierrath, S., Breitwieser, M., Holdcroft, S., 2022. On the stability of anion exchange membrane fuel cells incorporating polyimidazolium ionene (Aemion+®) membranes and ionomers. Sustain. Energy Fuels 6, 3351–3564. doi:10.1039/D2SE00690A.

Wierzbicki, S., Douglin, J.C., Kostuch, A., Dekel, D.R., Kruczała, K., 2020. Are radicals formed during anion-exchange membrane fuel cell operation? J. Phys. Chem. Lett. 19487185. 11 (18), 7630–7636. doi:10.1021/acs.jpclett.0c02349.

Wright, A.G., Fan, J., Britton, B., Weissbach, T., Lee, H.-F., Kitching, E.A., Peckham, T.J., Holdcroft, S., 2016. Hexamethyl-p-terphenyl poly(benzimidazolium): a universal hydroxide-conducting polymer for energy conversion devices. Energy Environ. Sci. 1754-5692. 9 (6), 2130–2142. doi:10.1039/c6ee00656f.

Wright, A.G., Holdcroft, S., 2014. Hydroxide-stable ionenes. ACS Macro Lett. 21611653. 3 (5), 444–447. doi:10.1021/mz500168d.

Wu, X., Chen, N., Klok, H., Lee, Y.M., Hu, X., Poly, B., 2022. (Aryl piperidinium) membranes for anion-exchange membrane fuel cells. Angew. Chem. Int. Ed. 1433-7851. 61 (7), doi:10.1002/anie.202114892.

Xing, B., Savadogo, O., 2000. Hydrogen/oxygen polymer electrolyte membrane fuel cells (PEMFCs) based on alkaline-doped polybenzimidazole (PBI). Electrochem. Commun. 13882481. 2 (10), 697–702. doi:10.1016/s1388-2481(00)00107-7.

Yan, J., Hickner, M.A., 2010. Anion exchange membranes by bromination of benzylmethyl-containing poly(sulfone)s. Macromolecules 00249297. 43 (5), 2349–2356. doi:10.1021/ma902430y.

Yanagi, H., Fukuta, K., 2008. Anion exchange membrane and ionomer for alkaline membrane fuel cells (AMFCs). ECS Trans. 1938-5862. 16 (2), 257–262. doi:10.1149/1.2981860.

Yassin, K., Rasin, I.G., Brandon, S., Dekel, D.R., 2024. How can we design anion-exchange membranes to achieve longer fuel cell lifetime? J. Membr. Sci. 18733123. 690. doi:10.1016/j.memsci.2023.122164.

Yu, E.H., Scott, K., 2004. Development of direct methanol alkaline fuel cells using anion exchange membranes. J. Power Sources 137 (2), 248–256. doi:10.1016/j.jpowsour.2004.06.004.

Zheng, Y., Serban, A., Zhang, H., Chen, N., Song, F., Hu, X., 2023. Anion exchange ionomers enable sustained pure-water electrolysis using platinum-group-metal-free electrocatalysts. ACS Energy Lett. 2380-8195. 8 (12), 5018–5024. doi:10.1021/acsenergylett.3c01866.

Žitka, J., Peter, J., Galajdová, B., Pavlovec, L., Pientka, Z., Paidar, M., Hnát, J., Bouzek, K., 2019. Anion exchange membranes and binders based on polystyrene-block-poly (Ethylene-ran-butylene)-block-polystyrene copolymer for alkaline water electrolysis. Desalination Water Treat. 19443986. 142, 90–97. doi:10.5004/dwt.2019.23411.

Ziv, N., Dekel, D.R., 2018. A practical method for measuring the true hydroxide conductivity of anion exchange membranes. Electrochem. Commun. 13882481. 88, 109–113. doi:10.1016/j.elecom.2018.01.021.

Chapter 5

Alkaline oxygen reduction reaction catalysts: Mechanisms, classification, design and challenges

Aoxin Ran[a] and Linhao Fan[a,b]
[a]*State Key Laboratory of Engines, Tianjin University, Tianjin, China*
[b]*School of Energy (National Industry-Education Platform for Energy Storage), Tianjin University, Tianjin, China*

1 Oxygen reduction reaction mechanisms

The oxygen reduction reaction (ORR) is a key process occurring at the complex three-phase interface formed by oxygen, electrolyte liquid, and catalyst solid, as shown in Fig. 5.1. At this interface, ORR primarily proceeds through two competitive reaction pathways: one is the two-electron reaction pathway, and the other is the four-electron reaction pathway (Yeager, 1984; Yan et al., 2023; Shi et al., 2019). In the two-electron pathway, oxygen is reduced to HO_2^- or H_2O_2 by obtaining two electrons and reacting with water at the cathode, with a relatively low positive potential. In the subsequent half-reaction, HO_2^- or H_2O_2 gains electrons and further decomposes into OH^- or O_2. In these two half-reactions, the generation of intermediate products reduces the energy conversion efficiency of the battery, and the oxidizing intermediate products corrode components like the catalyst and the electrolyte membrane. In the four-electron pathway, oxygen directly produces OH^- in an alkaline environment. Thus, catalysts that follow the direct four-electron pathway are more ideal cathode catalytic materials. Therefore, researchers have been dedicated to guiding ORR toward the four-electron pathway through meticulous design and regulation of the three phase interface characteristics, accelerating the reaction kinetics, and thereby effectively enhancing the catalytic efficiency of active sites (Zhao et al., 2024a; Li et al., 2021; Lin et al., 2021). Specifically, the electron conduction efficiency at the three phase interface of the catalyst, the mass transport rate of reactants such as oxygen, and the intrinsic activity of the catalyst's active sites collectively constitute the core factors influencing ORR kinetics and stability. Among them, the adsorption and desorption capabilities of the catalyst's

Alkaline Fuel Cells. https://doi.org/10.1016/B978-0-443-36638-3.00012-7

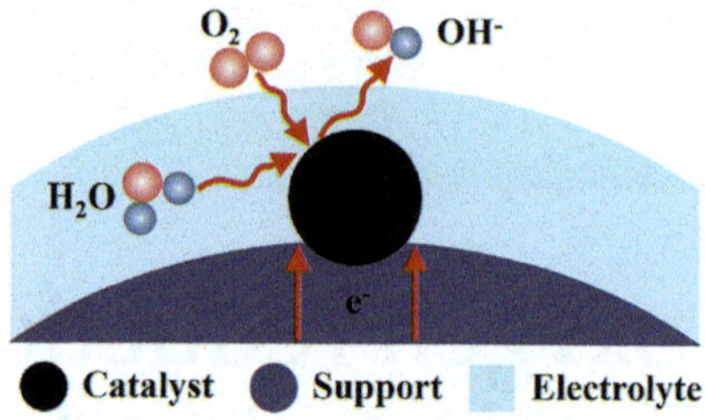

FIG. 5.1 Schematic representation of the three-phase reaction interface for ORR catalysts under alkaline conditions.

active sites for reaction species (including oxygen and oxygen containing intermediates) largely determine the kinetic rate of ORR (Pan and Yang, 2020). Therefore, researchers have utilized a series of advanced characterization techniques and computational materials science methods to deeply investigate the mechanisms of catalyst active sites in the processes of oxygen adsorption, activation, and adsorption/desorption of oxygen containing intermediates, thereby unraveling the catalytic mechanism of ORR.

The ORR mechanism is classified into the associative mechanism and dissociative mechanism based on the breaking pattern of the O-O bond in adsorbed oxygen molecules (Wang et al., 2022). In the associative mechanism, one or both ends of the oxygen molecule are adsorbed on the active sites. When electron acquisition triggers oxygen reduction, the O-O bond is typically not fully broken, forming OOH*. Subsequently, the oxygen containing intermediate OOH* continues to acquire electrons and is catalytically converted into O* and OH* intermediates. Finally, OH* desorbs from the active sites to complete the four-electron ORR catalytic process. The four-electron associative mechanism ORR pathway is as follows (Yeager, 1984; Yan et al., 2023):

$$O_2 + * \rightarrow OO* \tag{1}$$

$$OO* + H_2O + e^- \rightarrow OOH* + OH^- \tag{2}$$

$$OOH* + e^- \rightarrow O* + OH^- \tag{3}$$

$$O* + H_2O + e^- \rightarrow OH* + OH^- \tag{4}$$

$$OH* + e^- \rightarrow * + OH^- \tag{5}$$

In the dissociative mechanism, the oxygen molecule anchors at both ends onto adjacent active sites. When the oxygen molecule gains electrons, the O–O bond breaks directly, immediately generating two OH* intermediates. It is worth noting that the direct cleavage of the O–O bond effectively bypasses the formation stage of the OOH* intermediate. This transition not only significantly promotes selectivity toward the four-electron ORR pathway but also fundamentally eliminates the possibility of generating hydrogen peroxide via the two-electron ORR pathway. Subsequently, these OH* intermediates gain electrons and detach from the active site surface, converting into hydroxide ions. By reducing the complexity of reaction steps, the dissociative mechanism significantly optimizes the kinetic characteristics of the four-electron ORR, thereby drastically enhancing its reaction rate. The four-electron dissociative mechanism ORR pathway is as follows:

$$O_2 + ** \rightarrow *OO* \tag{6}$$

$$*OO* + 2H_2O + 2e^- \rightarrow 2OH* + 2OH^- \tag{7}$$

$$2OH* + 2e^- \rightarrow ** + 2OH^- \tag{8}$$

Under different reaction conditions, oxygen undergoes distinct reduction pathways. The feasibility of each reaction pathway can be estimated by calculating the Gibbs free energy of different intermediates in various reaction pathways. Density functional theory (DFT) calculations have shown that when the surface coverage of O_2 on the catalyst is low, the dissociative pathway represents the reaction with the lowest energy barrier (Nørskov et al., 2004). Conversely, when the surface coverage of O_2 is high, the associative pathway exhibits the lowest energy barrier and becomes the dominant pathway for ORR. In reality, the preference for different ORR pathways is primarily correlated with the electrode materials, and the specific scenarios vary significantly across different electrode materials.

According to the Sabatier principle, the key reaction intermediates must bind to the catalyst neither too strongly nor too weakly for the catalyst to achieve optimal performance (Nørskov et al., 2009). When the adsorption of intermediates is too weak, it becomes unfavorable for the electron and proton transfer in the reaction process. Conversely, when the adsorption of O* or OOH* on the sites is too strong, the desorption of H_2O becomes difficult, making it hard for the sites to be re-exposed and hindering the further progress of the reaction. Nørskov et al. constructed a volcano plot for ORR activity using the adsorption energies of oxygen containing species as descriptors (She et al., 2017). As shown in Fig. 5.2, the adsorption strength of different metal catalysts for the intermediate *O is displayed. Among all metals, Pt group metal catalysts (such as Pt and Pd) are closest to the peak of the activity volcano plot. Many design strategies, such as Pt alloys, have been proposed to reach the activity peak, but their high cost and limited resources restrict their commercial applications.

2 ORR catalysts classification

The exchange current density of ORR is generally extremely low. Even on the surface of Pt and Pt alloy catalysts with excellent catalytic performance, this value only reaches 10^{-5}–10^{-6} A m^{-2}, differing by approximately six orders of

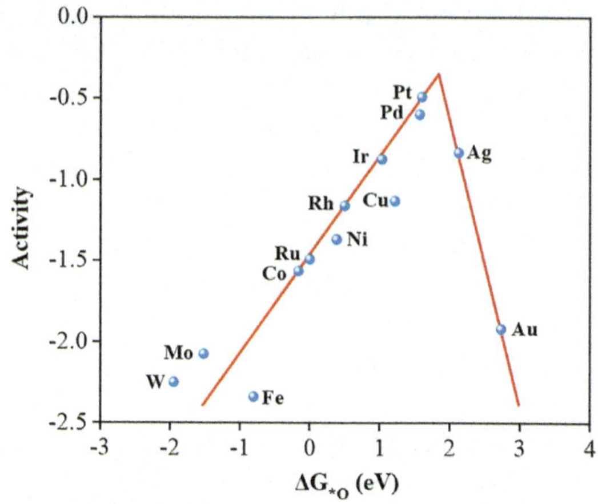

FIG. 5.2 The volcano plot of ORR activity versus *O binding free energy for different metals (Nørskov et al., 2004). *(Adapted from Nørskov, J.K., Rossmeisl, J., Logadottir, A., Lindqvist, L., Kitchin, J.R., Bligaard, T., Jónsson, H., 2004. Origin of the overpotential for oxygen reduction at a fuel-cell cathode. J. Phys. Chem. B. 1520-6106. 108 (46), 17886–17892. doi:10.1021/jp047349j, with permission of American Chemical Society.)*

magnitude from the hydrogen oxidation reaction (Xu et al., 2019). Thus, the ORR typically requires a high overpotential, necessitating reliance on highly active catalysts to overcome this barrier. Currently, commercial ORR electrocatalysts are predominantly high-activity Pt and its alloys, but their exorbitant costs significantly limit large-scale commercialization. Therefore, developing efficient and stable ORR electrocatalysts is critical for advancing the commercial application of AEMFCs (Radwan et al., 2021). In the field of ORR catalysis, alternative systems to Pt and Pt based alloy catalysts mainly include nonnoble metal catalysts such as metal-free carbon-based catalysts, transition metal-nitrogen-doped carbon materials, and transition metal compounds(Chen et al., 2020a), as shown in Fig. 5.3. The outstanding activity of these materials can be attributed to the synergistic effect between transition metals and nitrogen or carbon elements. This effect regulates the electronic structure of active centers, thereby optimizing the adsorption energy of active sites for reaction intermediates (e.g., $*O_2$, *OOH, *O, *OH, etc.) (Deng et al., 2022). Currently, there are two primary development pathways for ORR electrocatalysts: one is to develop efficient low-Pt content alloy catalysts to further reduce Pt usage, and the other is to research and develop nonnoble metal catalysts with high ORR activity (Xiaoxiao Li et al., 2022).

2.1 Metal-free carbon based catalysts

Metal-free carbon based ORR catalysts have emerged as a research hotspot in ORR catalysis in recent years, owing to their abundant raw material sources,

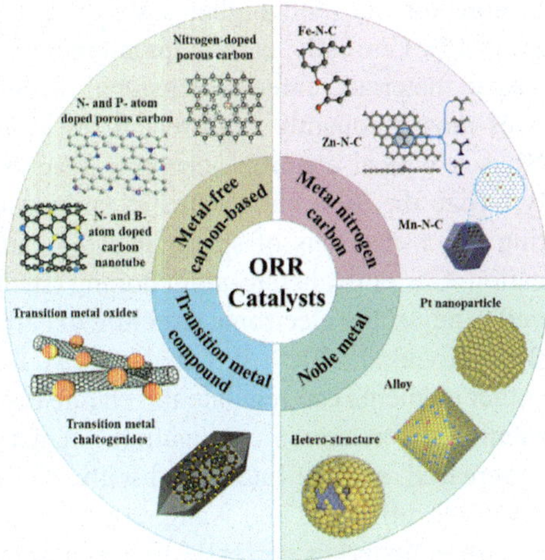

FIG. 5.3 ORR Catalysts classification: Metal-free carbon-based catalysts (Duan et al., 2021; Mao et al., 2018; Wang et al., 2011); Metal nitrogen carbon (M–N–C) catalysts (Alemany-Molina et al., 2025; Jin et al., 2024; Li et al., 2018); Transition metal compound catalysts (Ge et al., 2014; Shen et al., 2025); Noble metal catalysts (Jiao et al., 2021). *(Adapted from Duan, Z., Han, G., Huo, H., Lin, Z., Liping, G., Chunyu, D., Gao, Y., Yin, G., 2021. Monovacancy coupled pyridinic n site enables surging oxygen reduction activity of metal-free* CN_x *catalyst. ACS Sustain. Chem. Eng. 2168-0485. 9 (3), 1264–1271. doi:10.1021/acssuschemeng.0c07490; Mao, X., Cao, Z., Yin, Y., Wang, Z., Dong, H., Yang, S., 2018. Direct synthesis of nitrogen and phosphorus co-doped hierarchical porous carbon networks with biological materials as efficient electrocatalysts for oxygen reduction reaction. Int. J. Hydrogen Energy. 03603199. 43 (22), 10341–10350. doi:10.1016/j.ijhydene.2018.04.100; Wang, S., Iyyamperumal, E., Roy, A., Xue, Y., Yu, D., Dai, L., 2011. Vertically aligned BCN nanotubes as efficient metal-free electrocatalysts for the oxygen reduction reaction: a synergetic effect by co-doping with boron and nitrogen. Angew. Chem. Int. Ed. 15213773. 50 (49), 11756–11760. doi:10.1002/anie.201105204; Alemany-Molina, G., Vecchio, C.L., Baglio, V., Aricò, A.S., Morallón, E., Cazorla-Amorós, D., 2025. Pt nanoparticles for improving the performance and durability of Fe-N-C based materials towards oxygen reduction reaction in alkaline direct methanol fuel cells. J. Colloid Interface Sci. 00219797. 691, 137426. doi:10.1016/j.jcis.2025.137426; Jin, L., Kang, W., Liao, M., Wang, D., Jayaraman, B., Peera, S.G., Liu, C., 2024. Theoretical analysis of single Zn atoms with N/C on graphene promoting oxygen redox in alkaline medium. Mol. Catal. 24688231. 553, 113799. doi:10.1016/j.mcat.2023.113799; Li, J., Chen, M., Cullen, D.A., Hwang, S., Wang, M., Li, B., Liu, K., Karakalos, S., Lucero, M., Zhang, H., Lei, C., Xu, H., Sterbinsky, G.E., Feng, Z., Su, D., More, K.L., Wang, G., Wang, Z., Wu, G., 2018. Atomically dispersed manganese catalysts for oxygen reduction in proton-exchange membrane fuel cells. Nat. Catal. 25201158. 1 (12), 935–945. Available from: https://www.nature.com/natcatal/. doi:10.1038/s41929-018-0164-8; Ge, X., Liu, Y., Goh, F.W.T., Hor, T.S.A., Zong, Y., Xiao, P., Zhang, Z., Lim, S.H., Li, B., Wang, X., Liu, Z., 2014. Dual-phase spinel* $MnCo_2O_4$ *and spinel* $MnCo_2O_4$*/nanocarbon hybrids for electrocatalytic oxygen reduction and evolution. ACS Appl. Mater. Interfaces. 19448252. 6 (15), 12684–12691. Available from: http://pubs.acs.org/journal/aamick. doi:10.1021/am502675c; Jiao, K., Xuan, J., Qing, D., Bao, Z., Xie, B., Wang, B., Zhao, Y., Fan, L., Wang, H., Hou, Z., Huo, S., Brandon, N.P., Yin, Y., Guiver, M.D., 2021. Designing the next generation of proton-exchange membrane fuel cells. Nature. 0028-0836. 595 (7867), 361–369. doi:10.1038/s41586-021-03482-7, with permission of American Chemical Society, Elsevier, John Wiley and Sons, and Springer Nature.)*

excellent electrical conductivity, and higher stability compared to Pt-based catalysts (Lee et al., 2015). Currently, carbon quantum dots, carbon nanotubes (CNs), graphene-based materials, three-dimensional skeletal carbon materials, etc., are the ones most frequently explored in research (Li et al., 2023). However, the ORR is a typical electron accepting reaction, and it is rather difficult to directly utilize the inert πe lectrons in carbon materials for ORR. Therefore, activating the π electrons of sp^2 hybridized carbon is a key challenge for its application. Currently, hetero-atom doping (such as N, S, P, etc.) is commonly employed to adjust the electronic structure of carbon-based catalysts for achieving catalysis (Lv et al., 2020; Wu et al., 2019). Research has shown that incorporating other elements can regulate the charge distribution of carbon materials. By weakening the πelectron conjugation interaction, the number of active electrons is increased, and the interaction with adsorbents is improved, thus achieving the catalytic effect (Li and Wang, 2021).

Nitrogen doped carbon materials currently represent the most widely studied and applied category of metal-free carbon-based ORR catalysts. Their superior performance stems primarily from the unique electronic properties of nitrogen atoms. Compared to other ORR catalysts, nitrogen doped carbon materials exhibit excellent electrochemical performance, high specific surface area, well defined pore structures, and remarkable catalytic activity/selectivity. The N atom possesses lone pairs of electrons, and its introduction into carbon materials can alter the surface carbon arrangement, creating more defect sites (Shui et al., 2015). This structural modification significantly facilitates ORR by enhancing oxygen adsorption and electron transfer efficiency. As shown in Fig. 5.4A–D, Duan et al. (2021) fabricated a nitrogen doped hierarchically porous carbon (HPC) catalyst featuring single-vacancy-coupled pyridinic nitrogen sites by modifying ZIF-8 containing silica nanospheres. This catalyst exhibited a half wave potential of 0.855 V in an alkaline solution. The study revealed that the synergistic coupling between topological single-vacancy defects and edge-positioned pyridinic nitrogen sites significantly enhanced the turnover frequency of active sites, thereby achieving excellent ORR activity. Kong et al. (2022) designed a nitrogen-doped hollow porous carbon nanocage (N-HPCs) catalyst based on an interface-control strategy. The noncovalent bonding between nitrogen-deficient and nitrogen-rich heterostructures effectively promoted the adsorption and desorption of oxygen and intermediates, leading to half-wave potentials of 0.920 V (alkaline) and 0.780 V (acidic).

In addition to N doping, other heteroatoms such as P, B, and S are also frequently used to regulate the electronic structure and spin density within carbon materials. Furthermore, codoping with multiple different heteroatoms can induce more significant changes in defect density and electronic structure compared to single heteroatom doping. The synergistic effects among multiple heteroatoms effectively promote the ORR process. Wang et al. (2014) developed a facile method to prepare B and N isolate-doped graphitic nanosheets, as shown in Fig. 5.4E and verified its good ORR activity. In addition, Zheng et al. (2013)

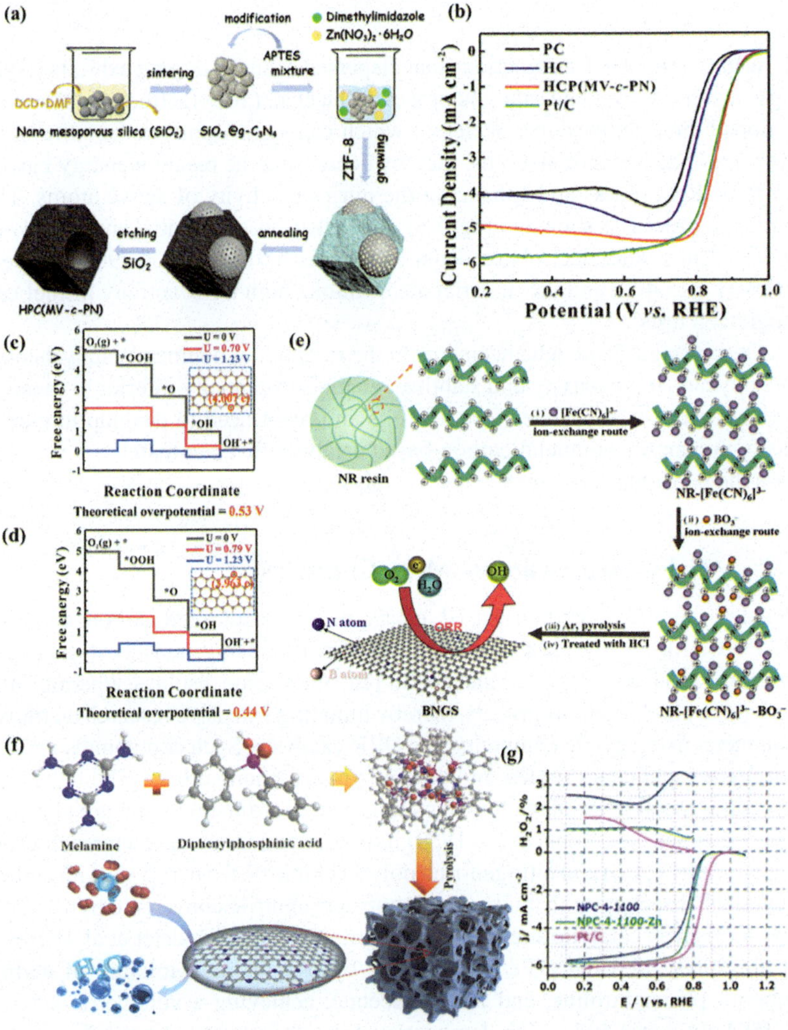

FIG. 5.4 Synthesis and performance characterization of N-doped carbon catalysts. (A) Schematic diagram of the synthesis process of N-doped HPC catalysts (Duan et al., 2021) (B) RDE polarization curves of PC, HC, HPC (MV-c-PN), and Pt/C; ORR free energy diagrams of (C) conventional pyridinic N site and (D) MV-c-PN site; (E) Synthesis process of B, N-codoped carbon catalysts (Wang et al., 2014); (F) Synthesis process of N, P dual-doped carbon nanosheet catalysts (Zhu et al., 2018); (G) RDE polarization curves and peroxide yield of NPC-4–1100, NPC-4–1100-Zn and commercial Pt/C. *(Adapted from Ref Duan, Z., Han, G., Huo, H., Lin, Z., Liping, G., Chunyu, D., Gao, Y., Yin, G., 2021. Monovacancy coupled pyridinic n site enables surging oxygen reduction activity of metal-free CN_x catalyst. ACS Sustain. Chem. Eng. 2168-0485. 9 (3), 1264–1271. doi:10.1021/acssuschemeng.0c07490; Wang, L., Peng, Y., Zhao, L., Tian, C., Zhao, D., Zhou, W., Yin, J., Wang, R., Honggang, F., 2014. B and N isolate-doped graphitic carbon nanosheets from nitrogen-containing ion-exchanged resins for enhanced oxygen reduction. Sci. Rep. 2045-2322. 4 (1). doi:10.1038/srep05184; Zhu, J., Xiao, M., Song, P., Jing, F., Jin, Z., Liang, M., Junjie, G., Liu, C., Chen, Z., Xing, W., 2018. Highly polarized carbon nano-architecture as robust metal-free catalyst for oxygen reduction in polymer electrolyte membrane fuel cells. Nano Energy. 22112855. 49, 23–30. doi:10.1016/j.nanoen.2018.04.021, with the permission of American Chemical Society, Springer Nature, and Elsevier.)*

showed through DFT calculations that dispersed N and B doping exhibits a synergistic effect during the catalytic process. Zhu et al. (2018) introduced P into an N doped three-dimensional skeleton, which can significantly reduce the work function of active carbon atoms, thereby decreasing the reaction energy barrier for the ORR process and enhancing the intrinsic activity of active atoms. The synthesis process is shown in Fig. 5.4F. The ORR onset potential and half-wave potential in acidic electrolytes reach as high as 0.910 V and 0.790 V, respectively (Fig. 5.4G), and the material also exhibits high ORR activity in fuel cell single-cell tests.

Despite significant advancements in the research of nonmetal catalysts over recent years, their activity and stability remain substantially inferior to those of metal-based catalysts. Therefore, there is an urgent need to develop low-cost, easily preparable nonmetal carbon-based catalysts that can match or even surpass the performance of metal-based catalysts.

2.2 Metal-nitrogen-carbon (M–N–C) catalyst

Besides metal-free carbon based catalysts, metal-nitrogen-carbon catalysts (M–N–C, M=Fe, Co, Mn, Ce, etc.) are also promising non-PGM catalysts. Jasinski (1964) first reported that cobalt phthalocyanine displays superior ORR activity in alkaline environments, thereby initiating extensive research on transition metal macrocyclic compounds as ORR catalysts. Such compounds include porphyrins, phthalocyanines, and their polymers, among others, where the type of central metal atom, ligand composition, and coordination structure all influence ORR activity. Gupta et al. (1989) introduced an innovative approach using metal salts (Fe, Co) and nitrogen-rich polyacrylonitrile for pyrolysis with carbon materials, discovering that nitrogen in polyacrylonitrile converts to pyridinic-N and forms new catalytic sites with transition metal ions. Dodelet et al. (Lefèvre et al., 2009) prepared a Fe/N/C catalyst by pyrolyzing a mixture of carbon support, phenanthroline, and ferrous acetate, achieving a volumetric activity of 99 A cm^{-3} at 0.8 V, which demonstrated the practicality of transition metal-nitrogen-doped carbon catalysts. They proposed that Fe-pyridinic-N coordination groups stored in micropores serve as active sites for ORR. Subsequently, replacing the carbon support with porous zeolitic imidazolate framework material (ZIF-8), their catalyst achieved a maximum power density of 750 MW cm^{-2} and a volumetric activity of 230 A cm^{-3} at 0.8 V.

Sun et al. (2019) investigated the relationship between the adsorption energy of reaction intermediates and ORR activity for different metal centers. They found that, in the four-electron pathway, the activity follows the order of $FeN_4 > CoN_4 > MnN_4 > CuN_4 > NiN_4$. The M–N–C catalysts with FeN_4 sites have exhibited ORR activity exceeding that of commercial Pt-based catalysts in alkaline electrolytes. Wang et al. (2018) prepared a single atom dispersed Fe–N–C catalyst by introducing ferrocene during the synthesis of ZIF-8, followed by pyrolysis, acid washing, and secondary pyrolysis. This catalyst achieved a half-wave potential of 0.904 V in 0.1 mol L^{-1} KOH. Although single-atom FeN_4 sites exhibit high ORR activity, their capability to catalyze the Fenton reaction is far greater than that of other transition metals. The generated free radicals attack the carbon support, significantly deteriorating stability. The ORR activity of Co is second only to Fe, and the milder Fenton reaction activity of Co-based single atom catalysts (SACs) also makes them attractive as ORR catalysts. Han et al. (2019b) precisely controlled the atomic dispersion of Co in bimetallic ZIF-8 by tuning the zinc content, synthesizing a series of cobalt nanoparticles (NPs), atomic clusters, and SACs, which are referred to as Co-NPs@NC, Co-ACs@NC, and Co-SAs@NC, respectively, as shown in Fig. 5.5. These catalysts exhibited comparable activity to commercial Pt based catalysts in alkaline solutions, providing critical insights into the size-activity

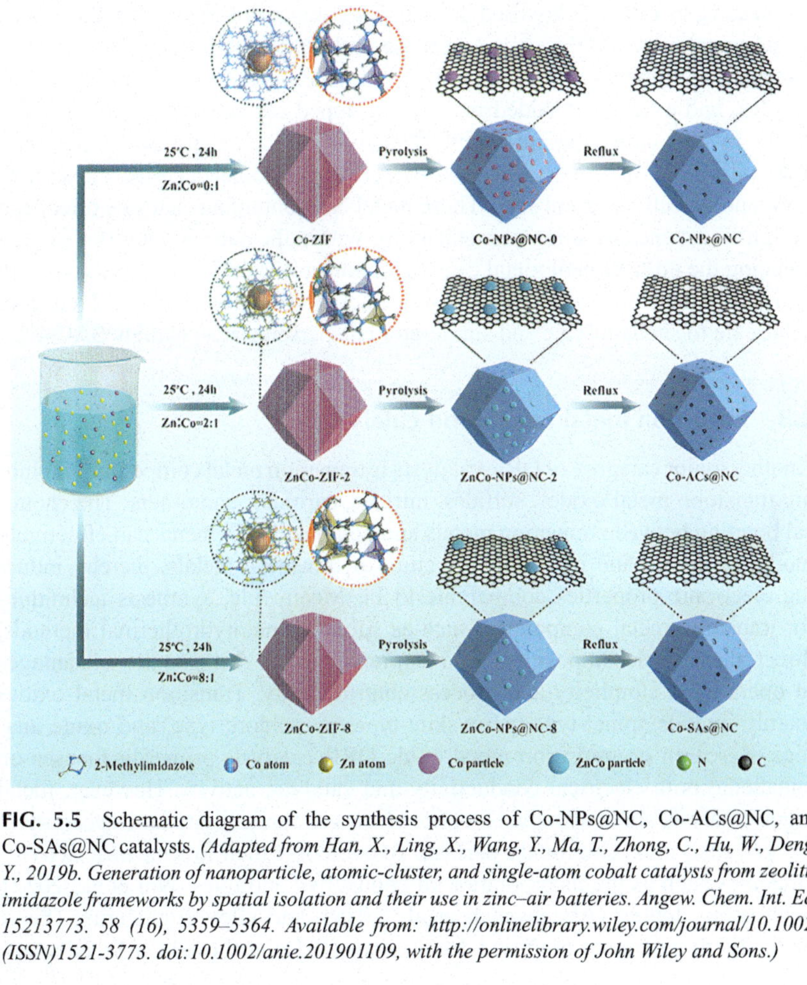

FIG. 5.5 Schematic diagram of the synthesis process of Co-NPs@NC, Co-ACs@NC, and Co-SAs@NC catalysts. *(Adapted from Han, X., Ling, X., Wang, Y., Ma, T., Zhong, C., Hu, W., Deng, Y., 2019b. Generation of nanoparticle, atomic-cluster, and single-atom cobalt catalysts from zeolitic imidazole frameworks by spatial isolation and their use in zinc–air batteries. Angew. Chem. Int. Ed. 15213773. 58 (16), 5359–5364. Available from: http://onlinelibrary.wiley.com/journal/10.1002/(ISSN)1521-3773. doi:10.1002/anie.201901109, with the permission of John Wiley and Sons.)*

relationship in ORR catalysts. Beyond Co, research has primarily focused on Mn and Cu. Li et al. (2018) employed DFT to analyze the ORR pathway on MnN_4C_{12} sites and subsequently synthesized a single atom dispersed Mn–N–C catalyst via combined chemical doping and postadsorption methods. This catalyst achieved a half-wave potential of 0.80 V in 0.1 mol L^{-1} $HClO_4$ solution. Han et al. (2019a) fabricated a high loading single atom Cu–N–C catalyst by directly anchoring Cu atoms into graphene using solid-phase Cu as the metal source via a self-initiated dispersion method under a nitrogen containing atmosphere. This catalyst exhibited a half wave potential of 0.847 V in 0.1 mol L^{-1} KOH. In addition, other transition metal SACs, such as Zn (Wang et al., 2020a), Ni (Noh et al., 2022), Ce (Zhu et al., 2021), and Cr (Luo et al., 2019) have demonstrated promising potential in the ORR.

Most current M–N–C materials require uncontrolled high-temperature treatment, resulting in catalysts with nonuniform compositions (Tylus et al., 2014). These can generally be classified into catalysts containing metal NPs, those with single metal-atom (MN_x) sites, and those with dual-metal atomic center sites. Catalysts with metal NP centers typically include metal NPs, metal carbides, nitrides, and oxides embedded in nitrogen doped carbon matrices (Cui et al., 2021). In such cases, the intrinsic ORR activity of the metal itself is generally low, with most of its effect exerted by influencing adjacent MN_x sites, N–C sites, and defective carbon sites. These metal NP-containing catalysts often lead to significant encapsulation of metal atoms within the particle interiors, greatly reducing the utilization of metal centers. To improve catalytic activity and atom utilization, transition metal–nitrogen-doped carbon catalysts have evolved from nanoscale to subnanoscale and single-atom dispersed states, namely SACs.

2.3 Transition metal compound catalysts

Another major category of ORR catalysts is transition metal compounds, including transition metal oxides, sulfides, nitrides, carbides, and others. The chemical bonding between transition metals and nonmetallic elements can effectively modulate the d-band electronic structure of transition metals, thereby inducing electronic properties comparable to Pt. Meanwhile, synthesis techniques for transition metal compounds, such as solvothermal/hydrothermal methods, direct thermal decomposition, and coprecipitation, demonstrate advantages in operational simplicity and process controllability. Transition metal oxides mainly include spinel-type, perovskite-type, pyrochlore-type, and oxide analogs. Research on transition metal oxide ORR catalysts primarily focuses on enhancing both electrical conductivity and catalytic activity. However, metal oxide catalysts are constrained by poor electrical conductivity and a scarcity of oxygen adsorption sites on the surface. To overcome their low intrinsic activity, strategies such as the incorporation of conductive additives (Niu et al., 2022), heteroatom doping (Abdelwahab et al., 2019), and alloy engineering (Peera et al., 2019) have been proposed. Liang et al. (2011) anchored Co_3O_4 NPs onto graphene oxide (GO) via a hydrothermal method and subsequently introduced nitrogen doping through ammoniation treatment. Their study revealed that the significantly enhanced electrical conductivity of Co_3O_4, coupled with the interfacial electronic coupling effect between the NPs and nitrogen-doped graphene, synergistically drove the improvement in ORR activity.

Transition metal chalcogenides (e.g., sulfides, selenides, tellurides) are predominantly characterized by two-dimensional layered structures. Their unique electron delocalization feature endows them with intrinsic electrochemical activity rivaling that of carbon based materials. By employing heteroatom doping, phase transition, defect engineering, and strain engineering, the d-band center position and M–X bond length (M = transition metal, X=S/Se/Te) of materials can be precisely modulated, which is conducive to oxygen reduction toward the four-electron pathway. Huang et al. (Cao et al., 2021) developed FeSe@NCs catalysts by engineering the crystal phase of ultrafine FeSe nanocrystals and strategically integrating them with a nitrogen doped carbon matrix. This hybrid structure demonstrated exceptional ORR activity in alkaline electrolytes, achieving a half-wave potential of 0.89 V and negligible activity decay over 10,000 cycles. Moreover, most transition metals can form nitrides with nitrogen. The introduction of nitrogen atoms enables the modulation of the catalyst's electronic structure, inducing a narrowing of the d-band filling states in transition metals, thereby exhibiting electronic characteristics analogous to noble metals. Zhang et al. (2019) prepared Ni-CoN@NC nitrogen doped nanospheres with the coexistence of nickel nitride and cobalt nitride NPs based on metal-organic frameworks. The onset potential and half-wave potential reached 0.97 V and 0.84 V, respectively. In transition metal carbides (such as Fe_3C, etc.), metal carbides generally coexist with nitrogen-doped carbon. For example, Liu et al. (2019) constructed a nitrogen doped carbon catalyst with Fe/Fe_3C NPs attached, which exhibited ORR performance close to that of the commercial Pt/C catalyst in both alkaline and neutral solutions. However, transition metal compounds still have drawbacks such as poor electrical conductivity, insufficient intrinsic activity, and easy dissolution due to reactions with electrolytes. In the future, more in-depth research needs to be carried out in aspects such as improving electrical conductivity, activity, and stability.

2.4 Noble metal catalyst

In recent years, numerous nonnoble metal catalysts, including carbon-based materials, M–N–C catalysts, and transition metal compound catalysts, have attracted extensive attention. Due to their inferior performance and stability, noble metal (e.g., Pt, Pd) catalysts still maintain an irreplaceable advantage as excellent ORR catalysts. However, the large scale consumption of noble metals severely restricts the commercialization process of AEMFC. Therefore, numerous efforts are devoted to reducing the use of noble metals. Currently, noble metal catalysts are mainly classified into three categories: Pt NP catalysts, alloy catalysts, and hetero structure catalysts. Pt NPs represent the most widely used noble metal catalyst system, with their activity highly dependent on size effects and crystal facet orientation. A previous study (Ma et al., 2021) has shown that Pt NPs with a particle size ranging from 2–4 nm exhibit the highest mass activity for ORR. Larger sized particles will hide most of the Pt atoms inside the particles, thus reducing the utilization rate of Pt atoms. In addition, high-index crystal planes (e.g., Pt (311)) exhibit oxygen-containing intermediate adsorption energies closer to the volcano plot peak, and their ORR performance is significantly superior to low-index crystal planes (e.g., Pt (111)) due to their abundance of edge sites (Zhu et al., 2019). Furthermore, introducing transition metals to form Pt–M alloys or intermetallic compound catalysts can synergistically modulate electronic structures and geometric effects, thereby enhancing ORR activity. Zhao et al. (2022) synthesized a ZIF-8-derived carbon-supported $Pt_3Co_{0.6}Ti_{0.4}$ intermetallic compound. The partial substitution of Co by Ti synergized ligand effects with the ordered atomic structure, endowing the catalyst with a mass activity of 1.84 A mg^{-1}Pt at 0.9 V versus RHE and exceptional stability with over 99% activity retention after 20,000 accelerated durability test cycles. Moreover, the construction of hetero-structured catalysts, such as core-shell architectures, enables dual enhancement in active site exposure and stability. Liu et al. (2018) synthesized a Co_2P@Pt core shell structure catalyst that utilizes nonprecious metals (Co_2P) as the core and an ultrathin Pt layer as the shell. This design reduces Pt loading while enhancing intrinsic activity via interfacial electron transfer effects, achieving a mass activity of 960 mA mg^{-1} Pt.

3 Catalyst design strategy

3.1 Atomic scale regulation: Active site coordination environment design

In the atomic-scale design of ORR catalysts, the coordination environment of SACs plays a crucial role in their electrochemical performance. Rational design of the coordination environment enables modulation of the electronic structure at active sites, including, but not limited to, the valence state distribution of central metal atoms, spin-state configuration, and the electron density distribution near the Fermi level (Shen et al., 2019). The improvement of the inherent catalytic performance of active sites can be achieved through designing a coordination environment by heteroatom doping, dual metal synergism, and carbon material defect engineering. These optimization strategies not only reshape the electronic structure of the central metal atom but also tune the adsorption and activation of oxygen and oxygenated intermediates at active sites, thereby significantly boosting the four-electron ORR activity.

Heteroatom doping is a highly delicate strategy that optimizes the intrinsic ORR activity of catalysts by incorporating nonmetallic elements beyond nitrogen (such as B, O, P, and S) into the coordination structure of active sites (Chen et al., 2023; Pan et al., 2022; Zhao et al., 2023). These nonmetallic elements can not only replace the nitrogen atoms closest to the central metal atom but also act on the next nearest carbon atoms, achieving precise tuning of the coordination environment through this substitution mechanism. In addition, introducing heteroatoms with diverse electronegativities alters the adsorption and activation of oxygen and oxygenated intermediates at active sites, thereby optimizing ORR catalytic performance. Shang et al. (2020) synthesized an S–Cu-ISA/SNC catalyst with an asymmetric Cu–S_1N_3 coordination structure by pyrolyzing a sulfur powder and Cu-ZIF-8 mixture. Extended X-ray absorption fine structure analysis revealed the coexistence of Cu–S bonds (2.32 Å) and Cu–N bonds (1.98 Å) around Cu atoms, confirming their asymmetric coordination environment, as shown in Fig. 5.6A and B. Volcano plot analysis based on oxygen intermediate adsorption free energy and overpotential demonstrated that the Cu–S1–N3 coordination structure lies closer to the activity peak than conventional Cu–N4 sites, as shown in Fig. 5.6C. This indicates optimized adsorption-desorption equilibrium for oxygen intermediates, resulting in superior ORR activity. Qu et al. (2025) introduced axial -SO_2 groups into the Fe–N_4 structure, regulating the spatial coordination environment of Fe single-atom sites to enhance the Lewis acidity of Fe sites (i.e., the ability of a substance to accept electron pairs). This effectively reduced the adsorption-free energy of *OOH and *OH intermediates, achieving a half-wave potential of 0.910 V for the ORR under alkaline conditions.

Besides heteroatom doping, some DACs with paired active centers (e.g., homonuclear Fe–Fe or Co–Co and heteronuclear Fe–Co-DACs) have demonstrated stronger oxygen adsorption capabilities than SACs, effectively lowering the energy barrier for O–O bond cleavage and showing higher ORR activity and greater selectivity for the four-electron mechanism (Li et al., 2022). Li et al. (2024) developed a series of Fe-metal dual-atom configuration catalysts, as shown in Fig. 5.6D. The introduction of different metal single atoms caused the Fe–N bond lengths in the coordination structure of iron single atoms to stretch or contract, thereby altering the electronic structure of Fe and regulating the ORR catalytic activity. The comparison of onset potential and half-wave potential between Fe–N–C and FeRu–N–C indicates that the Fe–Ru dual-atom configuration has the optimal ORR catalytic activity, as shown in Fig. 5.6E. In addition, in the stability test, after 5000 CV tests, the half-wave potential of Fe–Ru only decreased from 0.860 V to 0.846 V.

Carbon defect engineering, as an advanced strategy to precisely tailor the microstructure and properties of materials, has garnered significant attention (Zhao et al., 2024b; Wei et al., 2023). The core of this strategy lies in introducing multiple types of carbon defects into the next nearest neighbor carbon lattice around metal single atoms, enabling fine-tuning of the coordination environment of metal single atoms and thereby altering the electronic structure of metal atoms at active sites. As shown in Fig. 5.6F and G, when the next nearest neighbor of Fe–N_2 is a five-membered carbon ring, it forms the S1 site (FeN_4C_{10});

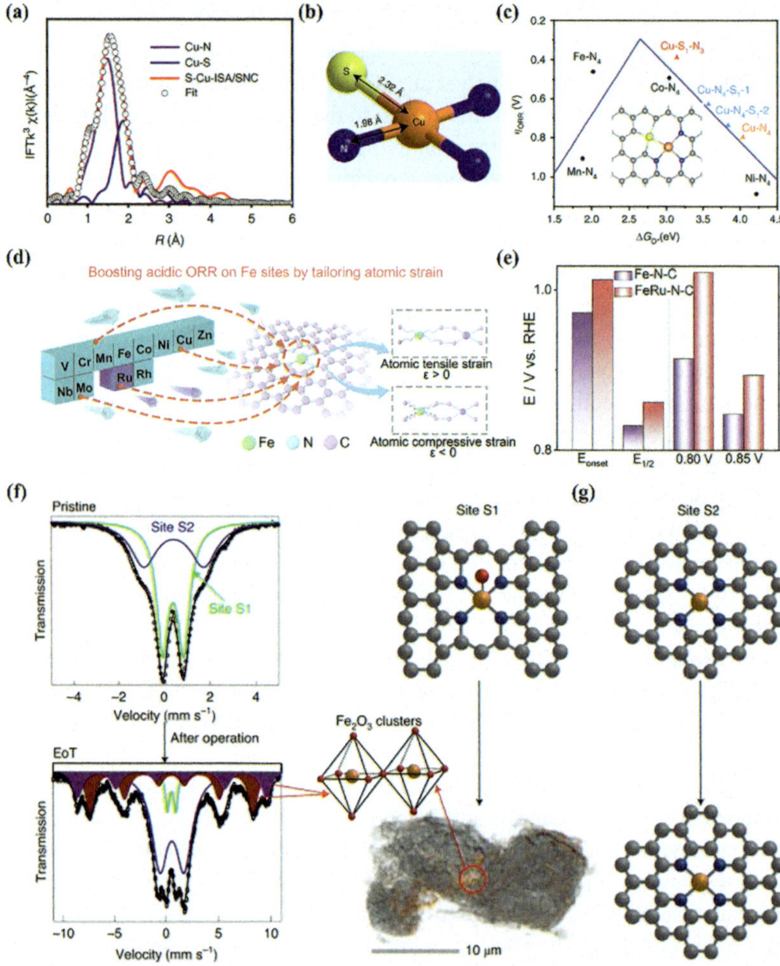

FIG. 5.6 Design strategies for the coordination environment of active sites. (A) FT Cu K-edge EXAFS fitting curves (Shang et al., 2020) and (B) the atomic model of S–Cu-ISA/SNC; (C) ORR overpotential (η_{ORR}) as a function of O* adsorption free energies (ΔG_{O*}) on different Cu-centered moieties; (D) schematic diagram of the modification of the Fe site by the second metal site in the Fe-metal diatomic coordination structure (Li et al., 2024); (E) comparison of onset potential and half-wave potential between Fe–N–C and FeRu–N–C; (F) structural schematic diagrams and Mössbauer spectra of the S1 site before and after cycling (Li et al., 2020); (G) structural schematic of the S2 site before and after cycling. *(Adapted from Shang, H., Zhou, X., Dong, J., Ang, L., Zhao, X., Liu, Q., Lin, Y., Pei, J., Zhi, L., Jiang, Z., Zhou, D., Zheng, L., Wang, Y., Zhou, J., Yang, Z., Cao, R., Sarangi, R., Sun, T., Yang, X., Zheng, X., Yan, W., Zhuang, Z., Jia, L., Chen, W., Wang, D., Zhang, J., Yadong, L., 2020. Engineering unsymmetrically coordinated Cu-S_1N_3 single atom sites with enhanced oxygen reduction activity. Nat. Commun. 2041-1723. 11 (1). doi:10.1038/s41467-020-16848-8; Li, L., Zhu, J., Kong, F., Wang, Y., Kang, C., Xu, M., Du, C., Yin, G., 2024. Tailoring atomic strain environment for high-performance acidic oxygen reduction by Fe-Ru dual atoms communicative effect. Matter. 25902385. 7 (4), 1517–1532. doi:10.1016/j.matt.2024.01.019.; Li, J., Sougrati, M.T., Zitolo, A., Ablett, J.M., Oğuz, I.C., Mineva, T., Matanovic, I., Atanassov, P., Cicco, A.D., Kumar, K., Dubau, L., Maillard, F., Jaouen, F., 2020. Identification of durable and non-durable FeNx sites in Fe-N-C materials for proton exchange membrane fuel cells. ChemRxiv 25732293. Available from: https://chemrxiv.org. doi:10.26434/chemrxiv.11842431.v1, with permission of Springer Nature and Elsevier.)*

when it is a six-membered carbon ring, it constitutes the S2 site (FeN_4C_{12}). The next nearest neighbor carbon ring structure regulates the electronic structure of the central Fe atom by influencing the neighboring N atoms (Li et al., 2020). Mössbauer spectroscopy reveals that the Fe atom in the S1 site exhibits a high-spin state, while the Fe atom in the S2 site is in a low-spin state. These different spin states affect the ORR catalytic performance at the three-phase interface: the high-spin state of the S1 site endows higher ORR activity, but its stability is relatively poor. It is prone to demetallation and tends to agglomerate into Fe_2O_3 clusters, reducing long term reliability. In contrast, the S2 site, under the low-spin state, although showing lower ORR activity, has an extremely stable structure.

3.2 Nanoscale optimization: Catalysts' surface modification and three-phase interfaces design

The nanoscale design strategy in this section mainly focuses on Pt based catalysts. The core of nanoscale design lies in optimizing the exposure of active sites, mass transport pathways, and interfacial interactions by controlling the size, morphology, surface structure, and composite architecture of catalysts. Specific strategies for surface structure modification of catalysts can be categorized into four aspects: regulation of Pt NP size and morphology (Li et al., 2016; Marković et al., 1994), combining Pt with another metal to generate heterostructure nanocrystals (Campbell, 1990; Rodriguez, 1996), surface modification (Ferrandez et al., 2013), and choosing appropriate catalyst supports. Beyond intrinsic surface modifications, the triple phase interface architecture at the nano level also needs to be optimized. This optimization ensures efficient reactant/product transport, thereby enhancing the ORR activity (Fan et al., 2023).

3.2.1 Nanoparticle size and morphology control

Catalyst activity is primarily governed by particle size, where reduced dimensions enhance performance through two synergistic mechanisms. Decreasing particle size to the nanoscale regime increases surface energy by exposing previously buried atoms, while simultaneously promoting the formation of high energy sites such as steps, edges, and kinks. These structural features exhibit elevated surface free energy and serve as preferential locations for catalytic reactions due to their optimized adsorption/desorption thermodynamics. In addition, reduced size enhances the interaction between the metal and the support, thereby improving catalytic activity and stability. Therefore, decreasing particle size is widely regarded as one of the most important methods to improve atomic utilization and catalytic activity. Previous studies (Ma et al., 2021; Gan et al., 2016) have shown that when the particle size decreases to the range of 2–5 nm, Pt exhibits optimal mass activity. The interior atoms of NPs are blocked by surface atoms and hardly participate in ORR processes. Therefore, hollow Pt NPs, created by removing interior Pt atoms from solid NPs, can maintain the original catalytic activity while reducing Pt usage. Chen et al. (2020c) prepared Pt–Co rhombic dodecahedra and then etched the inner Co-rich phases with nitric acid, leaving the Pt-rich phases at the edges, which in turn resulted in the Pt–Co nanoframe, as shown in Fig. 5.7A. The hollow internal structure of the Pt–Co nanoframe causes a decrease in the binding energy of oxygenated species by further shifting the d-band center downward, and its mass activity is 0.40 A mg^{-1} Pt. Meanwhile, the hollow structure increases active site exposure and optimizes internal mass transport, further boosting catalytic efficiency.

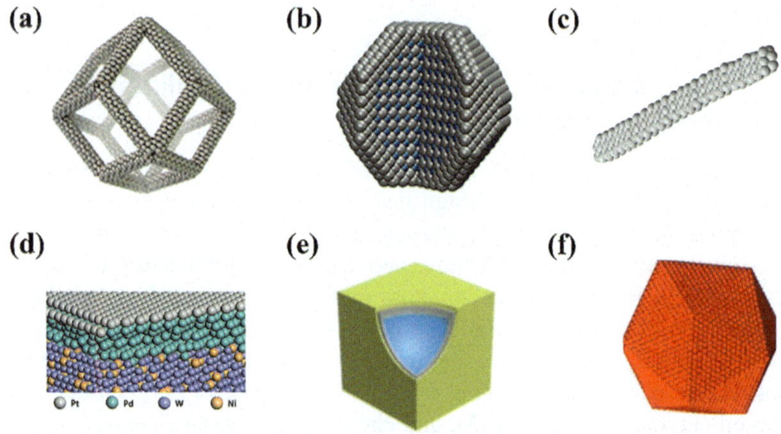

FIG. 5.7 New catalyst structure: (A) Nanoframe (Chen et al., 2020c), (B) Core-shell nanoparticle (Li et al., 2019), (C) Nanowire (Li et al., 2016), (D) Nanoplate (Song et al., 2020), (E) Nanocubes (He et al., 2021), (F) Nanopolyhedra (Zhu et al., 2016). *(Adapted from Chen, S., Li, M., Gao, M., Jin, J., Van Spronsen, M.A., Salmeron, M.B., Yang, P., 2020c. High-Performance Pt-Co Nanoframes for Fuel-Cell Electrocatalysis. Nano Lett. 15306992. 20 (3), 1974–1979. Available from: http://pubs.acs.org/journal/nalefd. doi:10.1021/acs.nanolett.9b05251; Li, M., Zhao, Z., Cheng, T., Fortunelli, A., Chen, C.Y., Yu, R., Zhang, Q., Gu, L., Merinov, B.V., Lin, Z., Zhu, E., Yu, T., Jia, Q., Guo, J., Zhang, L., Goddard, W.A., Huang, Y., Duan, X., 2016. Ultrafine jagged platinum nanowires enable ultrahigh mass activity for the oxygen reduction reaction. Science 10959203. 354 (6318), 1414–1419. Available from: http://science.sciencemag.org/content/sci/354/6318/1414.full.pdf. doi:10.1126/science.aaf9050; Li, J., Sharma, S., Liu, X., Pan, Y.-T., Spendelow, J.S., Chi, M., Jia, Y., Zhang, P., Cullen, D.A., Xi, Z., Lin, H., Yin, Z., Shen, B., Muzzio, M., Yu, C., Kim, Y.S., Peterson, A.A., More, K.L., Zhu, H., Sun, S., 2019. Hard-magnet L10-CoPt nanoparticles advance fuel cell catalysis. Joule. 25424351. 3 (1), 124–135. doi:10.1016/j.joule.2018.09.016.; Song, L., Liang, Z., Nagamori, K., Igarashi, H., Vukmirovic, M.B., Adzic, R.R., Sasaki, K., 2020. Enhancing Oxygen Reduction Performance of Pt Monolayer Catalysts by Pd(111) Nanosheets on WNi Substrates. ACS Catal. 21555435. 10 (7), 4290–4298. Available from: http://pubs.acs.org/page/accacs/about.html. doi:10.1021/acscatal.0c00040; He, T., Wang, W., Yang, X., Shi, F., Ye, Z., Zheng, Y., Li, F., Wu, J., Yin, Y., Jin, M., 2021. Deposition of atomically thin Pt shells on amorphous palladium phosphide cores for enhancing the electrocatalytic durability. ACS Nano. 1936-0851. 15 (4), 7348–7356. doi:10.1021/acsnano.1c00602.; Zhu, J., Xiao, M., Li, K., Liu, C., Zhao, X., Xing, W., 2016. Active Pt3Ni (111) Surface of Pt3Ni Icosahedron for Oxygen Reduction. ACS Appl. Mater. Interfaces. 19448252. 8 (44), 30066–30071. Available from: http://pubs.acs.org/journal/aamick. doi:10.1021/acsami.6b04237, with the permission of American Chemical Society and Elsevier.)*

In addition to ordinary Pt NPs, core-shell structures with special morphologies and structures, one-dimensional nanowires, two-dimensional nanoplates and nanosheets, and three-dimensional nanopolyhedrons and nanocages have been extensively studied (Li et al., 2016; Chen et al., 2024; Li et al., 2019; Chia and Pumera, 2018; Song et al., 2020; He et al., 2021; Zhu et al., 2016). Li et al. (2019) synthesized $L1_0$-CoPt core-shell catalyst structures with a size of 9 nm via annealing and etching under acidic conditions, as shown in Fig. 5.7B. In the MEA test, its mass activity reached as high as 0.56 A mg^{-1} Pt. Li et al. (2016) fabricated single-crystalline PtCo nanowires under mechanical stirring, as shown in Fig. 5.7C. The unique rhombic structure effectively enhances ORR catalytic activity and mass activity. The two-dimensional nanomaterials, nanosheets, and nanoplates exhibit remarkable advantages in catalytic efficiency and durability owing to their large specific surface area, excellent electrical conductivity, and uniform active site distribution (Chia and Pumera, 2018; Song et al., 2020). Song et al. (2020) reported a facile method for preparing Pt monolayer catalysts on Pd nanosheet/tungsten-nickel/carbon (PdNS/W–Ni/C) 2D substrates, as shown in Fig. 5.7D. Compared with commercial Pt/C, the oxygen-containing intermediate peak of PtML/PdNS/W–Ni/C shifted to a higher potential by approximately 75 mV, indicating that oxygen-containing species desorbed more easily on this catalyst and that it exhibited higher ORR catalytic activity. He et al. (2021) synthesized cubic Pd@Pt core-shell catalysts via glucose reduction, depositing high-density Pt octahedra on Pd crystals, as shown in Fig. 5.7E. Higher glucose concentration increased Pt deposition but reduced particle size. The catalysts with Pt octahedral shells showed enhanced ORR activity, benefiting from Pt (111) facets, achieving a mass activity of 0.164 A mg^{-1} Pt. Zhu et al. (2016) synthesized (111)-faceted icosahedral Pt_3M nanocrystals, as shown in Fig. 5.7F. Notably, the Pt_3Ni/C catalyst exhibited a significantly positive shift in redox potential due to the exposure of highly active Pt (111) facets, leading to an improvement in the ORR catalytic activity.

3.2.2 Nanoalloy catalysts construction

Integrating transition metals (such as Fe, Co, Ni, and Ru) into noble metal catalysts is a common method to regulate the electronic structure of such catalysts. The introduction of a second metal can generate a PtM alloy structure, as stated in Section 2.4. Due to the electronic interaction between the second metal and Pt, the d-band center of Pt can be lowered, leading to more filling of the Pt–O antibonding orbitals. This reduces the adsorption capacity for intermediate species, thereby enhancing the ORR activity of the alloy catalyst (Xiong et al., 2017). In addition, since the atomic size of the second metal is different from that of Pt, lattice strain will be generated inside the alloy particles after alloying. This strain increases the electron occupancy in the metal and causes the metal d-band center to decrease, an effect generally referred to as the strain effect. The atomic size of most transition metals is smaller than that of Pt, which causes compressive stress in alloy particles. Yang et al. (2021) synthesized 46 kinds of

Pt-based intermetallic compounds via universal high-temperature sulfur confinement and summarized a linear relationship between the intrinsic oxygen reduction activity of intermetallic compounds and the two-dimensional surface stress. It can be found that the catalytic activity monotonically increases with the enhancement of compressive strain. Among them, the PtNi intermetallic compound exhibits a mass activity of up to 1.84 $A\,mg^{-1}$ Pt in MEAs, demonstrating excellent application prospects.

3.2.3 Catalyst surface modification

Chemically modifying the surface via dopant introduction, which commonly involves substances such as metal clusters, molecules, ions, and organic or inorganic compounds to regulate surface electronic structures and interfacial interactions, thereby enhances the adsorption of oxygenated intermediates and oxygen and serves as an effective optimization strategy at the nanoscale (Ferrandez et al., 2013; Jiang et al., 2010; Chen et al., 2012). Different from atomic-level optimization, nanoscale surface optimization focuses on adjusting the overall surface structure and chemical state of NPs. Zhang et al. (2018) optimized the performance of the ORR by altering the structure of the cationic chains in ionic liquids to regulate their modification effects on the Pt catalyst surface. The modification by the ionic liquids changed the electronic properties and hydrophilicity/hydrophobicity of the Pt surface, thereby influencing the adsorption of reactants and reaction pathways on the surface. Recently, Lenne et al. (2023) regulated the interface of gold nanocatalysts through the chemical properties of calixarene ligand terminal groups and found that differences in terminal group hydrophilicity, oxygen affinity, etc., could modulate the interactions between the catalyst and electrolyte/reactants, significantly enhancing the efficiency, selectivity, and durability of the ORR. Asahi et al. (2019) utilized melamine-formaldehyde polymers to modify Pt NPs, enhancing their ORR activity. The polymer modification not only altered the surface chemical properties of the Pt NPs but also regulated their electronic structure, endowing the catalyst with higher catalytic activity toward the ORR.

3.2.4 Catalyst support optimization

In addition to structural improvements of catalysts, the properties of catalyst supports directly influence the particle size, distribution, and stability of catalysts, thereby affecting their catalytic activity and durability. Consequently, researchers have focused on utilizing novel materials and modifying support materials to enhance the performance of catalysts (Li et al., 2012). Carbon-based supports, such as CNs (Liu et al., 2017), graphene (Zeng et al., 2015), and carbon nanofibers (He et al., 2022), are promising due to their high specific surface area, suitable pore size distribution, and excellent electronic conductivity, as shown in Fig. 5.8A–C. To further enhance catalytic performance, many studies have conducted numerous nanoscale optimizations based on these novel carbon materials. For example, Li et al. demonstrated that intercalating carbon black between graphene sheets prevents the aggregation of GO, facilitating oxygen diffusion through the GO layers. Moreover, porous carbon supports can effectively enhance ORR performance. Yarlagadda et al. (2018) designed a porous carbon support, as shown in Fig. 5.8D, where catalyst particles are embedded within the micropores of the carbon matrix. This structure prevents the catalyst from being covered by the electrolyte while allowing ions and oxygen molecules to diffuse through the micropores and reach the catalyst surface, thereby improving ORR activity. In addition, the properties of carbon supports influence the distribution of catalysts, thereby affecting ORR performance. For example, Islam et al. (2022) synthesized a novel nitrogen-functionalized carbon

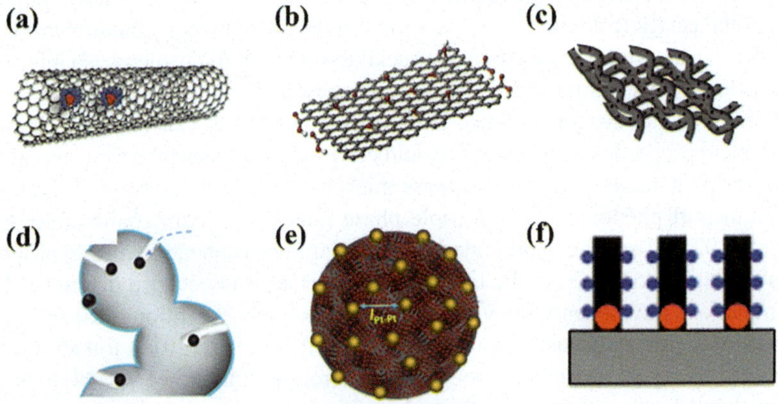

FIG. 5.8 New carbon support structure: (A) Carbon nanotube (Liu et al., 2017), (B) Graphene (Zeng et al., 2015), (C) Carbon nanofiber (He et al., 2022), (D) Porous carbon (Yarlagadda et al., 2018), (E) N-modified carbon (Islam et al., 2022), and (F) Vertically aligned carbon nanotubes (VANTs) (Murata et al., 2014). *(Adapted from Liu, S., Wang, Y., Liu, L., Mengli, L., Wenjie, L., Zhao, X., Qin, Z., Zhu, P., Wang, G., Long, Z., Huang, F., 2017. One-pot synthesis of Pd@PtNi core-shell nanoflowers supported on the multi-walled carbon nanotubes with boosting activity toward oxygen reduction in alkaline electrolyte. J. Power Sources. 03787753. 365, 26–33. doi:10.1016/j.jpowsour.2017.08.073; Zeng, L., Zhao, T.S., An, L., Zhao, G., Yan, X.H., Jung, C.Y., 2015. Graphene-supported platinum catalyst prepared with ionomer as surfactant for anion exchange membrane fuel cells. J. Power Sources. 03787753. 275, 506–515. doi:10.1016/j.jpowsour.2014.11.021; He, H., Lei, Y., Liu, S., Thummavichai, K., Zhu, Y., Wang, N., 2022. Tunable active-sites of Co-nanoparticles encapsulated in carbon nanofiber as high performance bifunctional OER/ORR electrocatalyst. J. Colloid Interface Sci. 630, 140–149; Yarlagadda, V., Carpenter, M.K., Moylan, T.E., Kukreja, R.S., Koestner, R., Gu, W., Thompson, L., Kongkanand, A., 2018. Boosting fuel cell performance with accessible carbon mesopores. ACS Energy Lett. 23808195. 3 (3), 618–621. Available from: http://pubs.acs.org/journal/aelccp. doi:10.1021/acsenergylett.8b00186; Islam, M.N., Basha, A.B.M., Kollath, V.O., Soleymani, A.P., Jankovic, J., Karan, K., 2022. Designing fuel cell catalyst support for superior catalytic activity and low mass-transport resistance. Nat. Commun. 20411723. 13 (1). Available from: https://www.nature.com/ncomms/. doi:10.1038/s41467-022-33892-8; Murata, S., Imanishi, M., Hasegawa, S., Namba, R., 2014. Vertically aligned carbon nanotube electrodes for high current density operating proton exchange membrane fuel cells. J. Power Sources. 03787753. 253, 104–113. doi:10.1016/j.jpowsour.2013.11.073, with the permission of Elsevier, American Chemical Society, and Springer Nature.)*

support, as depicted in Fig. 5.8E. This support enables uniform dispersion of Pt NPs, leading to enhanced ORR activity. Vertically aligned CNs have attracted attention due to their advantages over traditional carbon powders or porous carbon films, such as high specific surface area, excellent electrical conductivity, and unique ordered structure, as shown in Fig. 5.8F (Murata et al., 2014). This ordered carbon support structure accelerates the transport of reactants and products in the ORR, thereby improving catalytic performance.

3.2.5 *Three phase interface structures design*

The performance of catalysts is not only determined by their inherent active sites but is also heavily influenced by the efficient utilization at the triple phase interface, which determines the pathways for reactant access and product departure from the catalytic active sites. In addition to the advantageous modifications of surface structure for catalysts, it is crucial to optimize the triple phase interface structures. This optimization strategy is expected to substantially boost overall electrocatalytic performance. Although experiments can measure electrochemical properties such as activity and electrochemical surface area, as well as transport characteristics such as mass transport resistance, in-situ observations of transport phenomena at the triple phase interface remain challenging (Fan et al., 2025). To understand underlying mass transport phenomena and propose optimization strategies at the triple-phase interface, molecular dynamics (MD) simulations have been widely utilized owing to their exceptional capability in probing nanoscale transport mechanisms. Kurihara et al. (2019) found that the oxygen transport resistance through the electrolyte/Pt interface was higher than that through other regions in electrolyte films, meaning that the electrolyte/Pt interface played a dominant role in the oxygen transport resistance that hindered the ORR, as shown in Fig. 5.9A and B. Thus, suppressing the dense layer is crucial to reducing oxygen transport resistance. Zhou et al. (2024) also simulated the oxygen transport phenomenon through electrolyte films on different Pt particle morphologies and found that the electrolyte dense layer density is lower on 1D catalysts than on 3D ones, as shown in Fig. 5.9C. Finally, the oxygen flux to the 1D nanowire is approximately 2.3 times higher than that of 3D catalysts, as shown in Fig. 5.9D. Furthermore, Fan et al. (2022) conducted MD simulations to investigate the effect of ionic liquids on local oxygen transport, revealing that the addition of ionic liquids inhibits dense electrolyte packing on Pt particles, which reduces local oxygen transport resistance.

4 Future perspectives: Cross-scale simulation-guided design strategies

In the development of novel catalysts, computational simulation techniques have evolved into equally critical research tools alongside experimental characterization. The establishment of an iterative optimization paradigm encompassing computational design, performance prediction, and experimental

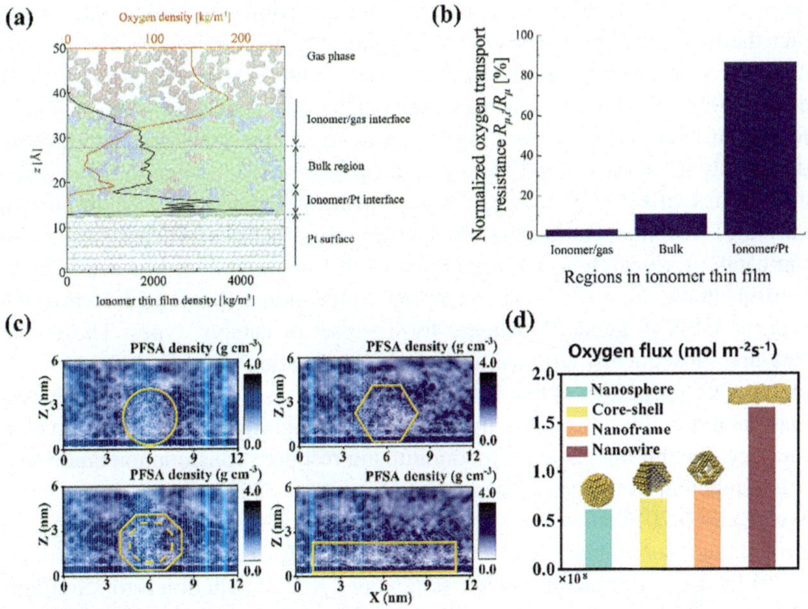

FIG. 5.9 Molecular dynamics (MD) simulation of oxygen transport at the three-phase interface. (A) Electrolyte film and oxygen density profiles in the thickness direction (Kurihara et al., 2019); (B) Oxygen transport resistances of three regions; (C) PFSA density distribution for different Pt particle structures (Zhou et al., 2024); (D) Oxygen flux for different Pt particle structures. *(Adapted from Kurihara, Y., Mabuchi, T., 2019. Takashi Tokumasu, Molecular dynamics study of oxygen transport resistance through ionomer thin film on Pt surface. J. Power Sources. 03787753. 414, 263–271. doi:10.1016/j.jpowsour.2019.01.011; Zhou, Y., Fan, L., Wang, J., Qing, D., Jiao, K., 2024. Revealing local oxygen transport in ionomer films on multidimensional nanoscale catalysts in fuel cells. J. Mater. Chem. A. 2050-7488. 12 (27), 16419–16426. doi:10.1039/d4ta01633e, with the permission of American Chemical Society and Royal Society of Chemistry.)*

verification can significantly reduce the temporal and economic costs associated with traditional trial-and-error approaches. Current mainstream catalytic simulation methodologies encompass DFT for electronic structure level analysis of active sites, MD simulations elucidating microscale dynamic interfacial reaction mechanisms, and the Lattice Boltzmann Method (LBM) for investigating mesoscale mass transport processes (Fan et al., 2023).

In recent years, the integration of machine learning (ML) with conventional theoretical approaches, such as DFT has been widely adopted for developing novel catalysts. Compared to traditional techniques, ML offers distinct advantages, including exceptional processing speeds for large datasets, enhanced precision in task execution, and irrelevant data filtering relative to laboratory-based manual approaches (Chen et al., 2020a; Li et al., 2021). ML algorithms can efficiently evaluate optimal environmental conditions, reaction parameters, and material formulations by simultaneously analyzing multiple variables, ultimately improving the accuracy of catalyst performance assessments and accelerating the development process. These capabilities enable ML to predict the theoretical properties of novel materials, facilitating the discovery of catalysts unattainable through conventional methods (Liu et al., 2012; Secor et al., 2023). Techniques such as neural networks, decision trees, and reinforcement learning have been implemented to simulate reaction kinetics, predict catalytic behavior, and recommend optimized reaction conditions. These approaches effectively manage multidimensional data (e.g., environmental variables, reaction parameters, and molecular structures), thereby streamlining catalytic research and development (Wang et al., 2020b; Wu et al., 2020). Current methodologies initially employ conventional theoretical approaches, such as DFT, to generate datasets for a subset of catalyst types. These datasets are subsequently utilized as training and testing data for ML models to extrapolate properties across all catalyst candidates, thereby identifying optimal catalytic solutions. This hybrid approach significantly reduces development cycles and lowers costs by minimizing resource consumption and energy waste inherent in conventional studies (Xing et al., 2022; Zhong et al., 2020). Wu et al. (2020) demonstrated a 130,000-fold acceleration in evaluating ORR performance for nonradioactive transition metal SACs using ML models trained on 14 DFT-derived metal species, achieving $<8.5\%$ prediction error. Similarly, Fang et al. (2024) combined DFT-derived thermodynamic data from 106 dual-metal-site catalysts with atomic properties to screen 623 candidates, establishing an efficient discovery pipeline. Zhang et al. (2024) employed crystal graph convolutional networks trained on 538 high-entropy intermetallic compounds to design $Pt(FeCoNiCu)_3$, which exhibited exceptional electrochemical stability with reduced computational complexity. These studies systematically replace a portion of conventional DFT computations with ML-trained datasets, thereby significantly reducing the required DFT calculations while enabling efficient screening of optimal catalytic solutions with maintained high prediction accuracy. The integration of traditional DFT methodologies with ML techniques creates a novel theoretical framework for catalyst development, which overcomes the computational efficiency and system size limitations inherent to conventional approaches. This hybrid strategy substantially decreases overall computational resource consumption throughout the entire workflow, rendering the development of novel catalysts more efficient and cost-effective.

MD simulations serve as a microscale modeling approach, while the LBM operates at the mesoscopic scale, collectively enabling in-depth investigations of phenomena and mass transport mechanisms across micro-mesoscopic regimes. These methodologies have been extensively employed to study the formation mechanisms of triplephase interfaces and mass transport characteristics within catalyst layers, thereby guiding structural optimizations (He et al., 2019; Liu et al., 2016; Yan et al., 2021; Yao et al., 2022). Murata et al. (2014) and Fan et al. (2022, 2020) utilized MD simulations to probe nanoscale oxygen transport near catalytic sites, revealing distinct oxygen transport resistance characteristics across different regions, which provided critical insights for optimizing oxygen-transport-efficient catalyst layer architectures. Furthermore, Chen et al. (2021), Shin et al. (2019), and Li et al. (2024) implemented LBM to investigate multiphase flow, oxygen diffusion, and electrochemical reaction processes within catalyst layers under varied fuel cell operating conditions, subsequently proposing corresponding structural design improvements. These multiscale simulation approaches effectively bridge microscopic mechanistic understanding with macroscopic performance optimization in catalyst development.

However, both the DFT-ML approach for developing novel catalyst architectures and the MD/LBM-based structural optimization strategies focus exclusively on single scale improvements, neglecting the synergistic effects of catalyst selection, design, and triple phase interface configuration on overall catalytic performance. We, therefore, propose a cross scale simulation-aided design strategy that integrates multiscale computational paradigms, as shown in Fig. 5.10. The cross scale simulation aided design strategy integrates multiscale computational paradigms. This methodology initially combines conventional DFT simulations with ML to rapidly develop novel nanocatalysts, subsequently constructs catalyst layer models based on these designed catalysts, and employs MD/LBM simulations to investigate triple phase interface formation mechanisms and mass transport characteristics within catalyst layers. This integrated workflow ultimately achieves full dimensional optimization of ORR catalyst design. The cross scale computational strategy establishes a comprehensive analytical framework bridging atomic level electronic structures with micro-mesoscopic transport phenomena through coupled multiscale investigations. It provides critical theoretical foundations for holistic catalyst design by correlating quantum scale catalytic activity with macroscopic transport efficiency. Compared to conventional single scale approaches, this strategy enables more efficient and cost effective development of high performance ORR catalysts with optimized structures. Future applications of this cross scale design paradigm are anticipated to become essential for achieving comprehensive ORR catalyst optimization. The methodology concurrently enhances catalytic activity, stability, and mass transport efficiency through systematic multiscale coordination, offering theoretical guidance and technical pathways for next generation

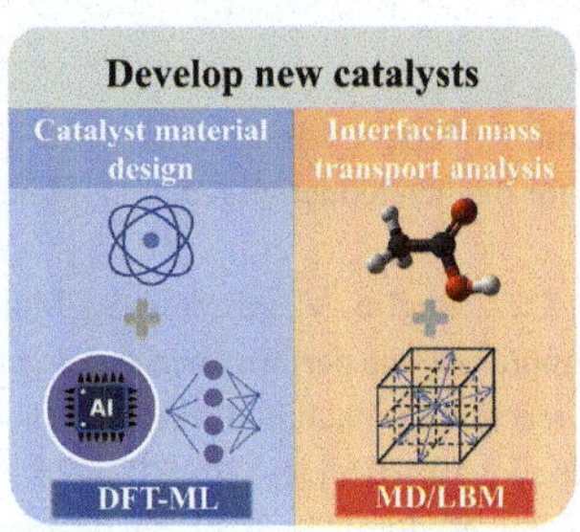

FIG. 5.10 The cross scale simulation aided design strategy that integrates multiscale computational paradigms.

fuel cell catalyst development. Its capacity to synergistically address electronic-structural requirements and transport limitations represents a transformative advancement in rational catalyst design methodologies.

5 Summary

ORR, as the core reaction of AEMFC, primarily proceeds via two-electron and four-electron pathways, with the latter being more desirable for high efficiency, as it directly generates OH^- without corrosive intermediates. Although Pt-based catalysts are the most favorable for the four-electron pathway of the ORR, their high cost hinders the commercialization of AEMFCs. Therefore, metal-free carbon based catalysts, metal-nitrogen-carbon catalysts, and transition metal compounds have emerged as potential alternatives to Pt based noble metal catalysts, especially for carbon based transition metal SACs. However, these catalysts face challenges of sluggish reaction kinetics and insufficient stability during the ORR process. Therefore, the atomic scale design focusing on coordination environment regulation is crucial, and its strategies mainly involve heteroatom doping, dual-metal synergism, and carbon material defect engineering. These optimization strategies not only reshape the electronic structure of the central metal atom but also tune the adsorption and activation of oxygen and oxygenated intermediates at active sites, thereby significantly boosting the four-electron ORR activity. Moreover, reducing the Pt usage by controlling the NP size and morphology and constructing nanoalloy catalysts is another way of promoting the commercialization of AEMFC. In addition to enhancing the intrinsic activity of the catalyst, the designs of catalyst support and the three phase interface also play a key role in the ORR performance, as they determine the transport efficiency of reactants and products of ORR. However, most theoretical calculation studies are devoted to enhancing the intrinsic activity of the catalyst but ignore the synergistic effect of mass transport. Therefore, we propose a cross scale simulation assisted design strategy: design novel nanocatalysts via DFT-ML integration, then investigate the three phase interface and mass transport in catalyst layers using MD/LBM simulations to achieve full-dimensional ORR catalyst optimization.

References

Abdelwahab, A., Carrasco-Marín, F., Pérez-Cadenas, A.F., 2019. Carbon xerogels hydrothermally doped with bimetal oxides for oxygen reduction reaction. Materials. 1996-1944. 12 (15), 2446. doi:10.3390/ma12152446.

Alemany-Molina, G., Vecchio, C.L., Baglio, V., Aricò, A.S., Morallón, E., Cazorla-Amorós, D., 2025. Pt nanoparticles for improving the performance and durability of Fe-N-C based materials towards oxygen reduction reaction in alkaline direct methanol fuel cells. J. Colloid Interface Sci. 00219797. 691, 137426. doi:10.1016/j.jcis.2025.137426.

Asahi, M., Yamazaki, S.-I., Taguchi, N., Ioroi, T., 2019. Facile approach to enhance oxygen reduction activity by modification of platinum nanoparticles by melamine-formaldehyde polymer. J. Electrochem. Soc. 0013-4651. 166 (8), F498. doi:10.1149/2.0641908jes.

Campbell, C.T., 1990. Bimetallic surface chemistry. Annu. Rev. Phys. Chem. 0066-426X. 41 (1), 775–837. doi:10.1146/annurev.pc.41.100190.004015.

Cao, Y., Huang, S., Peng, Z., Yao, F., Xiaohui, L., Liu, Y., Huang, H., Mingmei, W., 2021. Phase control of ultrafine FeSe nanocrystals in a N-doped carbon matrix for highly efficient and stable oxygen reduction reaction. J. Mater. Chem. A. 2050-7488. 9 (6), 3464–3471. doi:10.1039/d0ta09544c.

Chen, A., Zhang, X., Zhou, Z., 2020a. Machine learning: accelerating materials development for energy storage and conversion. InfoMat. 2567-3165. 2 (3), 553–576. doi:10.1002/inf2.12094.

Chen, H., Liang, X., Liu, Y., Xuan, A., Asefa, T., Zou, X., 2020b. Active site engineering in porous electrocatalysts. Adv. Mater. 0935-9648. 32 (44). doi:10.1002/adma.202002435.

Chen, L., Kang, Q., Tao, W., 2021. Pore-scale numerical study of multiphase reactive transport processes in cathode catalyst layers of proton exchange membrane fuel cells. Int. J. Hydrogen Energy. 03603199. 46 (24), 13283–13297. doi:10.1016/j.ijhydene.2021.01.146.

Chen, S., Li, M., Gao, M., Jin, J., Van Spronsen, M.A., Salmeron, M.B., Yang, P., 2020c. High-performance Pt-Co nanoframes for fuel-cell electrocatalysis. Nano Lett. 15306992. 20 (3), 1974–1979. Available from: http://pubs.acs.org/journal/nalefd. doi:10.1021/acs.nanolett.9b05251.

Chen, S., Liu, Y., Xiaogang, F., Wang, W., 2024. Recent advances in morphology control of platinum catalysts toward oxygen reduction reaction. Front. Energy. 2095-1701. 18 (3), 330–355. doi:10.1007/s11708-024-0929-5.

Chen, S., Wei, Z., Qi, X., Dong, L., Guo, Y.G., Wan, L., Shao, Z., Li, L., 2012. Nanostructured polyaniline-decorated Pt/C@PANI core-shell catalyst with enhanced durability and activity. J. Am. Chem. Soc. 15205126. 134 (32), 13252–13255. doi:10.1021/ja306501x.

Chen, Z., Peng, X., Chen, Z., Li, T., Zou, R., Shi, G., Huang, Y., Cui, P., Yu, J., Chen, Y., Chi, X., Loh, K.P., Liu, Z., Li, X., Zhong, L., Lu, J., 2023. Mass production of sulfur-tuned single-atom catalysts for Zn–air batteries. Adv. Mater. 15214095. 35 (15). Available from: http://onlinelibrary.wiley.com/journal/10.1002/(ISSN)1521-4095. doi:10.1002/adma.202209948.

Chia, X., Pumera, M., 2018. Characteristics and performance of two-dimensional materials for electrocatalysis. Nat. Catal. 2520-1158. 1 (12), 909–921. doi:10.1038/s41929-018-0181-7.

Cui, X., Gao, L., Lei, S., Liang, S., Zhang, J., Sewell, C.D., Xue, W., Liu, Q., Lin, Z., Yang, Y., 2021. Electrocatalysis: simultaneously crafting single-atomic fe sites and graphitic layer-wrapped Fe_3C nanoparticles encapsulated within mesoporous carbon tubes for oxygen reduction. Adv. Funct. Mater. 1616-301X. 31 (10). doi:10.1002/adfm.202170064.

Deng, L., Qiu, L., Rong, H., Yao, L., Zheng, Z., Ren, X., Yongliang, L., Chuanxin, H., 2022. Restricted diffusion preparation of fully-exposed Fe single-atom catalyst on carbon nanospheres for efficient oxygen reduction reaction. Appl. Catal. B Environ. 09263373. 305, 121058. doi:10.1016/j.apcatb.2021.121058.

Duan, Z., Han, G., Huo, H., Lin, Z., Liping, G., Chunyu, D., Gao, Y., Yin, G., 2021. Monovacancy coupled pyridinic n site enables surging oxygen reduction activity of metal-free CN_x catalyst. ACS Sustain. Chem. Eng. 2168-0485. 9 (3), 1264–1271. doi:10.1021/acssuschemeng.0c07490.

Fan, L., Deng, H., Zhang, Y., Du, Q., Leung, D.Y.C., Wang, Y., Jiao, K., 2023. Towards ultralow platinum loading proton exchange membrane fuel cells. Energy Environ. Sci. 17545706. 16 (4), 1466–1479. Available from: http://pubs.rsc.org/en/journals/journal/ee. doi:10.1039/d2ee03169h.

Fan, L., Wang, J., Diaz, D.F.R., Lincai, L., Wang, Y., Jiao, K., 2025. Molecular dynamics modeling in catalyst layer development for PEM fuel cell. Prog. Energy Combust. Sci. 03601285. 108, 101220. doi:10.1016/j.pecs.2025.101220.

Fan, L., Wang, Y., Jiao, K., 2020. Oxygen transport routes in ionomer film on polyhedral platinum nanoparticles. ACS Nano. 1936-0851. 14 (12), 17487–17495. doi:10.1021/acsnano.0c07856.

Fan, L., Wang, Y., Jiao, K., 2022. Enhancing oxygen transport in the ionomer film on platinum catalyst using ionic liquid additives. Fundam. Res. 26673258. 2 (2), 230–236. doi:10.1016/j.fmre.2021.09.004.

Fang, Z., Shuyuan, L., Zhang, Y., Wang, Y., Meng, K., Huang, C., Sun, S., 2024. The DFT and machine learning method accelerated the discovery of DMSCs with high ORR and OER catalytic activities. J. Phys. Chem. Lett.. 1948-7185. 15 (1), 281–289. doi:10.1021/acs.jpclett.3c02938.

Ferrandez, A.C., Baranton, S., Bigarré, J., Buvat, P., Coutanceau, C., 2013. Chemical functionalization of carbon supported metal nanoparticles by ionic conductive polymer via the "grafting From" method. Chem. Mater. 15205002. 25 (19), 3797–3807. doi:10.1021/cm401264f.

Gan, L., Rudi, S., Cui, C., Heggen, M., Strasser, P., 2016. Size-controlled synthesis of sub-10 nm $ptni_3$ alloy nanoparticles and their unusual volcano-shaped size effect on orr electrocatalysis. Small 16136829. 3189–3196. Available from: http://onlinelibrary.wiley.com/journal/10.1002/(ISSN)1613-6829. doi:10.1002/smll.201600027.

Ge, X., Liu, Y., Goh, F.W.T., Hor, T.S.A., Zong, Y., Xiao, P., Zhang, Z., Lim, S.H., Li, B., Wang, X., Liu, Z., 2014. Dual-phase spinel $MnCo_2O_4$ and spinel $MnCo_2O_4$/nanocarbon hybrids for electrocatalytic oxygen reduction and evolution. ACS Appl. Mater. Interfaces. 19448252. 6 (15), 12684–12691. Available from: http://pubs.acs.org/journal/aamick. doi:10.1021/am502675c.

Gupta, S., Tryk, D., Bae, I., Aldred, W., Yeager, E., 1989. Heat-treated polyacrylonitrile-based catalysts for oxygen electroreduction. J. Appl. Electrochem. 0021-891X. 19 (1), 19–27. doi:10.1007/bf01039385.

Han, G., Zheng, Y., Zhang, X., Wang, Z., Gong, Y., Chunyu, D., Banis, M.N., Yiu, Y.-M., Sham, T.-K., Lin, G., Sun, Y., Wang, Y., Wang, J., Gao, Y., Yin, G., Sun, X., 2019a. High loading single-atom Cu dispersed on graphene for efficient oxygen reduction reaction. Nano Energy. 22112855. 66, 104088. doi:10.1016/j.nanoen.2019.104088.

Han, X., Ling, X., Wang, Y., Ma, T., Zhong, C., Hu, W., Deng, Y., 2019b. Generation of nanoparticle, atomic-cluster, and single-atom cobalt catalysts from zeolitic imidazole frameworks by spatial isolation and their use in zinc–air batteries. Angew. Chem. Int. Ed. 15213773. 58 (16), 5359–5364. Available from: http://onlinelibrary.wiley.com/journal/10.1002/(ISSN)1521-3773. doi:10.1002/anie.201901109.

He, H., Lei, Y., Liu, S., Thummavichai, K., Zhu, Y., Wang, N., 2022. Tunable active-sites of Co-nanoparticles encapsulated in carbon nanofiber as high performance bifunctional OER/ORR electrocatalyst. J. Colloid Interface Sci. 630, 140–149.

He, T., Wang, W., Yang, X., Shi, F., Ye, Z., Zheng, Y., Li, F., Wu, J., Yin, Y., Jin, M., 2021. Deposition of atomically thin Pt shells on amorphous palladium phosphide cores for enhancing the electrocatalytic durability. ACS Nano. 1936-0851. 15 (4), 7348–7356. doi:10.1021/acsnano.1c00602.

He, Y.L., Liu, Q., Li, Q., Tao, W.Q., 2019. Lattice Boltzmann methods for single-phase and solid-liquid phase-change heat transfer in porous media: a review. Int. J. Heat Mass Transf. 00179310. 129, 160–197. Available from: http://www.journals.elsevier.com/international-journal-of-heat-and-mass-transfer/. doi:10.1016/j.ijheatmasstransfer.2018.08.135.

Islam, M.N., Basha, A.B.M., Kollath, V.O., Soleymani, A.P., Jankovic, J., Karan, K., 2022. Designing fuel cell catalyst support for superior catalytic activity and low mass-transport resistance. Nat. Commun. 20411723. 13 (1). Available from: https://www.nature.com/ncomms/. doi:10.1038/s41467-022-33892-8.

Jasinski, R., 1964. A new fuel cell cathode catalyst. Nature. 0028-0836. 201 (4925), 1212–1213. doi:10.1038/2011212a0.

Jiang, Z.Z., Wang, Z.B., Gu, D.M., Smotkin, E.S., 2010. Carbon riveted Pt/C catalyst with high stability prepared by in situ carbonized glucose. Chem. Commun. 13597345. 46 (37), 6998–7000. Available from: http://pubs.rsc.org/en/journals/journal/cc. doi:10.1039/c0cc02364g.

Jiao, K., Xuan, J., Qing, D., Bao, Z., Xie, B., Wang, B., Zhao, Y., Fan, L., Wang, H., Hou, Z., Huo, S., Brandon, N.P., Yin, Y., Guiver, M.D., 2021. Designing the next generation of proton-exchange membrane fuel cells. Nature. 0028-0836. 595 (7867), 361–369. doi:10.1038/s41586-021-03482-7.

Jin, L., Kang, W., Liao, M., Wang, D., Jayaraman, B., Peera, S.G., Liu, C., 2024. Theoretical analysis of single Zn atoms with N/C on graphene promoting oxygen redox in alkaline medium. Mol. Catal. 24688231. 553, 113799. doi:10.1016/j.mcat.2023.113799.

Kong, F., Cui, X., Huang, Y., Yao, H., Chen, Y., Tian, H., Meng, G., Chen, C., Chang, Z., Shi, J., 2022. N-doped carbon electrocatalyst: marked ORR activity in acidic media without the contribution from metal sites? Angew. Chem. Int. Ed. 15213773. 61 (15). Available from: http://onlinelibrary.wiley.com/journal/10.1002/(ISSN)1521-3773. doi:10.1002/anie.202116290.

Kurihara, Y., Mabuchi, T., Tokumasu, T., 2019. Molecular dynamics study of oxygen transport resistance through ionomer thin film on Pt surface. J. Power Sources. 03787753. 414, 263–271. doi:10.1016/j.jpowsour.2019.01.011.

Lee, S., Choun, M., Youngjin, Y., Lee, J., Mun, Y., Kang, E., Hwang, J., Lee, Y., Shin, C., Moon, S., Kim, S., Lee, E., Lee, J., 2015. Designing a highly active metal-free oxygen reduction catalyst in membrane electrode assemblies for alkaline fuel cells: effects of pore size and doping-site position. Angew. Chem. Int. Ed.. 1433-7851. 54 (32), 9230–9234. doi:10.1002/anie.201501590.

Lefèvre, M., Proietti, E., Jaouen, F., Dodelet, J.P., 2009. Iron-Based catalysts with improved oxygen reduction activity in polymer electrolyte fuel cells. Science. 10959203. 324 (5923), 71–74. doi:10.1126/science.1170051.

Lenne, Q., Mattiuzzi, A., Jabin, I., Hamon, J., Leroux, Y.R., Lagrost, C., 2023. Chemical tuning of metal nanocatalysts interface for ORR electrocatalysis. Adv. Mater. Interfaces. 21967350. 10 (6). Available from: http://onlinelibrary.wiley.com/journal/10.1002/(ISSN)2196-7350. doi:10.1002/admi.202202219.

Li, H., Zhang, X., Qin, Y., Liu, Y., Wang, J., Peng, L., Li, C., 2021. Crafting controllable Fe-based hierarchically organic-frameworks from bacterial cellulose nanofibers for efficient electrocatalysts in microbial fuel cells. J. Power Sources. 03787753. 512, 230522. doi:10.1016/j.jpowsour.2021.230522.

Li, J., Chen, M., Cullen, D.A., Hwang, S., Wang, M., Li, B., Liu, K., Karakalos, S., Lucero, M., Zhang, H., Lei, C., Xu, H., Sterbinsky, G.E., Feng, Z., Su, D., More, K.L., Wang, G., Wang, Z., Wu, G., 2018. Atomically dispersed manganese catalysts for oxygen reduction in proton-exchange membrane fuel cells. Nat. Catal. 25201158. 1 (12), 935–945. Available from: https://www.nature.com/natcatal/. doi:10.1038/s41929-018-0164-8.

Li, J., Sharma, S., Liu, X., Pan, Y.-T., Spendelow, J.S., Chi, M., Jia, Y., Zhang, P., Cullen, D.A., Xi, Z., Lin, H., Yin, Z., Shen, B., Muzzio, M., Yu, C., Kim, Y.S., Peterson, A.A., More, K.L., Zhu, H., Sun, S., 2019. Hard-magnet L10-CoPt nanoparticles advance fuel cell catalysis. Joule. 25424351. 3 (1), 124–135. doi:10.1016/j.joule.2018.09.016.

Li, J., Sougrati, M.T., Zitolo, A., Ablett, J.M., Oğuz, I.C., Mineva, T., Matanovic, I., Atanassov, P., Cicco, A.D., Kumar, K., Dubau, L., Maillard, F., Jaouen, F., 2020. Identification of durable and non-durable FeNx sites in Fe-N-C materials for proton exchange membrane fuel cells. ChemRxiv 25732293. Available from: https://chemrxiv.org. doi:10.26434/chemrxiv.11842431.v1.

Li, L., Zhu, J., Kong, F., Wang, Y., Kang, C., Xu, M., Du, C., Yin, G., 2024. Tailoring atomic strain environment for high-performance acidic oxygen reduction by Fe-Ru dual atoms communicative effect. Matter. 25902385. 7 (4), 1517–1532. doi:10.1016/j.matt.2024.01.019.

Li, M., Zhao, Z., Cheng, T., Fortunelli, A., Chen, C.Y., Yu, R., Zhang, Q., Gu, L., Merinov, B.V., Lin, Z., Zhu, E., Yu, T., Jia, Q., Guo, J., Zhang, L., Goddard, W.A., Huang, Y., Duan, X., 2016. Ultrafine jagged platinum nanowires enable ultrahigh mass activity for the oxygen reduction reaction. Science 10959203. 354 (6318), 1414–1419. Available from: http://science.sciencemag.org/content/sci/354/6318/1414.full.pdf. doi:10.1126/science.aaf9050.

Li, P., Wang, H.-L., 2021. Recent advances in carbon-supported iron group electrocatalysts for the oxygen reduction reaction. Carbon. 00086223. 183 (4), 1012. doi:10.1016/j.carbon.2021.07.084.

Li, Q., He, Y., Zhang, L., Pan, L.-M., Liu, H., Sun, W., Ma, Z., Zhu, L., Lian, Q., Tang, S., 2024. Pore-scale simulation of oxygen transport in a proton exchange membrane electrolysis cell: effect of the hydrophilia of porous transport layer and catalytic layer. J. Power Sources. 03787753. 595, 234048. doi:10.1016/j.jpowsour.2023.234048.

Li, S., Zhang, Y., Hu, Y., Wang, B., Sun, S., Yang, X., He, H., 2021. Predicting metal-organic frameworks as catalysts to fix carbon dioxide to cyclic carbonate by machine learning. J. Materiomics. 23528478. 7 (5), 1029–1038. doi:10.1016/j.jmat.2021.02.005.

Li, Y., Li, Y., Zhu, E., McLouth, T., Chiu, C.Y., Huang, X., Huang, Y., 2012. Stabilization of high-performance oxygen reduction reaction Pt electrocatalyst supported on reduced graphene oxide/carbon black composite. J. Am. Chem. Soc. 15205126. 134 (30), 12326–12329. doi:10.1021/ja3031449.

Li, Y., Nong, W., Zeng, Z., Wang, C., 2023. Degradation mechanism and enhancing strategies of oxygen reduction reaction catalyzed by carbon-based metal free catalysts in acidic solution. Adv. Energy Mater. 1614-6832. 13 (1). doi:10.1002/aenm.202203159.

Li, Z., Li, B., Hu, Y., Wang, S., Yu, C., 2022. Highly-dispersed and high-metal-density electrocatalysts on carbon supports for the oxygen reduction reaction: from nanoparticles to atomic-level architectures. Mater. Adv. 2633-5409. 3 (2), 779–809. doi:10.1039/d1ma00858g.

Liang, Y., Yanguang, L., Wang, H., Zhou, J., Wang, J., Regier, T., Dai, H., 2011. Co3O4 nanocrystals on graphene as a synergistic catalyst for oxygen reduction reaction. Nat. Mater. 1476-1122. 10 (10), 780–786. doi:10.1038/nmat3087.

Lin, L., Xing, D., Wang, J., Chen, H., Xuan, H., Fang, W., Yang, L., Chen, Z., Zhao, L., 2021. High-pressure vapor-solid transformation of 2D template to derive 2D Co–N–C electrocatalysts for oxygen reduction reaction. Nanotechnology. 0957-4484. 33 (10), 105601. doi:10.1088/1361-6528/ac3d64.

Liu, C., Zhong, M., Cui, M., Zhang, Z., Zhang, X., Dong, S., Murray, C.B., Wang, J.X., Zhang, S., 2018. Favorable core/shell interface within Co_2P/Pt nanorods for oxygen reduction electrocatalysis. Nano Lett. 1530-6984. 18 (12), 7870–7875. doi:10.1021/acs.nanolett.8b03666.

Liu, H., Kang, Q., Leonardi, C.R., Schmieschek, S., Narváez, A., Jones, B.D., Williams, J.R., Valocchi, A.J., Harting, J., 2016. Multiphase lattice Boltzmann simulations for porous media applications. Comput. Geosci. 1420-0597. 20 (4), 777–805. doi:10.1007/s10596-015-9542-3.

Liu, J., Xue, Y., Zhang, M., Dai, L., 2012. Graphene-based materials for energy applications. MRS Bull. 0883-7694. 37 (12), 1265–1272. doi:10.1557/mrs.2012.179.

Liu, S., Wang, Y., Liu, L., Mengli, L., Wenjie, L., Zhao, X., Qin, Z., Zhu, P., Wang, G., Long, Z., Huang, F., 2017. One-pot synthesis of Pd@PtNi core-shell nanoflowers supported on the multi-walled carbon nanotubes with boosting activity toward oxygen reduction in alkaline electrolyte. J. Power Sources. 03787753. 365, 26–33. doi:10.1016/j.jpowsour.2017.08.073.

Liu, Y., Wang, X., Zhao, B., Shao, X., Huang, M., 2019. Fe/Fe_3C nanoparticles encapsulated in N-doped hollow carbon spheres as efficient electrocatalysts for the oxygen reduction reaction over a wide pH range. Chem. Eur. J. 0947-6539. 25 (41), 9650–9657. doi:10.1002/chem.201806111.

Luo, Z., Wang, G., Gong, Z., et al., 2019. Single-atom Cr–N4 sites designed for durable oxygen reduction catalysis in acid media. Angew. Chem. Int. Ed. 58 (36), 12469–12475.

Lv, Q., Wang, N., Si, W., Hou, Z., Li, X., Wang, X., Zhao, F., Yang, Z., Zhang, Y., Huang, C., 2020. Pyridinic nitrogen exclusively doped carbon materials as efficient oxygen reduction electrocatalysts for Zn-air batteries. Appl. Catal. B Environ. 09263373. 261, 118234. doi:10.1016/j.apcatb.2019.118234.

Ma, Y., Kuhn, A. N., Gao, W., Al-Zoubi, T., Du, H., Pan, X., Yang, H., 2021. Strong electrostatic adsorption approach to the synthesis of sub-three nanometer intermetallic platinum–cobalt oxygen reduction catalysts. Nano Energy. 22112855. 79, 105465. doi:10.1016/j.nanoen.2020.105465.

Mao, X., Cao, Z., Yin, Y., Wang, Z., Dong, H., Yang, S., 2018. Direct synthesis of nitrogen and phosphorus co-doped hierarchical porous carbon networks with biological materials as efficient electrocatalysts for oxygen reduction reaction. Int. J. Hydrogen Energy. 03603199. 43 (22), 10341–10350. doi:10.1016/j.ijhydene.2018.04.100.

Marković, N.M., Adžić, R.R., Cahan, B.D., Yeager, E.B., 1994. Structural effects in electrocatalysis: oxygen reduction on platinum low index single-crystal surfaces in perchloric acid solutions. J. Electroanal. Chem. 15726657. 377 (1-2), 249–259. doi:10.1016/0022-0728(94)03467-2.

Murata, S., Imanishi, M., Hasegawa, S., Namba, R., 2014. Vertically aligned carbon nanotube electrodes for high current density operating proton exchange membrane fuel cells. J. Power Sources. 03787753. 253, 104–113. doi:10.1016/j.jpowsour.2013.11.073.

Niu, W., Luo, Y., Lanlan, W., Liu, G., Xiaohang, D., Wang, Y., Jingde, L., Chen, Z., 2022. Conductive oxide support design and synergistic engineering of bimetallic high-performance electrocatalyst for oxygen reduction reaction. Chem. Eng. J. 13858947. 442, 136266. doi:10.1016/j.cej.2022.136266.

Noh, W.Y., Mun, J., Lee, Y., Kim, E.M., Kim, Y.K., Kim, K.Y., Jeong, H.Y., Lee, J.H., Song, H.-K., Lee, G., Lee, J.S., 2022. Molecularly engineered carbon platform to anchor edge-hosted single-atomic M–N/C (M=Fe, Co, Ni, Cu) electrocatalysts of outstanding durability. ACS Catal. 2155-5435. 12 (13), 7994–8006. doi:10.1021/acscatal.2c00697.

Nørskov, J.K., Bligaard, T., Rossmeisl, J., Christensen, C.H., 2009. Towards the computational design of solid catalysts. Nat. Chem. 17554349. 1 (1), 37–46. doi:10.1038/nchem.121.

Nørskov, J.K., Rossmeisl, J., Logadottir, A., Lindqvist, L., Kitchin, J.R., Bligaard, T., Jónsson, H., 2004. Origin of the overpotential for oxygen reduction at a fuel-cell cathode. J. Phys. Chem. B. 1520-6106. 108 (46), 17886–17892. doi:10.1021/jp047349j.

Pan, F., Yang, Y., 2020. Designing CO_2 reduction electrode materials by morphology and interface engineering. Energy Environ. Sci. 13 (8), 2275–2309.

Pan, Y., Xuelu, M., Wang, M., Yang, X., Liu, S., Chen, H., Zhuang, Z., Zhang, Y., Cheong, W., Zhang, C., Cao, X., Shen, R., Qian, X., Zhu, W., Liu, Y., Wang, X., Zhang, X., Yan, W., Jun, L., Chen, H.M., Chen, C., Yadong, L., 2022. Construction of N, P Co-doped carbon frames anchored with Fe single atoms and Fe_2P nanoparticles as a robust coupling catalyst for electrocatalytic oxygen reduction. Adv. Mater. 0935-9648. 34 (29). doi:10.1002/adma.202203621.

Peera, S.G., Lee, T.G., Sahu, A.K., 2019. Pt-rare earth metal alloy/metal oxide catalysts for oxygen reduction and alcohol oxidation reactions: an overview. Sustain. Energy Fuels. 2398-4902. 3 (8), 1866–1891. doi:10.1039/c9se00082h.

Qu, Q., Mao, Y., Ji, S., Liao, J., Dong, J., Wang, L., Wang, Q., Liang, X., Zhang, Z., Yang, J., Li, H., Zhou, Y., Wang, Z., Waterhouse, G.I.N., Wang, D., Li, Y., 2025. Engineering the lewis acidity of fe single-atom sites via atomic-level tuning of spatial coordination configuration for enhanced oxygen reduction. J. Am. Chem. Soc. 0002-7863. 147 (8), 6914–6924. doi:10.1021/jacs.4c17444.

Radwan, A., Jin, H., Liu, B., Chen, Z., Qian, W., Zhao, X., Daping, H., Shichun, M., 2021. 3D-ZIF scaffold derived carbon encapsulated iron nitride as a synergistic catalyst for ORR and zinc-air battery cathodes. Carbon. 00086223. 171, 368–375. doi:10.1016/j.carbon.2020.09.024.

Rodriguez, J., 1996. Physical and chemical properties of bimetallic surfaces. Surf. Sci. Rep. 01675729. 24 (7-8), 223–287. doi:10.1016/0167-5729(96)00004-0.

Secor, M., Soudackov, A.V., Hammes-Schiffer, S., 2023. Density matrix-based features as descriptors for oxygen reduction and evolution catalysts. J. Phys. Chem. C. 127 (31), 15246–15256.

Shang, H., Zhou, X., Dong, J., Ang, L., Zhao, X., Liu, Q., Lin, Y., Pei, J., Zhi, L., Jiang, Z., Zhou, D., Zheng, L., Wang, Y., Zhou, J., Yang, Z., Cao, R., Sarangi, R., Sun, T., Yang, X., Zheng, X., Yan, W., Zhuang, Z., Jia, L., Chen, W., Wang, D., Zhang, J., Yadong, L., 2020. Engineering unsymmetrically coordinated Cu-S_1N_3 single atom sites with enhanced oxygen reduction activity. Nat. Commun. 2041-1723. 11 (1). doi:10.1038/s41467-020-16848-8.

She, Z.W., Kibsgaard, J., Dickens, C.F., Chorkendorff, I., Nørskov, J.K., Jaramillo, T.F., 2017. Combining theory and experiment in electrocatalysis: insights into materials design. Science 10959203. 355 (6321). Available from: http://science.sciencemag.org/content/sci/355/6321/eaad4998.full.pdf. doi:10.1126/science.aad4998.

Shen, H., Thomas, T., Rasaki, S.A., Saad, A., Chun, H., Wang, J., Yang, M., 2019. Oxygen reduction reactions of Fe-N-C catalysts: current status and the way forward. Electrochem. Energy Rev. 2520-8489. 2 (2), 252–276. doi:10.1007/s41918-019-00030-w.

Shen, L., Gong, Y., Yang, Q., Deng, J., Lian-Hua, X., Shan, D., 2025. Metal selenization strategy to construct porous heterostructures for oxygen electrocatalysis: selenite coordination at metal sites of NH_2-MIL-88(Fe). J. Alloys Compd. 09258388. 1031, 181015. doi:10.1016/j.jallcom.2025.181015.

Shi, Q., Zhu, C., Dan, D., 2019. Yuehe Lin, Robust noble metal-based electrocatalysts for oxygen evolution reaction. Chem. Soc. Rev. 0306-0012. 48 (12), 3181–3192. doi:10.1039/c8cs00671g.

Shin, S., Liu, J., Akbar, A., Sukkee, U., 2019. Nanoscale transport characteristics and catalyst utilization of vertically aligned carbon nanotube catalyst layers for fuel cell applications: comprehensive stochastic modeling of composite morphological structures. J. Catal. 00219517. 377, 465–479. doi:10.1016/j.jcat.2019.07.053.

Shui, J., Wang, M., Feng, D., Dai, L., 2015. N-doped carbon nanomaterials are durable catalysts for oxygen reduction reaction in acidic fuel cells. Sci. Adv. 2375-2548. 1 (1). doi:10.1126/sciadv.1400129.

Song, L., Liang, Z., Nagamori, K., Igarashi, H., Vukmirovic, M.B., Adzic, R.R., Sasaki, K., 2020. Enhancing oxygen reduction performance of Pt monolayer catalysts by Pd(111) nanosheets on WNi substrates. ACS Catal. 21555435. 10 (7), 4290–4298. Available from: http://pubs.acs.org/page/accacs/about.html. doi:10.1021/acscatal.0c00040.

Sun, Y., Silvioli, L., Sahraie, N.R., Ju, W., Li, J., Zitolo, A., Li, S., Bagger, A., Arnarson, L., Wang, X., Moeller, T., Bernsmeier, D., Rossmeisl, J., Jaouen, F., Strasser, P., 2019. Activity-selectivity trends in the electrochemical production of hydrogen peroxide over single-site metal-nitrogen-carbon catalysts. J. Am. Chem. Soc. 15205126. 141 (31), 12372–12381. Available from: http://pubs.acs.org/journal/jacsat. doi:10.1021/jacs.9b05576.

Tylus, U., Jia, Q., Strickland, K., Ramaswamy, N., Serov, A., Atanassov, P., Mukerjee, S., 2014. Elucidating oxygen reduction active sites in pyrolyzed metal–nitrogen coordinated non-precious-metal electrocatalyst systems. J. Phys. Chem. C. 1932-7447. 118 (17), 8999–9008. doi:10.1021/jp500781v.

Wang, J., Han, G., Wang, L., Du, L., Chen, G., Gao, Y., Ma, Y., Du, C., Cheng, X., Zuo, P., Yin, G., 2018. ZIF-8 with ferrocene encapsulated: a promising precursor to single-atom Fe embedded nitrogen-doped carbon as highly efficient catalyst for oxygen electroreduction. Small 16136829. 14 (15). Available from: http://onlinelibrary.wiley.com/journal/10.1002/(ISSN)1613-6829. doi:10.1002/smll.201704282.

Wang, J., Zhao, C.X., Liu, J.N., Song, Y.W., Huang, J.Q., Li, B.Q., 2022. Dual-atom catalysts for oxygen electrocatalysis. Nano Energy. 22112855. 104. Available from: http://www.journals.elsevier.com/nano-energy/. doi:10.1016/j.nanoen.2022.107927.

Wang, L., Peng, Y., Zhao, L., Tian, C., Zhao, D., Zhou, W., Yin, J., Wang, R., Honggang, F., 2014. B and N isolate-doped graphitic carbon nanosheets from nitrogen-containing ion-exchanged resins for enhanced oxygen reduction. Sci. Rep. 2045-2322. 4 (1). doi:10.1038/srep05184.

Wang, Q., Ina, T., Chen, W.T., Shang, L., Sun, F., Wei, S., Sun-Waterhouse, D., Telfer, S.G., Zhang, T., Waterhouse, G.I.N., 2020a. Evolution of Zn(II) single atom catalyst sites during the pyrolysis-induced transformation of ZIF-8 to N-doped carbons. Sci. Bull. 20959281. 65 (20), 1743–1751. Available from: http://link.springer.com/journal/11434. doi:10.1016/j.scib.2020.06.020.

Wang, S., Iyyamperumal, E., Roy, A., Xue, Y., Yu, D., Dai, L., 2011. Vertically aligned BCN nanotubes as efficient metal-free electrocatalysts for the oxygen reduction reaction: a synergetic effect by co-doping with boron and nitrogen. Angew. Chem. Int. Ed. 15213773. 50 (49), 11756–11760. doi:10.1002/anie.201105204.

Wang, X., Xiao, B., Yihang, L., Tang, Y., Liu, F., Chen, J., Liu, Y., 2020b. First-principles based machine learning study of oxygen evolution reactions of perovskite oxides using a surface center-environment feature model. Appl. Surf. Sci.. 01694332. 531, 147323. doi:10.1016/j.apsusc.2020.147323.

Wei, S., Sun, Y., Qiu, Y.Z., Li, A., Chiang, C.Y., Xiao, H., Qian, J., Li, Y., 2023. Self-carbon-thermal-reduction strategy for boosting the Fenton-like activity of single Fe-N4 sites by carbon-defect engineering. Nat. Commun. 20411723. 14 (1). Available from: https://www.nature.com/ncomms/. doi:10.1038/s41467-023-43040-5.

Wu, L., Guo, T., Li, T., 2020. Rational design of transition metal single-atom electrocatalysts: a simulation-based, machine learning-accelerated study. J. Mater. Chem. A. 2050-7488. 8 (37), 19290–19299. doi:10.1039/d0ta06207c.

Wu, Q., Wang, Z., Li, Y., Wang, Z., Liang, Y., Hu, Y., Ou, R., Wang, J., Lu, T., Wang, J., Wang, H., Zhao, C., Han, L., 2019. Nitrogen-rich, well-dispersed nanoporous carbon materials for super-efficient oxygen reduction reaction. ChemElectroChem 21960216. 6 (6), 1894–1900. Available from: http://onlinelibrary.wiley.com/journal/10.1002/(ISSN)2196-0216. doi:10.1002/celc.201900198.

Xiaoxiao, L., Yan, Y., Zheng, X., Yao, Y., Liu, Y., 2022. Atomically dispersed V-N-C catalyst with saturated coordination effect for boosting electrochemical oxygen reduction. Chem. Eng. J. 13858947. 444, 136363. doi:10.1016/j.cej.2022.136363.

Xing, M., Zhang, Y., Shuyuan, L., Hong, H., Sun, S., 2022. Prediction of carbon dioxide reduction catalyst using machine learning with a few-feature model: WLEDZ. J. Phys. Chem. C. 1932-7447. 126 (40), 17025–17035. doi:10.1021/acs.jpcc.2c02161.

Xiong, Y., Shan, H., Zhou, Z., Yan, Y., Chen, W., Yang, Y., Liu, Y., Tian, H., Jianbo, W., Zhang, H., Yang, D., 2017. Tuning surface structure and strain in Pd–Pt core–shell nanocrystals for enhanced electrocatalytic oxygen reduction. Small. 1613-6810. 13 (7). doi:10.1002/smll.201603423.

Xu, X., Xia, Z., Zhang, X., Sun, R., Sun, X., Li, H., Wu, C., Wang, J., Wang, S., Sun, G., 2019. Atomically dispersed Fe-N-C derived from dual metal-organic frameworks as efficient oxygen reduction electrocatalysts in direct methanol fuel cells. Appl. Catal. B Environ. 09263373. 259, 118042. doi:10.1016/j.apcatb.2019.118042.

Yan, L., Panpan, L., Zhu, Q., Kumar, A., Sun, K., Tian, S., Sun, X., 2023. Atomically precise electrocatalysts for oxygen reduction reaction. Chem. 24519294. 9 (2), 280–342. doi:10.1016/j.chempr.2023.01.003.

Yan, M., Chenlu, X., Sun, Y., Pan, H., Hong, L., 2021. Manipulating Zn anode reactions through salt anion involving hydrogen bonding network in aqueous electrolytes with PEO additive. Nano Energy. 22112855. 82, 105739. doi:10.1016/j.nanoen.2020.105739.

Yang, C.L., Wang, L.N., Yin, P., Liu, J., Chen, M.X., Yan, Q.Q., Wang, Z.S., Xu, S.L., Chu, S.Q., Cui, C., Ju, H., Zhu, J., Lin, Y., Shui, J., Liang, H.W., 2021. Sulfur-anchoring synthesis of platinum intermetallic nanoparticle catalysts for fuel cells. Science 10959203. 374 (6566), 459–464. Available from: https://www.science.org/doi/10.1126/science.abj9980. doi:10.1126/science.abj9980.

Yao, N., Chen, X., Zhong-Heng, F., Zhang, Q., 2022. Applying classical, ab initio, and machine-learning molecular dynamics simulations to the liquid electrolyte for rechargeable batteries. Chem. Rev. 0009-2665. 122 (12), 10970–11021. doi:10.1021/acs.chemrev.1c00904.

Yarlagadda, V., Carpenter, M.K., Moylan, T.E., Kukreja, R.S., Koestner, R., Gu, W., Thompson, L., Kongkanand, A., 2018. Boosting fuel cell performance with accessible carbon mesopores. ACS Energy Lett. 23808195. 3 (3), 618–621. Available from: http://pubs.acs.org/journal/aelccp. doi:10.1021/acsenergylett.8b00186.

Yeager, E., 1984. Electrocatalysts for O_2 reduction. Electrochim. Acta. 00134686. 29 (11), 1527–1537. doi:10.1016/0013-4686(84)85006-9.

Zeng, L., Zhao, T.S., An, L., Zhao, G., Yan, X.H., Jung, C.Y., 2015. Graphene-supported platinum catalyst prepared with ionomer as surfactant for anion exchange membrane fuel cells. J. Power Sources. 03787753. 275, 506–515. doi:10.1016/j.jpowsour.2014.11.021.

Zhang, G.R., Wolker, T., Sandbeck, D.J.S., Munoz, M., Mayrhofer, K.J.J., Cherevko, S., Etzold, B.J.M., 2018. Tuning the electrocatalytic performance of ionic liquid modified Pt catalysts for the oxygen reduction reaction via cationic chain engineering. ACS Catal. 21555435. 8 (9), 8244–8254. Available from: http://pubs.acs.org/page/accacs/about.html. doi:10.1021/acscatal.8b02018.

Zhang, J., Chen, J., Luo, Y., 2019. MOFs-assisted synthesis of hierarchical porous nicke-cobalt nitride heterostructure for oxygen reduction reaction and supercapacitor. ACS Sustain. Chem. Eng. 8, 382–392.

Zhang, L., Zhang, X., Chen, C., Zhang, J., Tan, W., Xu, Z., Zhong, Z., Du, L., Song, H., Liao, S., Zhu, Y., Zhou, Z., Cui, Z., 2024. Machine learning-aided discovery of low-Pt high entropy intermetallic compounds for electrochemical oxygen reduction reaction. Angew. Chem. Int. Ed. 15213773. 63 (51). Available from: http://onlinelibrary.wiley.com/journal/10.1002/(ISSN)1521-3773. doi:10.1002/anie.202411123.

Zhao, C., Liu, X., Liu, J., Wang, J., Wan, X., Wang, C., Xi-Yao, L., Shui, J., Song, L., Peng, H., Bo-Quan, L., Zhang, Q., 2023. Molecular recognition regulates coordination structure of single-atom sites. Angewandte Chemie. 0044-8249. 135 (48). doi:10.1002/ange.202313028.

Zhao, H., Ren, J.T., Yuan, Z.Y., 2024a. Microenvironment engineering of gas-involving energy electrocatalysis and device applications. Coord. Chem. Rev. 00108545. 514. Available from: https://www.sciencedirect.com/science/journal/00108545. doi:10.1016/j.ccr.2024.215901.

Zhao, W., Chi, B., Liang, L., Yang, P., Zhang, W., Xin, G., Wang, L., Cui, Z., Liao, S., 2022. Optimizing the electronic structure of ordered Pt–Co–Ti ternary intermetallic catalyst to boost acidic oxygen reduction. ACS Catal. 2155-5435. 12 (13), 7571–7578. doi:10.1021/acscatal.2c00554.

Zhao, Y., Chen, H., Xuelu, M., Jiaye, L., Yuan, Q., Zhang, P., Wang, M., Junxi, L., Min, L., Wang, S., Guo, H., Ruanbo, H., Kun-Hua, T., Zhu, W., Xuning, L., Yang, X., Pan, Y., 2024b. Vacancy defects inductive effect of asymmetrically coordinated single-atom $Fe{-}N_3S_1$ active sites for robust electrocatalytic oxygen reduction with high turnover frequency and mass activity. Adv. Mater. 0935-9648. 36 (11). doi:10.1002/adma.202308243.

Zheng, Y., Jiao, Y., Ge, L., Jaroniec, M., Qiao, S.Z., 2013. Two-step boron and nitrogen doping in graphene for enhanced synergistic catalysis. Angew. Chem. Int. Ed. 15213773. 52 (11), 3110–3116. doi:10.1002/anie.201209548.

Zhong, M., Tran, K., Min, Y., Wang, C., Wang, Z., Dinh, C.T., De Luna, P., Yu, Z., Rasouli, A.S., Brodersen, P., Sun, S., Voznyy, O., Tan, C.S., Askerka, M., Che, F., Liu, M., Seifitokaldani, A., Pang, Y., Lo, S.C., Ip, A., Ulissi, Z., Sargent, E.H., 2020. Accelerated discovery of CO_2 electrocatalysts using active machine learning. Nature 14764687. 581 (7807), 178–183. Available from: http://www.nature.com/nature/index.html. doi:10.1038/s41586-020-2242-8.

Zhou, Y., Fan, L., Wang, J., Qing, D., Jiao, K., 2024. Revealing local oxygen transport in ionomer films on multidimensional nanoscale catalysts in fuel cells. J. Mater. Chem. A. 2050-7488. 12 (27), 16419–16426. doi:10.1039/d4ta01633e.

Zhu, J., Elnabawy, A.O., Lyu, Z., Xie, M., Murray, E.A., Chen, Z., et al., 2019. Facet-controlled Pt–Ir nanocrystals with substantially enhanced activity and durability towards oxygen reduction. Mater. Today. 35, 69–77.

Zhu, J., Xiao, M., Li, K., Liu, C., Zhao, X., Xing, W., 2016. Active Pt3Ni (111) Surface of Pt3Ni Icosahedron for Oxygen Reduction. ACS Appl. Mater. Interfaces. 19448252. 8 (44), 30066–30071. Available from: http://pubs.acs.org/journal/aamick. doi:10.1021/acsami.6b04237.

Zhu, J., Xiao, M., Song, P., Jing, F., Jin, Z., Liang, M., Junjie, G., Liu, C., Chen, Z., Xing, W., 2018. Highly polarized carbon nano-architecture as robust metal-free catalyst for oxygen reduction in polymer electrolyte membrane fuel cells. Nano Energy. 22112855. 49, 23–30. doi:10.1016/j.nanoen.2018.04.021.

Zhu, M., Zhao, C., Liu, X., Wang, X., Zhou, F., Wang, J., Yanmin, H., Zhao, Y., Yao, T., Yang, L.-M., Yuen, W., 2021. Single atomic cerium sites with a high coordination number for efficient oxygen reduction in proton-exchange membrane fuel cells. ACS Catal. 2155-5435. 11 (7), 3923–3929. doi:10.1021/acscatal.0c05503.

Chapter 6

Cell design and stack engineering: Theory, integration and perspectives

Haoran Du and Zhiming Bao
Tianjin University, Tianjin, China

The fuel cell is a one of the most promising power sources that converts the chemical energy of fuel and oxidant directly into electric energy through electrochemical reaction (Jiao and Li, 2011; Wang et al., 2020). There are various types of fuel cells, and the proton exchange membrane fuel cell (PEMFC) has become the focus of scholars' research at this stage as a kind of fuel cell whose by-products are only water, and its advanced characteristics such as high efficiency, zero-emission, and fast dynamic response (Wang et al., 2016; Huang et al., 2021). However, this type of fuel cell typically requires an acidic environment and utilizes expensive platinum metal as a catalyst. The alkaline fuel cell (AFC) is the most mature technology among various fuel cells, offering advantages such as high efficiency and low cost. However, due to the use of corrosive liquid electrolytes in AFC, it also has certain drawbacks, including a degree of danger and the potential to cause environmental pollution. In contrast, alkaline anion exchange membrane fuel cell (AEMFC) work in an alkaline environment, greatly avoiding the use of precious metal as catalysts and making them more conducive to the oxygen reduction reaction (ORR), thereby reducing the need for catalysts. It also avoids the problem of electrolyte leakage, significantly improves the safety and environmental friendliness of the system, and has great potential for cost reduction and performance improvement (Arunchander et al., 2017; Ziv et al., 2018).

1 Fundamentals of alkaline fuel cells and anion exchange membrane fuel cells

The traditional AFC, which was used in the 1960s for space and military missions in various countries, such as the United States Apollo space program, was the first fuel cell to be put into use (Li, 2005). Typically, a 35% ~ 45%

Alkaline Fuel Cells. https://doi.org/10.1016/B978-0-443-36638-3.00016-4

KOH or NaOH solution is used as the electrolyte, and precious metals (e.g., Pt, Pd) and excess metals (e.g., Ni, Co) and their combined alloys are employed as catalysts. Early AFC systems used an asbestos membrane adsorbed with a saturated KOH solution as the electrolyte diaphragm. The asbestos membrane serves to keep the electrolyte in place, in addition to separating the fuel from the oxidizing agent, but the electrolyte can also corrode it. Allis-Chalmers was the first to develop the asbestos mesh matrix type of AFC, as shown in Fig. 6.1. However, asbestos has a carcinogenic effect, and many countries have proposed banning its use in AFC. Rosa et al. explored materials such as polyphenylene sulfide (PPS) and polytetrafluoroethylene (PTFE). They found that PPS and PTFE exhibit properties very similar to those of asbestos in alkaline solutions, as well as improved corrosion resistance and lower electrical resistance, with PPS being even superior to asbestos. Since AFC adopts alkaline solutions such as KOH or NaOH as an electrolyte, which leads to its easy reaction with carbon dioxide in the air, producing the phenomenon of CO_2 poisoning and drastically reducing the performance and life span of AFC, many scholars believe that AFC is not suitable for automobile power, and its use scenarios are limited.

Fig. 6.1 presents an early AFC structure, featuring key components such as oxygen/hydrogen electrodes, chambers, support plates, and an asbestos mesh matrix.

In recent years, due to the emergence of anion-exchange membranes (Zhuo et al. 2015), they have become the optimal choice to replace liquid electrolytes, which can utilize air as the reaction gas and avoid the phenomenon of CO_2 toxicity that occurs in traditional AFCs. Due to the many advantages of AEMFC (Jianjun, 2020), more and more scholars have begun to invest in the research of AEMFC (Gutru et al., 2020). They compared the working principles of PEMFC

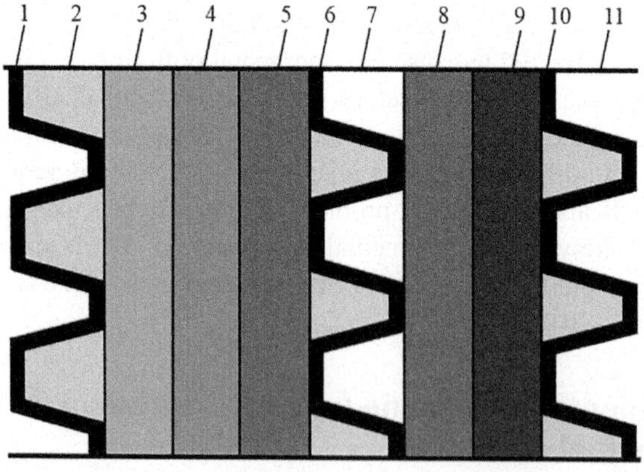

FIG. 6.1 Schematic diagram of an early AFC structure (where 1-oxygen support plate, 2-oxygen chamber, 3-oxygen electrode, 4-asbestos mesh, 5-hydrogen electrode, 6-hydrogen chamber, 7-hydrogen support plate, 8-drainage feeler, 9-drainage membrane support plate, 10-evaporation chamber, and 11-water-removal plate).

and AEMFC in the review literature (shown in Fig. 6.2). Although AEMFC has considerable advantages and application value, there are still many challenges in the performance of membranes and water management because AEMFC has the advantages and application value of AEMFC (Krewer et al., 2018). It is not only the best choice for replacing liquid electrolytes but also for preventing the CO_2 poisoning phenomenon associated with traditional AFCs. As AEMFC works in an alkaline environment, the anode hydrogen reacts with hydroxide ions to produce water, and the cathode water reacts with oxygen and electrons to produce hydroxide ions. Meanwhile, due to its unique electroosmotic drag (EOD) effect, water will move from the cathode side to the anode side with the transfer of ions, so it is easy to lead to water flooding at the anode, while the cathode side will experience the phenomenon of membrane drying, which is significantly different from PEMFCs. Therefore, for AEMFCs, water management is a key factor affecting cell performance (Zhang and Jiao, 2018).

Fig. 6.2 compares the working principles of PEMFCs (A) and alkaline AEMFCs (B). In PEMFCs, hydrogen (H_2) oxidizes at the anode, releasing protons (H^+) that migrate through the membrane to form water at the cathode. In AEMFCs, hydroxide ions (OH^-) transport across the membrane, enabling hydrogen oxidation at the anode.

The basic structure of AEMFC is shown in Fig. 6.3, which mainly consists of a bipolar plate (BP), flow field or flow channel (FF), gas diffusion layer (GDL), microporous layer (MPL), catalyst layer (CL), alkaline anion exchange membrane (AAEM). Among these, the AAEM is positioned between the cathode and anode, and together with the CL, they form the

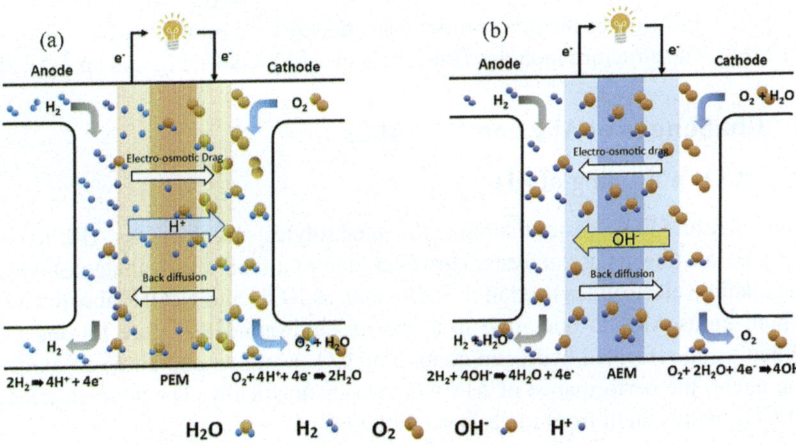

FIG. 6.2 Comparison of the working principles of PEMFC and AEMFC (Gutru et al., 2020) (a represents PEMFC, and b represents AEMFC). *(Gutru, R., Turtayeva, Z., Xu, F., Maranzana, G., Vigolo, B. & Desforges, A. (2020). A comprehensive review on water management strategies and developments in anion exchange membrane fuel cells. Int. J. Hydrogen Energy 45(38), 19642–19663.)*

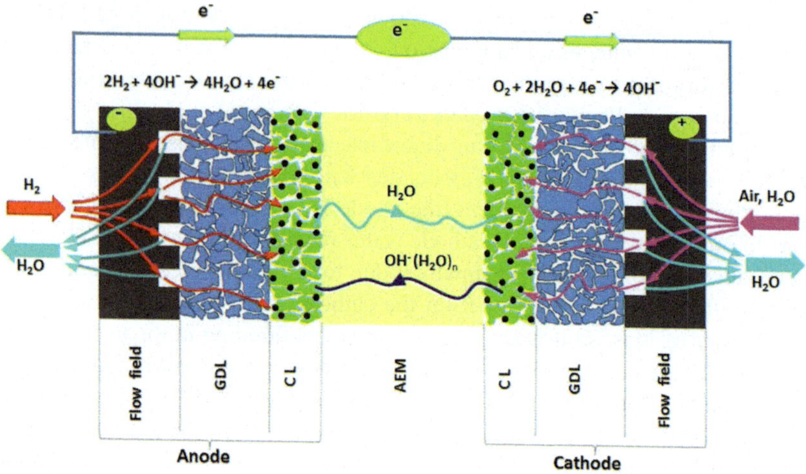

FIG. 6.3 Component structure and working principle of AAMFC (Dekel, 2018). *(Dekel, D. R. (2018). Review of cell performance in anion exchange membrane fuel cells. J. Power Sources 375, 158–169.)*

Membrane-Electrode Assembly (MEA) (Gutru et al., 2020). The two sides of the MEA are supported by the GDL, and as part of the cell mechanism, the GDL and the CL are also separated by the MPL. The permeabilities of the three components are ranked from largest to smallest as follows: GDL > MPL > CL. The reactive gases are transported from the flow channel through the GDL and MPL to the CL, where they participate in the electrochemical reaction.

Fig. 6.3 illustrates the key components and working principle of an alkaline AEMFC. The structure includes flow channels (FF), GDL, CL, and an AAEM.

2 Challenges of AFCs and AEMFCs

2.1 CO_2 poisoning of AFC

The ORR of AFC is faster than in acidic electrolytes, which allows AFC to use nonprecious metals as catalysts. However, due to a fatal disadvantage of AFC, the alkaline electrolytes used in the cell (such as KOH and NaOH) absorb CO_2, which limits AFC to operating in a special environment of pure H_2 and O_2. When using reformed hydrogen containing CO_2 as a fuel and air as an oxidizing agent, the performance of AFCs decreases drastically. The adverse effects of CO_2 mainly stem from its reaction with OH^-.

$$CO_2 + 2OH^- \rightarrow CO_3^{2-} + H_2O \tag{1}$$

The CO_2 poisoning phenomenon causes the following reasons for the reduced performance of AFC:

1. Decreased OH^- concentration affects the reaction kinetics
2. Increase in the viscosity of the electrolyte, resulting in a lower diffusion coefficient and limiting current density
3. Precipitation of carbonates in the pores of porous electrodes
4. Decreased solubility of O_2
5. Decrease in the conductivity of the electrolyte

Kordesch et al., 1983 investigated the effect of whether CO_2 is involved in the reaction on the performance and lifetime of AFC. The study on the performance impact showed that in the short term, the use of air or air without CO_2 at the cathode had a negligible impact on the performance of AFC. However, as the reaction continued, the AFC with air as the cathode reaction gas showed a significant decrease in performance. In the investigation of the effect of lifetime, it was found that the lifetime of AFC using air as the reaction gas ranged from 1600 to 3400 h when the current density was 65 mA/cm^2, while the lifetime of AFC with the cathode reaction gas without CO_2 under the same experimental conditions could reach 4000 to 5000 h. In conclusion, the poisoning phenomenon of CO_2 can negatively affect the performance and lifetime of AFCs.

The toxicity of CO_2 to AFCs is gradually being addressed as scholars continue to explore the issue in depth. Ahuja et al. proposed the use of liquid hydrogen to remove CO_2. When liquid hydrogen absorbs heat and gasifies, a heat exchanger is used to achieve the condensation of CO_2, thus reducing the CO_2 content to <0.001%. However, the fact that hydrogen tends to be stored in the form of compressed gas rather than liquid limits the use of this method. Cifrain and Kordesch (2004) proposed the method of circulating electrolytes. By renewing the electrolyte, removing the carbonates generated in the electrolyte, and continuously adding OH^- as carriers, the mechanical damage to the electrode caused by carbonate precipitation is attenuated, ensuring an adequate number of carriers and also facilitating the management of water and heat, which allows the AFC to work for an extended period and with high efficiency; however, this method has undoubtedly increased the complexity of the AFC. Gülzow and Schulze (2004) found that when the catalyst material was mixed with PTFE fine particles at high speed during electrode preparation, the small PTFE particles with a particle size of less than 1 μm covered the catalyst surface, which increased the electrode strength while avoiding the electrode being completely flooded by the electrolyte and reduced the possibility of carbonate precipitation blocking the micropores as well as mechanical damage to the electrode. Sleem-ur-Rahman and Al-Zakri (1998) combined the dry and wet methods for electrode preparation, proposing a filtration method to optimize the electrode's performance by controlling the PTFE content and grinding time. By using the new electrode preparation method, the tolerance of AFC to CO_2 was improved.

In addition, scholars have developed a free electrolyte-type AFC, which contains an electrolyte circulation system inside the cell, can cool the electrolyte, evaporate water outside the cell, and is easy to adapt to changes in the volume of the liquid and electrolyte exchange. The single-cell structure of the free

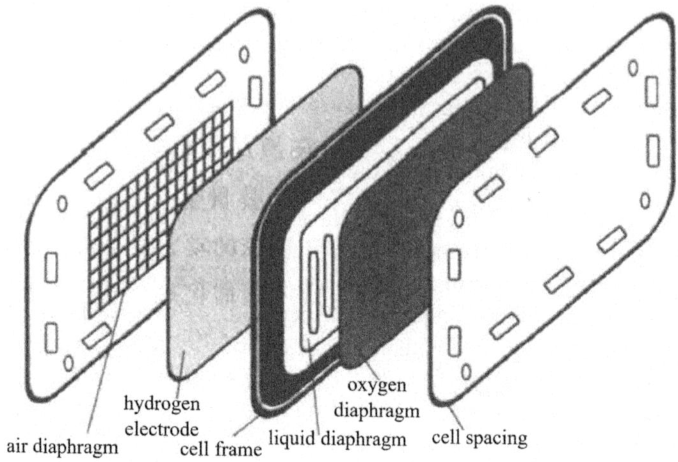

FIG. 6.4 Free electrolyte type alkaline fuel cell.

electrolyte-type AFC is shown in Fig. 6.4. The sealing material for the gas and electrolyte channels is a rubber gasket. When the hydrogen gas cycle method is used to remove water, the porous nickel partition at the back of the hydrogen electrode can function as an electrolyte storage tank to compensate for the electrolyte volume changes caused by temperature and concentration fluctuations. To meet the application's requirements, several single cells can be connected in series to form a power stack (Chaolin et al., 2022).

Fig. 6.4 illustrates the single-cell structure of a free electrolyte-type AFC. This design incorporates an internal electrolyte circulation system, which facilitates electrolyte cooling and external water evaporation. Additionally, the system allows for efficient adaptation to liquid volume changes and enables straightforward electrolyte exchange, enhancing operational flexibility and performance.

2.2 Water management in AEMFC

The working principle and water transfer mechanism of the AEMFC are generally the same as those of the PEMFC (shown in Fig. 6.5), which is that the H_2 energized at the anode reacts with the O_2 energized at the cathode to produce H_2O, and at the same time generates electrical and thermal energy. However, the specific process of the reaction is quite different.

Fig. 6.5 compares the operating principles and water transfer mechanisms of PEMFC and AEMFC. While both systems generate water, electrical, and thermal energy through the reaction of hydrogen (H_2) and oxygen (O_2), PEMFC relies on proton (H^+) transport across the membrane, whereas AEMFC involves hydroxide ion (OH^-) migration. The water formation pathways and transport directions differ significantly between the two technologies.

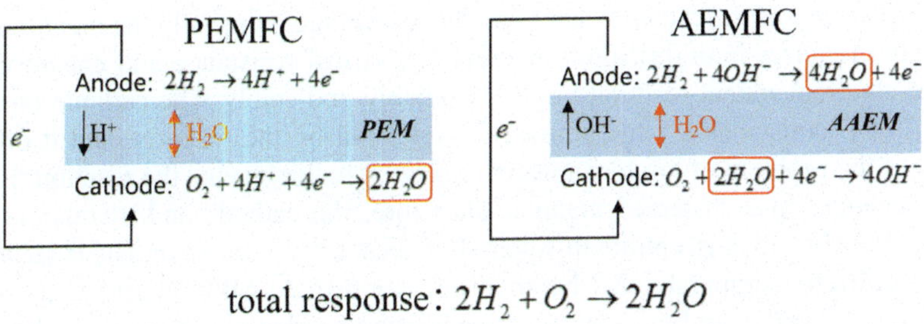

FIG. 6.5 Comparison of operating principle and water transfer mechanism between PEMFC and AEMFC.

The electrochemical reactions at the anode and cathode of the PEMFC are:

$$H_2 \rightarrow 2H^+ + 2e^-$$

$$\frac{1}{2}O_2 + 2H^+ + 2e^- \rightarrow H_2O$$

Anodic and cathodic electrochemical reactions in AEMFC:

$$H_2 + 2OH^- \rightarrow 2H_2O + 2e^-$$

$$\frac{1}{2}O_2 + H_2O + 2e^- \rightarrow 2OH^-$$

A comparison of the PEMFC and AEMFC reaction processes can be seen: the catalytic layer of PEMFC is designed for an acidic environment, while the catalytic layer of AEMFC is suited for an alkaline environment. The catalytic layer's compatibility with different working environments determines that the catalyst can be selected from a broader range of applications and can be more widely used in the market. In the acidic working environment of PEMFC, a stable catalytic reaction can be achieved using precious metals like platinum and its alloys, which leads to the cost of PEMFC being difficult to reduce. However, in the alkaline working environment of AEMFC, nonprecious metal catalysts such as nickel and its alloys can be used, and the corrosion problem of the catalyst is less severe, which is conducive to promoting the commercialization of AEMFC. In addition, the two types of fuel cells have different locations of water generation. In PEMFC, water is generated at the cathode (Lee et al., 2015), whereas in AEMFC, water is consumed at the cathode and generated at the anode. The different mechanisms of water consumption, generation, and transport determine that reasonable and effective water management is extremely important for the performance of AEMFC.

Specifically, hydrogen combines with OH^- at the anode of the AEMFC to form water and e^-, while water and oxygen react at the cathode, and the resulting OH^- migrates from the cathode to the anode through the membrane. Water also migrates along with the OH^- to the anode due to the electroosmotic drag (EOD) effect. Appropriate water content is essential to maintain the ionic conductivity of the catalytic layer and the membrane, especially the membrane, which requires humidification to reduce electrical resistance and ensure that the electrochemical reaction proceeds reliably and stably. The cathode of the AEMFC consumes a large amount of water during the reaction, and it often faces the problem of too low a water content, which prevents the reaction from proceeding in a sustained manner. Therefore, the cathode and membrane of the AEMFC are susceptible to water shortages, and the reaction gas is usually humidified. Comparatively, the anodic reaction in AEMFC generates a large amount of water, and the operating temperature inside the cell is usually below 100°C, so the anodic porous layer in AEMFC is prone to the phenomenon of water flooding (water flooding) (Yan et al., 2022). In addition, the hydraulic pressure on both sides of the cell differs, which also drives water transport across the membrane (permeation). In summary, the water transport process in AEMFC is highly complex, and water management becomes one of the primary factors limiting cell performance (Huo et al., 2012).

Xue et al. (2024) designed a lightly branched PAP anion exchange membrane based on a polyarylpiperidine-type anion exchange membrane. This new anion exchange membrane was able to maintain good chemical structure stability as well as excellent mechanical properties under a high-temperature processing environment. The mass transfer problem in the high-current region was solved, and an excellent "water balance" was achieved, which enabled excellent durability to be maintained while ensuring high power density output of the cell. Deng et al. (2016b) optimized water management by establishing a three-dimensional multiphase nonisothermal model of the whole cell in AEMFC, considering the interfacial effects of two-phase transport in porous electrodes to carry out a study related to the optimization of water management. The results showed that slight pressurization of the anode could effectively address the problems of anode flooding and membrane drying. Deng et al. (2016a) investigated the effects of electrode structure and operating conditions on the performance of AEMFC. The results showed that increasing the ionomer content as well as increasing the humidity of the inlet gas could keep the membrane wet and thus improve the overall output performance of the cell.

Similar to PEMFC, in AEMFC, water exists in three forms: liquid water, gaseous water, and membrane water, where membrane water refers to water adsorbed on hydrophilic functional groups in the membrane and catalytic layer. Under certain conditions, these three forms of water can be interconverted. The content of membrane-state water is directly related to key properties such as ionic conductivity and membrane-state water diffusivity (Jiao et al., 2014a), with lower ionic conductivity leading to higher ohmic losses. In addition, the lower ionic conductivity in AEMFC compared to the higher proton conductivity in PEMFC, as well as the higher sensitivity of cell performance to ionic conductivity, makes the adequate humidification of the membrane and catalytic layer critical for improving cell performance. Meanwhile, the content of liquid water in the porous layer has a significant effect on cell performance. In fuel cell research, the liquid water volume fraction is typically used to characterize the liquid water content of the porous layer. The higher the liquid water volume fraction, the smaller the available activation area for electrochemical reactions and the lower the electrochemical reaction rate of the cell. Excessive liquid water content may even hinder the electrochemical reaction, leading to water flooding; therefore, it is crucial to improve the drainage capacity of the anode or enhance water transfer to the cathode.

Water transport across the membrane on both sides of the AEMFC is achieved in three ways: electroosmotic drag, diffusion due to concentration gradients, and osmosis due to hydraulic gradients (Jiao et al., 2014b). Among them, electroosmotic drag is positively correlated with OH^- transport, which enhances water transport from the cathode to the anode and exacerbates flooding and membrane drying, while diffusion and osmosis increase water transport to the cathode, which mitigates flooding and membrane drying to a certain extent, and is an important mechanism to enhance the performance of AEMFC. The use of anode-generated water to humidify the AEMFC and participate in the electrochemical reaction at the cathode, as well as the enhancement of anode drainage to mitigate the adverse effects of liquid water on reactive gas transport and the electrochemical reaction, is key to improving the performance of the AEMFC. Effective water management is essential (Li et al., 2010).

3 Cell design of AFCs and AEMFCs

3.1 Flow field design

The BP and FF are crucial components of a fuel cell, serving as the transport and distribution channels for reaction gases, discharging the generated water, conducting electricity and heat, and providing mechanical support (Chang et al., 2007; Tseng et al., 2012). It occupies 70% of the cell stack's volume and 30%–40% of the cost (Tsuchiya and Kobayashi, 2004; Jiao et al., 2021). Similar to the PEMFC, the FF is created by cutting grooves in the pole plate, resulting in a groove-ridge structure; thus, the conventional FF is also referred to as a flow channel (Fan et al., 2018). Conventional flow channels are most commonly known as serpentine channels (Jeon et al., 2008; Choi et al., 2011), parallel channels (Wang et al., 2011), and interdigitated channels (Cooper et al., 2016, 2017), etc. Domestic and foreign researchers have investigated the characteristics of different flow channels and their comparisons (Su et al., 2005; Hashemi et al., 2012; Lim et al., 2017) (shown in Fig. 6.6), explored the influence of the ridge ratio and other factors on battery performance (Kerkoub et al., 2018), and improved the flow channel design by changing the cross-section shape (Li et al., 2017; Wang et al., 2017; Lee et al., 2020) and adding baffles in the flow channel (Yin et al., 2018; Shen et al., 2019) to enhance battery performance (Zhang and Jiao, 2018; Yan et al., 2019). However, the trench-ridge FF structure will inevitably bring about uneven distribution of reactive gases and generated film-state water and liquid water within the porous layer structure, such as the GDL under the trench and the ridge, presenting different heat and mass transfer

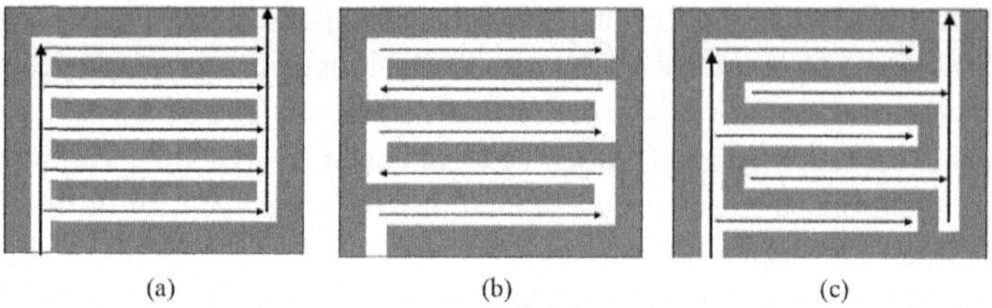

FIG. 6.6 Common flow field types. (A) parallel flow field; (B) serpentine flow field; (C) cross-finger flow field (Wang et al., 2008). *(Wang, Y., Basu, S., Wang, C.-Y., 2008. Modeling two-phase flow in PEM fuel cell channels. J. Power Sources 179 (2), 603–617.)*

characteristics, leading to uneven distribution of temperature and exacerbating the problems of water flooding and film drying (Park et al., 2018).

Fig. 6.6 illustrates three common FF configurations in fuel cells: (A) parallel FF, (B) serpentine FF, and (C) cross-finger FF. Each design offers distinct advantages in reactant distribution and water management.

A series of studies have been conducted on the mechanism by which the FF influences cell performance and the optimization of the FF structure (Nishida et al., 2015). Suresh et al. (2011) pioneered an open-fork multi-fluid channel scheme and verified it through simulation and experiments. The results show that this scheme can increase the transverse flow of reactant gases in the GDL and distribute the reactant gases more uniformly throughout the entire reactive zone. It is worth noting that this structure effectively alleviates the phenomenon of undergassing and water flooding at the U-bend in the serpentine flow channel. Bachman et al. (2012) noted that transverse flow plays a crucial role in altering the internal transport characteristics of the fuel cell, as well as its performance. Although the transverse flow of reactant gases occurs in traditional serpentine and forked finger channels, the loss of cell performance is significant. It is challenging to utilize transverse flow to promote the transport of reactant gases and the discharge of water flooding in the cell. In contrast, for parallel flow channels, the lateral flow between different channels can be enhanced by setting different gas pressures in different channels. The two-fluid model established by Jiao et al. (2014a) investigated the specific mechanism of the lateral flow in the parallel flow channels affecting the gas transport characteristics and battery performance. The results show that the lateral flow of reactive gases in the porous layer can be enhanced by setting different gas pressures between the flow channels. As a result, the drainage capacity and gas transport capacity of the cell are improved, leading to a more uniform distribution of reactive gases and water. Kerkoub et al. (2018) investigated the effect of the trench-ridge ratio on the performance of the cell in different flow channel structures. The results show that in the high output voltage range, the ridge ratio has a negligible effect on cell performance, whereas in the low output voltage range, narrowing the ridge ratio is beneficial to cell performance. Through in-depth investigation of the transport mechanism inside the cell, it is known that the reduction of the ridge ratio will increase the pressure drop of the cell and strengthen the gas convection under the ridge, which will enhance the transport speed of the reactive gases under the ridge and make the gas distribution in the catalytic layer more uniform, which is conducive to the improvement of the cell performance. Yin et al. (2018) and Shen et al. (2019) investigated the effect of adding baffles and other structures in the flow channel on the performance of the cell. The results show that the addition of baffles can significantly improve the mass transfer efficiency of the cell, as reflected in the acceleration of the local gas flow rate, the increase in the concentration of reactive gases in the GDL, and the enhancement of liquid water discharge efficiency. Moreover, in the high electric density region, the addition of baffles can significantly improve the cell's performance. Zhang and Jiao (2018) also investigated the effects of the number of added baffles and the arrangement of baffles on the mass transfer characteristics of the cell. The results indicated that the higher the number of baffles, the more uniform the gas velocity distribution in the flow channel, which can effectively promote the transport of reactive gases.

With the gradual deepening and diversification of research on novel flow channels, metal foam (MF) FF design and three-dimensional FF design forms have been studied by an increasing number of scholars and applied in fuel cells. MF has garnered increasing attention due to its porous, breathable, lightweight, and excellent electrical and thermal conductivity properties. Scholars have investigated its structural characteristics (Jung and Diebels, 2016) and heat and mass transfer properties (Afshari and Houreh, 2014; Hossain and Shabani, 2015) through a combination of experiments and simulations. MF, as a high-strength foam material (Bao et al., 2019), offers the potential to overcome the shortcomings of the groove-ridge flow channel structure and has become a promising solution for the FF of fuel cells (Carton and Olabi, 2015). The physical structure of the MF material is shown in Fig. 6.7 (Huo, 2018). The MF material exhibits excellent electrical and thermal conductivity, a compressible porous structure (Banhart, 2001), and controllable porosity as well as permeability (Yuan et al., 2012). The use of a MF FF instead of the traditional trench-ridge FF can significantly reduce the weight of the cell and its associated costs, promoting the uniform distribution of reactive gases and enhancing the cell's liquid water discharge ability (Huo et al., 2017). In recent years, three-dimensional FF design has garnered increasing attention from scholars. Li et al. (2017) proposed a new wave-shaped three-dimensional FF. They carried out experimental and simulation studies, which showed that the periodic wave structure could bring about periodic changes in the flow direction, flow velocity and local pressure of the local reactive gases and enhance the forced convection in the vertical direction, which strengthens the transport of reactive gases from the runner to the reactive region as well as the liquid water discharge. The wave structure can induce periodic changes in the local flow direction, flow velocity, and local pressure,

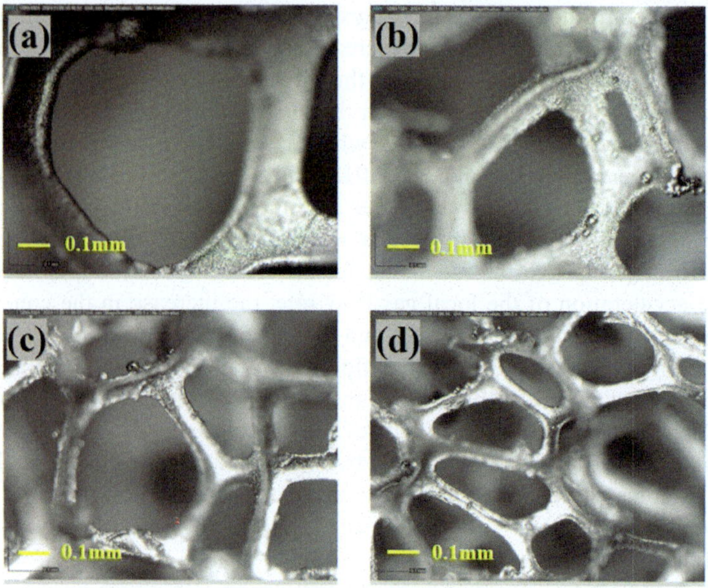

FIG. 6.7 Basic physical structure of metal foam materials (Shen et al., 2025). *(Shen, Z., Wang, Q., Xu, F., Zhou, X., Ni, L., Cheng, Z., Jiang, J., 2025. Investigation on strengthening effect of metal foams in enlarged microchannels for heterogeneous liquid-liquid mass transfer. Chem. Eng. Process.-Process Intensif. 0255-2701. 110379.)*

thereby enhancing vertical forced convection. This strengthens the transmission of reaction gas from the flow channel to the reaction area and facilitates the discharge of liquid water. The new FF can significantly enhance battery performance in the high current density range, resulting in a 17.8% improvement in the maximum net power density of the battery. Yan et al. (2019), Shen et al. (2019) proposed a tapered 3D wave FF perpendicular to the direction of the FF based on the aforementioned 3D wave FF. The results show that the new tapering wave 3D FF further increases the gas flow velocity in both the horizontal and vertical directions, enhancing the efficiency of reaction gas transport as well as liquid water discharge. Toyota in Japan has designed the famous 3D grid FF, which consists of 3D spatial cells arranged in a shape similar to a fish scale, increasing the ultimate current density of the reactor by 2.4 times.

Fig. 6.7 presents the basic physical structure of MF materials at different scales (a–d). The images reveal the characteristic porous architecture with interconnected cells.

Additionally, scholars have proposed a series of bionic FF structures inspired by natural and human body structures. Badduri et al. (2020) proposed and validated the design of lung slice-type and leaf-type FFs based on the structure of tree leaves and human lung slices (shown in Fig. 6.8). Compared to the traditional serpentine FF, the bionic FFs shown can significantly improve

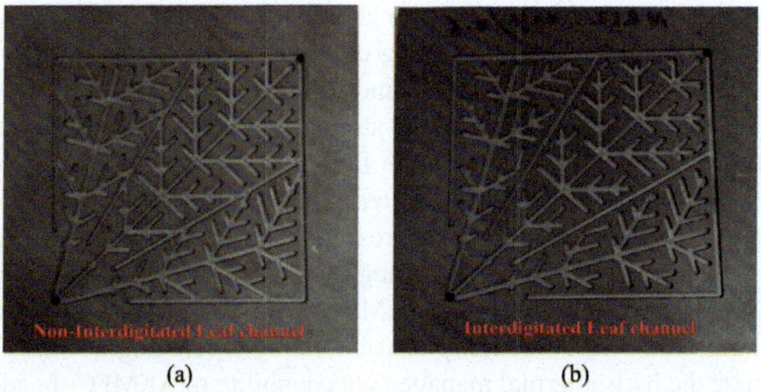

FIG. 6.8 Bionic flow field design. (A) non-interdigitated leaf channel; (B) Interdigitated leaf channel (Badduri et al., 2020). *(Badduri, S.R., Srinivasulu, G.N., Rao, S.S., 2020. Influence of bio-inspired flow channel designs on the performance of a PEM fuel cell. Chin. J. Chem. Eng. 1004-9541. 28 (3), 824–831.)*

the performance of the battery. The performance of the fork-and-finger and nonfork-and-finger FFs in the dendritic FF is compared experimentally. The results show that the fork-and-finger FF performs better, increasing the gross output power density and net output power density by 6.72% and 5.58%, respectively.

Fig. 6.8 showcases bionic flow field designs inspired by biological structures: (A) non-interdigitated leaf channel mimicking natural leaf venation, and (B) interdigitated leaf channel combining biomimicry with forced convection principles.

In fuel cells with conventional trench-ridge FFs, the GDL plays multiple roles, including support, electrical conductivity, thermal conductivity, and gas distribution, and is a key component that connects the FF and the MEA. Since the porous structure of the MF is similar to that of the GDL, scholars have proposed an integrated FF (GDL-less) structure. Unlike the traditional fuel cell composition, the integrated structure features fewer fuel cell parts, a more compact design, and significantly reduced volume and weight, which significantly reduces cell cost and increases power density. At the same time, due to the significant shortening of the substance and charge transport paths, the resistance and mass transfer resistance of the cell are significantly reduced, and the drainage capacity is enhanced. Tanaka and Shudo (2014) investigated the integrated flow structure using experimental methods based on a mesh FF. The results showed that the novel structural design could reduce the risk of flooding in fuel cells in high-current density regions. However, due to the reduction of the contact area between the FF and the MPL after GDL removal, its high-frequency resistance increases, thereby increasing the ohmic resistance of the cell. Park et al. (2019) found that the integrated FF design form can reduce the cell thickness by 82%, which can significantly improve the cell's volumetric power density.

In summary, MF, bionic, and integrated FF structures have great potential and commercial value in reducing cell volume, minimizing mass transfer losses, enhancing cell performance, and lowering cell costs. However, based on the open literature, these new FF structures have not yet been applied to AEMFC, which may be related to the fact that AEMFC technology is still in the development stage, with high material compatibility requirements and an immature preparation process. Future research could focus on exploring the following directions: firstly, developing highly conductive and corrosion-resistant MF materials suitable for alkaline AEMFC; secondly, improving the homogeneity of gas-water-heat-electricity distribution by leveraging novel FF designs to optimize the hydrothermal management capability of AEMFC. In addition, it is also necessary to systematically evaluate the long-term operational stability and economic viability of these new FFs in AEMFC to promote its transition from laboratory research to industrial application. With the continuous progress of AEMFC technology, these innovative FF structures are expected to provide a new technological path to solve the complex and costly hydrothermal management problems currently faced.

3.2 Distribution area design

The manifold is the key structure connecting the inlet and outlet to the FF. The structure of the distribution zone has a significant impact on the distribution of reactive gases, and the uniform distribution of reactive gases in the electrochemical reaction is a key factor affecting the battery's performance. The design of the manifold facilitates the discharge of liquid water generated by the electrochemical reaction, preventing water from accumulating in the FF and thereby minimizing its adverse impact on the battery's water management. The role of the manifold is vital, especially for batteries such as AFC, which are prone to flooding. The PEMFC pole plate features a distribution zone (Lu et al., 2023) that connects the gas inlet and the FF in the reaction zone, thereby enhancing the uniformity of gas distribution from the inlet to the FF. The distribution of reactants is crucial for the efficiency of the PEMFC. The structure of the distribution zone has a significant impact on the gas supply and drainage of the fuel cell (Chen et al., 2021). Jiao et al. (2006) numerically investigated the air-water flow in the micro-parallel channel at the inlet and outlet of the fuel cell. They proposed that the inlet collector with a progressively decreasing cross-sectional area could solve the water alignment problem. Liu et al. (2018) constructed a micro-distributor in order to improve the homogeneity of the reactant gases. Moreover, they investigated the effect of micro-distributors on the performance of parallel fuel cells. They found that the micro-distributor area could maximize the power density. Pistoresi et al. (2015) designed trapezoidal and stepped distribution zones to improve the uniformity of the gas flow distribution. Kumar and Singh (2019) explored the location of the inlet and outlet of the fuel cell, as well as the angle of inclination of the distribution zone, to minimize the inhomogeneities of the flow distribution and improve the performance of the fuel cell. The uniformity of the flow distribution is described by the inhomogeneity distribution factor (Pistoresi et al., 2015; Liu et al., 2018; Kumar and Singh, 2019), which is described in detail in the subsequent sections. Hossain et al. (2017) and Carton and Olabi (2010) proposed the design of an inclined inlet collector and stepped collector, respectively, both of which can improve the uniformity of the gas distribution in the FF of a fuel cell.

Siddiqui and Zubair (2017) proposed shunt-type and continuous-type distribution zones. The shunt-type distribution zone is complex to design and fabricate, has poor mechanical strength, and is sensitive to inlet conditions. Therefore, it is primarily used in small fuel cells. In contrast, the continuous type is more commonly used in large industrial fuel cells (Wang, 2015). Distribution zones play a crucial role in connecting the inlet and outlet to the FF. They are typically designed with large area dimensions to accommodate parallel FF configurations in the reaction zone. Currently, distribution zones in polar plates are typically classified into two types: dot matrix distribution zones and forced shunt distribution zones. Dabiri et al. (2018) designed a novel distributor with a cylindrical dot matrix design, increasing fluid distribution uniformity through shunt and diffusion effects. However, the use of a global dot matrix led to an increase in manufacturing costs. Lu et al. (2023) proposed a local dot matrix design with different dot matrix sizes and positions to achieve a more uniform flow. Yong et al. (2022a) proposed a "one part in two, two parts in four" structure for commercial large-scale fuel cells to improve the distribution of the gas before it enters the reaction zone. Yu et al. (2021) proposed the design of a point distribution zone and a shunt distribution zone, and the distribution becomes more uniform when the steering angle of the path in the distribution zone reaches a specific value. They found that the homogeneity of gas distribution depends on the diversion and dispersion effects of the fractional array. Although the addition of a distribution zone may increase the manufacturing cost of the fuel cell due to the reduction in the reaction zone area, it is still considered an important component of commercial, large-size fuel cells. It has been adopted by many manufacturers and researchers (Tsukamoto et al., 2021; Yong et al., 2022a, 2022b).

Uniform distribution of the electrode area not only facilitates the electrochemical reaction but also reduces the polarization effect and improves the efficiency and stability of the cell. Uneven gas distribution may lead to excessive stress or loss of the anion exchange membrane or pole plate electrodes in some areas, thus accelerating the decline of the battery. Uniform distribution helps reduce loading in these areas, thereby extending the battery's life. Designing a reasonable distribution area structure is crucial for improving AFC performance and service life.

4 Stack engineering of AFCs and AEMFCs

In response to AFC, the earliest use of AFC stacks dates back to the 1970s. The AFCs used in the US Apollo space program could reach an average power of 0.6 kW using pure hydrogen, pure oxygen and a concentrated electrolyte (85% KOH), which allows the cell to operate at a low pressure (around 60 psi). The electrolyte does not boil, but due to the use of a highly concentrated electrolyte, the performance (150 mA/cm^2 at 0.85 V) is not as good as that of the low-concentration electrolyte, requiring an increase in the operating overtemperature to 260°C. The AFC in the Shuttle orbiter, with an average power of 7 kW and the same operating pressure as in the Apollo program, can have a high current density at lower temperatures (470 mA/cm^2 at 0.86 V) due to the high content of precious metals on its electrodes [anode: 10 mg/cm^2 (80% Pt and 20% Pd); cathode: 20 mg/cm^2 (90% Au and 10% Pd). mA/cm^2]. In the 21st century, AFC reactors are increasingly utilized in various engineering applications. In 2015, AFC Energy (United Kingdom) successfully tested the first 51-cell reactor (model: Beta cartridge, half the number of commercial 101-cell designs), which operated stably for a fortnight, reaching the benchmark performance for large-scale reactor operation and reaching the set operating temperature, better than the planned 12-hour target (Barrett, 2015). In the same year, AFC Energy completed initial testing of the first 101-cell electro stack in the United Kingdom, which has continued to operate stably since 3 March 2015, with the thermal management of the electrostack and the consistency of the 101 individual cells meeting expectations and signed a commercialization agreement with a Korean company for a 50 MW class. In addition, the company has proposed a modular design for the electric stack, which allows the applicant to flexibly compose different capacities (e.g., multiples of the 10 kW class) by stacking 101 single cells, thereby expanding the low-power market (Barrett, 2015). In 2019, AFC Energy introduced the HydroX-Cell (L) AFC with a liquid electrolyte for large industrial scenarios. The device tolerates low-purity hydrogen and is planned for use in off-grid chargers for electric vehicles (Anon, 2019).

For AEMFC, as scholars continue to deepen their research, the power density of AEMFC stacks, driven by the emergence of new materials and structures, has led to the initial performance (BOL) of AEMFC stacks being very close to that of PEMFC stacks. In 2007, Varcoe et al. (2007) demonstrated the BOL of AEMFC stacks by employing ethylene-co-tetrafluoroethylene (ETFE) as an anion-exchange membrane, as well as using a carrier-free metal catalyst (anode: 4 mg/cm^2 of Pt–Ru, cathode: Pt), which enabled the cell to reach a peak power of 0.13 W/cm^2 In 2010, Gu et al. (2010) used a carbon-loaded platinum catalyst (anode: 0.2 mg/cm^2) for AEMFC stacks to achieve a peak power density of 0.26 W/cm^2. In 2013, Yu et al. (2014) optimized the structure of membrane electrodes in anion-exchange membrane fuel cells, as well as the preparation process of membrane electrode assemblies, to design a kilowatt-scale AFC stack. The power of a single cell reached 0.52 W/cm^2, and the power density of the electro stack reached 0.188 W/cm^2 for over 500 hours of continuous, stable operation. Meanwhile, the prototype of the electrostack operated continuously for more than 500 hours, achieving a power output of 1074 W. In 2015, Wang et al. (2015) used a 50 μm-thick quaternary ammonium sulfonated polysulfone AEM, an ion-aggregating polysulfone binder, and a carbon-loaded catalyst (anode: 0.4 mg Pt–Ru/cm^2 of Pt–Ru/C catalyst, cathode: 0.4 mg/cm^2 of Pt/C catalyst) to achieve a peak power density of 0.6 W/cm^2. In the same year, Pan et al., 2014)further increased the peak power density to 1.0 W/cm^2 by replacing the commercial Pt–Ru/C catalyst at the anode with a Pt/C catalyst. In 2017, Varcoe et al. (2007) and Kaspar et al. (2015) used a radiation-grafted AEM of 47 μm-thick ETFE, powder-ground radiation-grafted ETFE electrodes, and carbon-supported catalysts (anode: 0.4 mg Pt–Ru/cm^2 of Pt–Ru/C catalyst, cathode: 0.4 mg/cm^2 of Pt/C catalyst), achieving an even higher peak power density of 1.16 W/cm^2. In 2019, AFC Energy launched the HydroX-Cell(S) AFC, equipped with a solid anion-exchange membrane, with a current density of over 1 A/cm^2 in a single cell and a performance close to that of a PEMFC. This technology covered stationary and mobile scenarios such as transportation, marine, and defense. The cell is also ammonia fuel compatible, significantly reducing fuel costs and dependence on purity. As the power density of the AEMFC electro stack continues to increase, successive attempts have been made to integrate the electro stack with other electrochemical devices to meet the needs of more complex applications. In 2025, Kale et al. (2025) integrated an AEMFC, AEMWE (Anion Exchange Membrane Water Electrolyzer), and PV (solar photovoltaic) systems for uninterrupted hydrogen generation and power generation. The AEMFC, by consuming hydrogen generated by the AEMWE, can produce a total output of up to 2436.6 kW, which accounts for 35% of the power demand of the entire integrated system.

AFCs are gradually approaching the PEMFCs in terms of performance and scale-up for commercial applications. They are expected to achieve wider application and unleash greater development potential in the future due to their inherent advantages, such as low cost and strong fuel compatibility, as well as continuous optimization through technological iteration.

5 Perspectives

Traditional AFC, as a high-performance, low-cost, and mature technology for transportation, has specific development and application prospects. Especially as the problem of CO_2 poisoning is gradually being solved, the practical application of AFC presents a promising outlook. AFC and PEMFC each have their strengths, and while focusing on the development of PEMFC, AFC should also be explored more deeply to further uncover its potential. Compared to PEMFC, AEMFC has significant advantages in terms of catalyst application and battery cost but also faces shortcomings and challenges, such as low power density and poor battery stability. However, as scholars continue to conduct research in the field of AEMFC, the power density of AEMFC is gradually increasing. The anode side of nonplatinum catalysts, rather than platinum, is also a typical area of research. Once the anode of the AEMFC can be used with ultra-low platinum loading or nonplatinum catalysts, the cost of constructing commercial, large-size power stacks will be significantly reduced. The ideal goal of "platinum-free anion-exchange membrane fuel cells" can be achieved, but even if the use of platinum cannot be entirely eliminated, the cost advantage of adopting ultra-low-loaded platinum is still substantial. While PEMFCs' high CO_2 tolerance makes them well-suited to carbon-containing fuels, AEMFCs are better suited to nitrogen-based fuels such as hydrazine and ammonia, which cannot be used in PEMFCs, making AEMFCs an excellent option for transport applications.

References

Afshari, E., Houreh, N.B., 2014. Performance analysis of a membrane humidifier containing porous metal foam as flow distributor in a PEM fuel cell system. Energy Convers. Manag. 0196-8904. 88, 612–621.

Anon, 2019. AFC energy unveils branding of inaugural technology platforms. Fuel Cells Bull. 2019 (12), 12. doi:10.1016/S1464-2859(19)30529-2.

Arunchander, A., Peera, S.G., Sahu, A.K., 2017. Self-assembled manganese sulfide nanostructures on graphene as an oxygen reduction catalyst for anion exchange membrane fuel cells. ChemElectroChem. 2196-0216. 4 (6), 1544–1553.

Bachman, J., Santamaria, A., Tang, H.-Y., Park, J.W., 2012. Investigation of polymer electrolyte membrane fuel cell parallel flow field with induced cross flow. J. Power Sources. 0378-7753. 198, 143–148.

Badduri, S.R., Srinivasulu, G.N., Rao, S.S., 2020. Influence of bio-inspired flow channel designs on the performance of a PEM fuel cell. Chin. J. Chem. Eng. 1004-9541. 28 (3), 824–831.

Banhart, J., 2001. Manufacture, characterisation and application of cellular metals and metal foams. Prog. Mater. Sci. 0079-6425. 46 (6), 559–632.

Bao, Z., Niu, Z., Jiao, K., 2019. Numerical simulation for metal foam two-phase flow field of proton exchange membrane fuel cell. Int. J. Hydrog. Energy. 0360-3199. 44 (12), 6229–6244.

Barrett, S., 2015. AFC Energy doubles alkaline stack size in German industrial trial. Fuel Cells Bull. 2, 6.

Carton, J.G., Olabi, A.-G., 2010. Design of experiment study of the parameters that affect performance of three flow plate configurations of a proton exchange membrane fuel cell. Energy. 0360-5442. 35 (7), 2796–2806.

Carton, J.G., Olabi, A.-G., 2015. Representative model and flow characteristics of open pore cellular foam and potential use in proton exchange membrane fuel cells. Int. J. Hydrog. Energy. 0360-3199. 40 (16), 5726–5738.

Chang, H.-P., Chou, C.-L., Chen, Y.-S., Hou, T.-I., Weng, B.-J., 2007. The design and cost analysis of a portable PEMFC UPS system. Int. J. Hydrog. Energy. 0360-3199. 32 (3), 316–322.

Chaolin, W., Guang, W., Qian, W., 2022. Hydrogen Energy and Fuel Cell, vol. 6.

Chen, W.-H., Tsai, Z.-L., Chang, M.-H., You, S., Kuo, P.-C., 2021. Geometry optimization and pressure analysis of a proton exchange membrane fuel cell stack. Int. J. Hydrog. Energy. 0360-3199. 46 (31), 16717–16733.

Choi, K.-S., Kim, H.-M., Moon, S.-M., 2011. Numerical studies on the geometrical characterization of serpentine flow-field for efficient PEMFC. Int. J. Hydrog. Energy. 0360-3199. 36 (2), 1613–1627.

Cifrain, M., Kordesch, K.V., 2004. Advances, aging mechanism and lifetime in AFCs with circulating electrolytes. J. Power Sources. 0378-7753. 127 (1-2), 234–242.

Cooper, N.J., Santamaria, A.D., Becton, M.K., Park, J.W., 2017. Investigation of the performance improvement in decreasing aspect ratio interdigitated flow field PEMFCs. Energy Convers. Manag. 0196-8904. 136, 307–317.

Cooper, N.J., Smith, T., Santamaria, A.D., Park, J.W., 2016. Experimental optimization of parallel and interdigitated PEMFC flow-field channel geometry. Int. J. Hydrog. Energy. 0360-3199. 41 (2), 1213–1223.

Dabiri, S., Hashemi, M., Rahimi, M., Bahiraei, M., Khodabandeh, E., 2018. Design of an innovative distributor to improve flow uniformity using cylindrical obstacles in header of a fuel cell. Energy. 0360-5442. 152, 719–731.

Dekel, D.R., 2018. Review of cell performance in anion exchange membrane fuel cells. J. Power Sources. 0378-7753. 375, 158–169.

Deng, H., Wang, D., Wang, R., Xie, X., Yin, Y., Qing, D., Jiao, K., 2016a. Effect of electrode design and operating condition on performance of hydrogen alkaline membrane fuel cell. Appl. Energy. 0306-2619. 183, 1272–1278.

Deng, H., Wang, D., Xie, X., Zhou, Y., Yin, Y., Qing, D., Jiao, K., 2016b. Modeling of hydrogen alkaline membrane fuel cell with interfacial effect and water management optimization. Renew. Energy. 0960-1481. 91, 166–177.

Fan, L., Niu, Z., Zhang, G., Jiao, K., 2018. Optimization design of the cathode flow channel for proton exchange membrane fuel cells. Energy Convers. Manag. 0196-8904. 171, 1813–1821.

Gu, S., Cai, R., Luo, T., Jensen, K., Contreras, C., Yan, Y., 2010. Quaternary phosphonium-based polymers as hydroxide exchange membranes. ChemSusChem. 1864-5631. 3 (5), 555–558.

Gülzow, E., Schulze, M., 2004. Long-term operation of AFC electrodes with CO_2 containing gases. J. Power Sources. 0378-7753. 127 (1-2), 243–251.

Gutru, R., Turtayeva, Z., Xu, F., Maranzana, G., Vigolo, B., Desforges, A., 2020. A comprehensive review on water management strategies and developments in anion exchange membrane fuel cells. Int. J. Hydrog. Energy. 0360-3199. 45 (38), 19642–19663.

Hashemi, F., Rowshanzamir, S., Rezakazemi, M., 2012. CFD simulation of PEM fuel cell performance: effect of straight and serpentine flow fields. Math. Comput. Model. 0895-7177. 55 (3-4), 1540–1557.

Hossain, M.S., Shabani, B., 2015. Metal foams application to enhance cooling of open cathode polymer electrolyte membrane fuel cells. J. Power Sources. 0378-7753. 295, 275–291.

Hossain, M.S., Shabani, B., Cheung, C.P., 2017. Enhanced gas flow uniformity across parallel channel cathode flow field of proton exchange membrane fuel cells. Int. J. Hydrog. Energy. 0360-3199. 42 (8), 5272–5283.

Huang, Z., Jian, Q., Luo, L., Huang, B., Bai, X., Li, D.,2021. Rapid thermal response and sensitivity analysis of proton exchange membrane fuel cell stack with ultra-thin vapor chambers. Appl. Therm. Eng. 1359-4311. 199, 117526.

Huo, S., 2018. Theoretical and experimental study on cold start and ice formation mechanism of metal foam PEM fuel cell. doi:10.27356/d.cnki.gtjdu.2018.000370.

Huo, S., Cooper, N.J., Smith, T.L., Park, J.W., Jiao, K., 2017. Experimental investigation on PEM fuel cell cold start behavior containing porous metal foam as cathode flow distributor. Appl. Energy. 0306-2619. 203, 101–114.

Huo, S., Deng, H., Chang, Y., Jiao, K., 2012. Water management in alkaline anion exchange membrane fuel cell anode. Int. J. Hydrog. Energy. 0360-3199. 37 (23), 18389–18402.

Jeon, D.H., Greenway, S., Shimpalee, S., Van Zee, J.W., 2008. The effect of serpentine flow-field designs on PEM fuel cell performance. Int. J. Hydrog. Energy. 0360-3199. 33 (3), 1052–1066.

Jiao, K., Bachman, J., Zhou, Y., Park, J.W., 2014a. Effect of induced cross flow on flow pattern and performance of proton exchange membrane fuel cell. Appl. Energy. 0306-2619. 115, 75–82.

Jiao, K., He, P., Du, Q., Yin, Y., 2014b. Three-dimensional multiphase modeling of alkaline anion exchange membrane fuel cell. Int. J. Hydrog. Energy. 0360-3199. 39 (11), 5981–5995.

Jiao, K., Li, X., 2011. Water transport in polymer electrolyte membrane fuel cells. Prog. Energy Combust. Sci. 0360-1285. 37 (3), 221–291.

Jiao, K., Xuan, J., Du, Q., Bao, Z., Xie, B., Wang, B., Zhao, Y., Fan, L., Wang, H., Hou, Z., 2021. Designing the next generation of proton-exchange membrane fuel cells. Nature. 0028-0836. 595 (7867), 361–369.

Jiao, K., Zhou, B., Quan, P., 2006. Liquid water transport in straight micro-parallel-channels with manifolds for PEM fuel cell cathode. J. Power Sources. 0378-7753. 157 (1), 226–243.

Jung, A., Diebels, S., 2016. Modelling of metal foams by a modified elastic law. Mech. Mater. 0167-6636. 101, 61–70.

Kale, R.K., Vairakannu, P., Rajaraman, P.V., 2025. Solar powered integration of anion exchange membrane based water electrolyzer and fuel cell for uninterrupted hydrogen and power generation. Int. J. Hydrog. Energy. 0360-3199. 128, 771–785.

Kaspar, R.B., Letterio, M.P., Wittkopf, J.A., Gong, K., Gu, S., Yan, Y., 2015. Manipulating water in high-performance hydroxide exchange membrane fuel cells through asymmetric humidification and wetproofing. J. Electrochem. Soc. 1945-7111. 162 (6), F483.

Kerkoub, Y., Benzaoui, A., Haddad, F., Ziari, Y.K., 2018. Channel to rib width ratio influence with various flow field designs on performance of PEM fuel cell. Energy Convers. Manag. 0196-8904. 174, 260–275.

Kordesch, K., Gsellmann, J., Kraetschmer, B., 1983. AFC performance. Power Sources 9, 21.

Krewer, U., Weinzierl, C., Ziv, N., Dekel, D.R., 2018. Impact of carbonation processes in anion exchange membrane fuel cells. Electrochim. Acta. 0013-4686. 263, 433–446.

Kumar, S., Singh, P.K., 2019. Effects of flow inlet angle on flow maldistribution and thermal performance of water cooled mini-channel heat sink. Int. J. Therm. Sci. 1290-0729. 138, 504–511.

Lee, J., Gundu, M.H., Lee, N., Lim, K., Lee, S.W., Jang, S.S., Kim, J.Y., Ju, H., 2020. Innovative cathode flow-field design for passive air-cooled polymer electrolyte membrane (PEM) fuel cell stacks. Int. J. Hydrog. Energy. 0360-3199. 45 (20), 11704–11713.

Lee, N.W., Kim, S.I., Kim, Y.S., Kim, S.H., Ahn, B.K., Kim, M.S., 2015. An effective discharge method for condensed water inside the GDL using pressure gradient of a PEM fuel cell. Int. J. Heat Mass Transf. 0017-9310. 85, 703–710.

Li, W., Zhang, Q., Wang, C., Yan, X., Shen, S., Xia, G., Zhu, F., Zhang, J., 2017. Experimental and numerical analysis of a three-dimensional flow field for PEMFCs. Appl. Energy. 0306-2619. 195, 278–288.

Li, X., 2005. Principles of Fuel Cells, CRC press.

Li, Y.S., Zhao, T.S., Yang, W.W., 2010. Measurements of water uptake and transport properties in anion-exchange membranes. Int. J. Hydrog. Energy. 0360-3199. 35 (11), 5656–5665.

Lim, B.H., Majlan, E.H., Daud, W.R.W., Rosli, M.I., Husaini, T., 2017. Numerical analysis of modified parallel flow field designs for fuel cells. Int. J. Hydrog. Energy. 0360-3199. 42 (14), 9210–9218.

Liu, H., Yang, W., Tan, J., An, Y., Cheng, L., 2018. Numerical analysis of parallel flow fields improved by micro-distributor in proton exchange membrane fuel cells. Energy Convers. Manag. 0196-8904. 176, 99–109.

Lu, W., Pan, W., Chen, Z., Gao, Y., Ding, L., Chen, X., Wang, F.,2023. Structural optimization of distribution zone for large-sized PEMFC with high power density. Chem. Eng. Sci. 0009-2509. 276, 118803.

Nishida, K., Taniguchi, R., Ishizaki, Y., Tsushima, S., Hirai, S., 2015. Impacts of channel wettability and flow direction on liquid water transport in the serpentine flow field of a polymer electrolyte fuel cell. J. Power Sources. 0378-7753. 275, 447–457.

Pan, J., Chen, C., Li, Y., Wang, L., Tan, L., Li, G., Tang, X., Xiao, L., Lu, J. and Zhuang L.., 2014. Constructing ionic highway in alkaline polymer electrolytes. Energy Environ. Sci. 7 (1), 354–360. doi:10.1039/C3EE43275K.

Park, J.E., Lim, J., Kim, S., Choi, I., Ahn, C.-Y., Hwang, W., Lim, M.S., Cho, Y.-H., Sung, Y.-E., 2018. Enhancement of mass transport in fuel cells using three-dimensional graphene foam as flow field. Electrochim. Acta. 0013-4686. 265, 488–496.

Park, J.E., Lim, J., Lim, M.S., Kim, S., Kim, O.-H., Lee, D.W., Lee, J.H., Cho, Y.-H., Sung, Y.-E.,2019. Gas diffusion layer/flow-field unified membrane-electrode assembly in fuel cell using graphene foam. Electrochim. Acta. 0013-4686. 323, 134808.

Pistoresi, C., Fan, Y., Luo, L., 2015. Numerical study on the improvement of flow distribution uniformity among parallel mini-channels. Chem. Eng. Process. Process Intensif. 0255-2701. 95, 63–71.

Shen, J., Tu, Z., Chan, S.H., 2019. Enhancement of mass transfer in a proton exchange membrane fuel cell with blockage in the flow channel. Appl. Therm. Eng. 1359-4311. 149, 1408–1418.

Shen, Z., Wang, Q., Xu, F., Zhou, X., Ni, L., Cheng, Z., Jiang, J., 2025. Investigation on strengthening effect of metal foams in enlarged microchannels for heterogeneous liquid-liquid mass transfer. Chem. Eng. Process. Process Intensif. 0255-2701. 215, 110379.

Siddiqui, O.K., Zubair, S.M., 2017. Efficient energy utilization through proper design of microchannel heat exchanger manifolds: a comprehensive review. Renew. Sustain. Energy Rev. 1364-0321. 74, 969–1002.

Sleem-ur-Rahman, M.A., Al-Zakri, A.S., 1998. Parametric study of the preparation of gas-diffusion electrodes for alkaline fuel cells by a filtration method. J. Power Sources 72 (1), 71–76.

Su, A., Chiu, Y.C., Weng, F.B., 2005. The impact of flow field pattern on concentration and performance in PEMFC. Int. J. Energy Res. 0363-907X. 29 (5), 409–425.

Suresh, P.V., Jayanti, S., Deshpande, A.P., Haridoss, P., 2011. An improved serpentine flow field with enhanced cross-flow for fuel cell applications. Int. J. Hydrog. Energy. 0360-3199. 36 (10), 6067–6072.

Tanaka, S., Shudo, T., 2014. Corrugated mesh flow channel and novel microporous layers for reducing flooding and resistance in gas diffusion layer-less polymer electrolyte fuel cells. J. Power Sources. 0378-7753. 268, 183–193.

Tseng, C.-J., Tsai, B.T., Liu, Z.-S., Cheng, T.-C., Chang, W.-C., Lo, S.-K., 2012. A PEM fuel cell with metal foam as flow distributor. Energy Convers. Manag. 0196-8904. 62, 14–21.

Tsuchiya, H., Kobayashi, O., 2004. Mass production cost of PEM fuel cell by learning curve. Int. J. Hydrog. Energy 0360-3199. 29 (10), 985–990.

Tsukamoto, T., Aoki, T., Kanesaka, H., Taniguchi, T., Takayama, T., Motegi, H., Takayama, R., Tanaka, S., Komiyama, K., Yoneda, M.,2021. Three-dimensional numerical simulation of full-scale proton exchange membrane fuel cells at high current densities. J. Power Sources. 0378-7753. 488, 229412.

Varcoe, J.R., Slade, R.C.T., Yee, E.L.H., Poynton, S.D., Driscoll, D.J., Apperley, D.C., 2007. Poly (ethylene-co-tetrafluoroethylene)-derived radiation-grafted anion-exchange membrane with properties specifically tailored for application in metal-cation-free alkaline polymer electrolyte fuel cells. Chem. Mater. 0897-4756. 19 (10), 2686–2693.

Wang, B., Zhou, Y., Du, Q., Yin, Y., Jiao, K., 2016. Transient investigation of passive alkaline membrane direct methanol fuel cell. Appl. Therm. Eng. 1359-4311. 100, 1245–1258.

Wang, J., 2015. Theory and practice of flow field designs for fuel cell scaling-up: a critical review. Appl. Energy. 0306-2619. 157, 640–663.

Wang, X.-D., Xu, J.-L., Yan, W.-M., Lee, D.-J., Su, A., 2011. Transient response of PEM fuel cells with parallel and interdigitated flow field designs. Int. J. Heat Mass Transf. 0017-9310. 54 (11-12), 2375–2386.

Wang, Y., Basu, S., Wang, C.-Y., 2008. Modeling two-phase flow in PEM fuel cell channels. J. Power Sources. 0378-7753. 179 (2), 603–617.

Wang, Y., Seo, B., Wang, B., Zamel, N., Jiao, K., Adroher, X.C.,2020. Fundamentals, materials, and machine learning of polymer electrolyte membrane fuel cell technology. Energy AI. 2666-5468. 1, 100014.

Wang, Y., Wang, G., Li, G., Huang, B., Pan, J., Liu, Q., Han, J., Xiao, L., Lu, J., Zhuang, L., 2015. Pt–Ru catalyzed hydrogen oxidation in alkaline media: oxophilic effect or electronic effect? Energy Environ. Sci. 8 (1), 177–181.

Wang, Y., Yue, L., Wang, S., 2017. New design of a cathode flow-field with a sub-channel to improve the polymer electrolyte membrane fuel cell performance. J. Power Sources. 0378-7753. 344, 32–38.

Xue, J., Douglin, J.C., Yassin, K., Huang, T., Jiang, H., Zhang, J., Yin, Y., Dekel, D.R., Guiver, M.D., 2024. High-temperature anion-exchange membrane fuel cells with balanced water management and enhanced stability. Joule. 2542-4785. 8 (5), 1457–1477.

Yan, X., Guan, C., Zhang, Y., Jiang, K., Wei, G., Cheng, X., Shen, S., Zhang, J., 2019. Flow field design with 3D geometry for proton exchange membrane fuel cells. Appl. Therm. Eng. 1359-4311. 147, 1107–1114.

Yan, Y.Y., Weikang, L., Junfeng, Z., 2022. Structure and performance of anode dual catalyst layer for anion-exchange membrane fuel cell. J. Tianjin Univ. 55 (01), 40–45.

Yin, Y., Wang, X., Shangguan, X., Zhang, J., Qin, Y., 2018. Numerical investigation on the characteristics of mass transport and performance of PEMFC with baffle plates installed in the flow channel. Int. J. Hydrog. Energy. 0360-3199. 43 (16), 8048–8062.

Yong, Z., Shirong, H., Xiaohui, J., Yuntao, Y., Mu, X., Xi, Y.,2022a. Characteristics of proton exchange membrane fuel cell considering “dot matrix” gas distribution zones and waveform staggered flow field with cooling channels. Energy Convers. Manag. 0196-8904. 267, 115881.

Yong, Z., Shirong, H., Xiaohui, J., Yuntao, Y., Mu, X., Xi, Y., 2022b. Performance study on a large-scale proton exchange membrane fuel cell with cooling. Int. J. Hydrog. Energy. 0360-3199. 47 (18), 10381–10394.

Yu, Y., Zhan, Z., He, L., Yang, X., Wan, X., Sui, P.-C., Pan, M., 2021. Effects of distribution zone design on flow uniformity and pressure drop in PEMFC. J. Electrochem. Soc. 1945-7111. 168 (9), 094505.

Yuan, W., Tang, Y., Yang, X., Wan, Z., 2012. Porous metal materials for polymer electrolyte membrane fuel cells–A review. Appl. Energy. 0306-2619. 94, 309–329.

Zhang, G., Jiao, K., 2018. Three-dimensional multi-phase simulation of PEMFC at high current density utilizing Eulerian–Eulerian model and two-fluid model. Energy Convers. Manag. 0196-8904. 176, 409–421.

Ziv, N., Mustain, W.E., Dekel, D.R., 2018. The effect of ambient carbon dioxide on anion-exchange membrane fuel cells. ChemSusChem. 1864-5631. 11 (7), 1136–1150.

Further Reading

Hongmei, Y., et al., 2014. Alkaline Anion Exchange Membrane Fuel Cell, 27, 5. SNAD000001596053.

Jianjun, Z., 2020. Preparation and applied research of anion exchange membranes for alkaline fuel cell. doi:10.27517/d.cnki.gzkju.2020.000691.

Zhuo, Z., Aihua, W., Shengji, L., Guanghe, Y., Chunyu, Z., 2015. Research and development of anion exchange membrane for fuel cell. Polym. Bull. (08), 52–62. doi:10.14028/j.cnki.1003-3726.2015.08.007.

Chapter 7

Performance testing and characterization techniques

Xingyi Shi[a], Dixian Wu[a], and Qiang Zheng[b]
[a]*City University of Hong Kong (Dongguan), Dongguan, China*
[b]*Eastern Institute of Technology, Zhejiang Key Laboratory of Industrial Intelligence and Digital Twin, Ningbo, China*

1 Introduction

The alkaline fuel cell is regarded as one of the most promising systems in the field of energy conversion, offering a sustainable and efficient method for transforming chemical energy into electricity (McLean et al., 2002; Kordesch et al., 2000). It produces electricity through electrochemical reactions, making it a viable and eco-friendly solution for various energy needs (Verhaert et al., 2009). Unlike conventional proton exchange membrane fuel cells, which rely on the transport of protons across the membrane, the operation of the alkaline fuel cell depends on the movement of hydroxide ions (Ferriday and Middleton, 2021). The core functionality of the alkaline fuel cell is determined by three key components: the membrane, cathode, and anode (Tuan and Kim, 2016). Each of these components plays a critical role in the overall performance and functionality of the fuel cell, and a thorough analysis of their individual behaviors is essential for optimizing the overall cell performance.

1. Membrane: The membrane serves as a passage for the movement of hydroxide ions during electrochemical reactions (Maegawa et al., 2022). It must allow hydroxide ions to migrate freely while maintaining electrical insulation between the anode and cathode. Representatively, anion exchange membranes (AEMs) are widely used in alkaline fuel cells. The choice of membrane material plays a significant role in determining the membrane properties and further influences the fuel cell's output power, durability, and efficiency. To ensure long-term stability, it is also vital for the membrane to withstand the operating conditions required for fast electrochemical reactions.
2. Cathode: The cathode, or positive electrode, is where the reduction reaction of the oxidant occurs. It is typically composed of a bilayered structure

Alkaline Fuel Cells. https://doi.org/10.1016/B978-0-443-36638-3.00010-3

comprising a diffusion layer and a catalyst layer. The diffusion layer facilitates the transport of oxidants to enable even distribution across the catalyst layer (Bidault et al., 2009; Jo et al., 2000). The catalytic activity of the catalyst layer, on the other hand, determines the activation loss, while the catalyst's stability is critical for ensuring the long-term performance of the fuel cell. Additionally, the cathode is required to withstand varied operating conditions, such as high temperatures and humidity, to ensure stable and efficient power output.

3. Anode: The anode, or negative electrode, is where the oxidation reaction of fuels takes place. Like the cathode, it is also usually made of two layers to promote the transport of reactants while ensuring a low fabrication cost. The anode's performance is equally important to that of the cathode, as it directly influences the cell's electrochemical performance and overall power output. Additionally, the selection of anode material, coupled with its ability to withstand the electrochemical reactions and harsh environmental impacts, is also a key consideration during the anode manufacturing process.

A comprehensive understanding of the roles of these components is essential for the optimization of alkaline fuel cell performance. The efficiency, durability, and operational capability of each component under varied operating conditions directly impact the fuel cell's overall performance. In addition to the key components, the performance of an alkaline fuel cell is also highly sensitive to operating conditions such as temperature, humidity, and the concentration of reactants (Verhaert et al., 2012; Ergül et al., 1997; Verhaert et al., 2011; Zhang et al., 2011; Verma and Basu, 2007). These factors can all significantly impact the fuel cell. For example, higher temperatures can increase the fuel cell's output performance by promoting the reactions, but they can also accelerate the deterioration of the membrane, resulting in performance degradation over time. Similarly, fluctuations in humidity can affect conductivity and the movement of reactants, further impacting fuel cell performance. The availability of reactants is another crucial factor affecting fuel cell performance, as low concentrations can significantly limit its power output.

Thus, to ensure a comprehensive understanding of fuel cell behavior and optimize its performance, it is essential to conduct thorough performance analyses and characterizations. Employing a diverse range of testing techniques is of paramount importance to identify bottlenecks, improve component reliability, and ensure the fuel cell meets the required standards for real-world applications. Techniques such as polarization curve tests, constant-current discharging behavior tests, and durability tests provide insights into operational efficacy across diverse conditions, thereby serving as a cornerstone for technological advancement and commercial viability.

In conclusion, the development of alkaline fuel cells is fundamentally governed by the careful optimization of constituent materials—including membranes, cathodes, and anodes—coupled with rigorous characterization techniques. This chapter serves as a comprehensive guide to the essential testing and characterization techniques vital for the development and application of alkaline fuel cells, emphasizing their importance in driving innovation and ensuring optimal performance in various industrial and consumer settings.

2 Electrode characterization techniques

The performance of fuel cells relies heavily on the interaction between the electrodes and the electrolyte to convert chemical energy into electrical energy. Therefore, understanding the electrochemical behavior, as well as the physical and structural properties of the electrodes, is essential for optimizing the overall performance of the fuel cell system. To date, a variety of characterization techniques have been developed and employed, providing unique insights into different aspects of the electrodes. In this section, some typical methods used for analyzing the electrodes are summarized and discussed.

In terms of physical and structural properties, scanning electron microscopy (SEM) is one of the most important techniques for providing insights into the surface topography and defects of electrodes. It provides detailed surface maps, allowing for the identification of specific structural features. The high-resolution images of the electrode surfaces obtained can help reveal the factors that may impact electrochemical reactions as well as mass transfer processes. Such structural analyses are crucial for understanding the relationship between electrode structure and its performance. For example, Yang et al. employed SEM to demonstrate a notable increase in catalyst agglomerate size with elevated ionomer content (Fig. 7.1), which could negatively impact the surface area and catalyst utilization (Yang et al., 2014).

With further assistance from energy dispersive spectroscopy (EDS) mapping, complementary information could be obtained, including detailed chemical composition and distribution data. This technique is indispensable for elucidating elemental dispersion and quantifying compositional variations, thereby helping to understand the interaction between the electrode materials and the electrolyte. Using the EDS technique, Bhandary et al. examined the AgCd electrocatalyst coated on the carbon paper (Bhandary et al., 2016), confirming a uniform distribution of Ag and Cd particles (Fig. 7.2) across the surface of the carbon paper.

X-ray diffraction (XRD) is another powerful technique widely used to study the crystallographic structure of materials. In the context of analyzing catalysts in the electrodes, XRD provides valuable insights into the electrode composition. Furthermore, it can also help reveal the presence of impurities or defects in the electrode material. Thermal analysis techniques, such as differential scanning calorimetry and thermogravimetric analysis (TGA), are critical techniques for understanding the thermal stability and decomposition behavior of the electrodes. They can help in identifying the onset temperature and progression of electrode degradation, which is essential in determining the operational conditions and durability of electrodes.

FIG. 7.1 SEM images. SEM images of surface morphology at different ionomer contents: (A) 10 wt.%, (B) 15 wt.%, (C) 20 wt.%, and (D) 25 wt.%. *(Reproduced with permission from Yang, D., Yu, H., Li, G., Zhao, Y., Liu, Y., Zhang, C., Song, W., Shao, Z., 2014. Fine microstructure of high performance electrode in alkaline anion exchange membrane fuel cells. J. Power Sources, 267, 39–47. https://doi.org/10.1016/j.jpowsour.2014.04.053.)*

To comprehensively examine the electrochemical performance of electrodes, cyclic voltammetry (CV) stands as one of the most widely used and powerful characterization techniques. It provides dynamic insights into redox processes and reaction kinetics. By linearly scanning back and forth between set electrode potentials and monitoring the current response, CV reveals characteristic oxidation/reduction peaks that elucidate reaction mechanisms and catalytic activity. Analysis of the shape, position, and intensity of the peaks can reveal critical information, such as reaction reversibility, intermediate species, and degradation behaviors. Additionally, CV is highly effective for assessing electrode behavior under varied operating conditions, such as different scan rates or electrolyte compositions, making it a versatile tool for performance analysis. Utilizing CV, Fazil et al. first evaluated the performance of catalysts with varied loadings of Ag/CNT, as illustrated in Fig. 7.3A, proving CV to be an effective technique for assessing electrochemical performance. Subsequently, the stability of these catalysts was further investigated over fifteen testing cycles (Fig. 7.3B), revealing the influence of metal loading on durability (Fazil and Chetty, 2014).

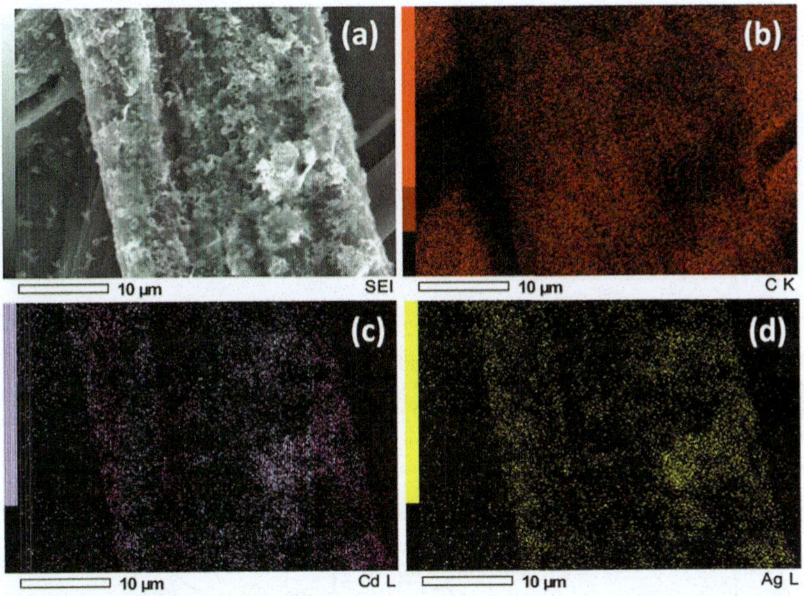

FIG. 7.2 SEM image and EDS mapping. (A) SEM image, and (B) to (D) EDS mapping of the AgCd electro-catalyst on the carbon paper surface. *(Reproduced with permission from Bhandary, N., Basu, S., Ingole, P.P., 2016. Rudimentary simple, single step fabrication of nano-flakes like AgCd alloy electro-catalyst for oxygen reduction reaction in alkaline fuel cell. Electrochim. Acta, 212, 122–129. https://doi.org/10.1016/j.electacta.2016.06.143.)*

Furthermore, CV could also serve as a quantitative method for determining the electrochemical active surface area (ECSA) of catalytic materials. Unlike physical surface area measurements obtained through techniques like BET adsorption, which quantify total exposed surface area, ECSA specifically measures the catalyst surface area that is electrochemically accessible and participates in charge transfer reactions. This distinction makes it critical for evaluating catalyst utilization efficiency, as it accounts for factors such as inactive sites, pore accessibility, and surface passivation. Ganesh et al. conducted a comprehensive analysis of the ECSA across catalysts with diverse shapes, providing quantitative evidence that enhanced surface area is a key factor contributing to the superior electrochemical performance of the catalyst (Ganesh et al., 2021).

Linear sweep voltammetry (LSV) is also an essential electrochemical technique, particularly useful for studying reaction kinetics and mass transport phenomena. Unlike CV, which employs a cyclic potential scan, LSV applies a single, unidirectional voltage sweep at a controlled rate, allowing for the direct observation of the current response as a function of the applied potential. This method is especially valuable for evaluating onset potentials for electrocatalytic processes (e.g., oxygen evolution or hydrogen evolution reactions) and assessing Tafel slopes. Using a rotating ring-disk electrode, Chai et al. obtained LSV

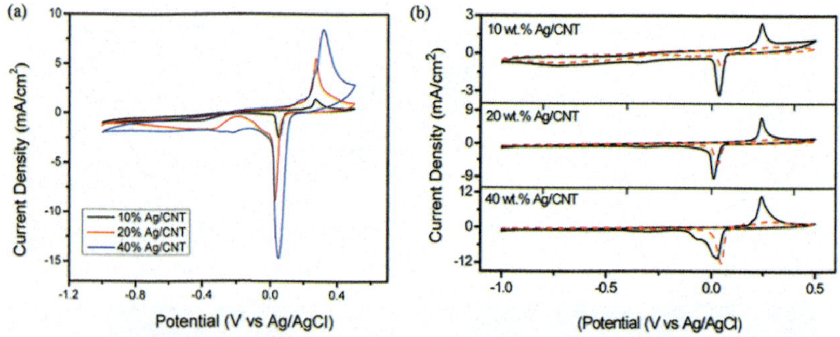

FIG. 7.3 CV curves. CV of Ag/CNT with different weights (A) at a scan rate of 100 mV s^{-1}, and (B) a comparative analysis of the first and 50th cycles (Fazil and Chetty, 2014). *(Reproduced with permission from Fazil, A., Chetty, R., 2014. Synthesis and Evaluation of Carbon Nanotubes Supported Silver Catalyst for Alkaline Fuel Cell. Electroanalysis, 26 (11), 2380–2387. https://doi.org/10.1002/elan.201400246.)*

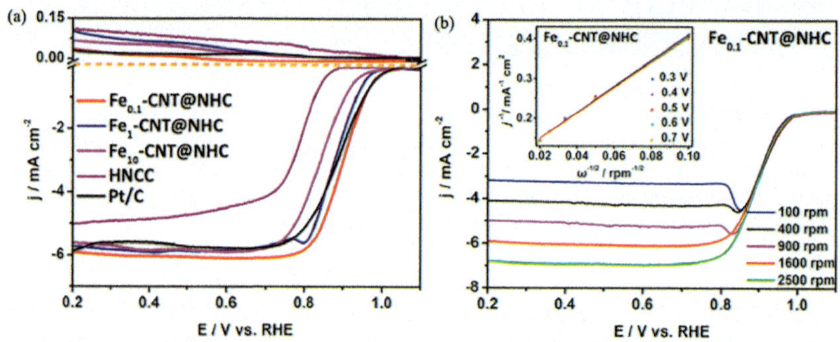

FIG. 7.4 LSV curves. LSV curves: (A) at 1600 rpm for Fe_x-CNT@NHC and Pt/C, and (B) at 100–2500 rpm. *(Reproduced with permission from Chai, L., Hu, Z., Wang, X., Zhang, L., Li, T.T., Hu, Y., Pan, J., Qian, J., Huang, S., 2021. Fe7C3 nanoparticles with in situ grown CNT on nitrogen doped hollow carbon cube with greatly enhanced conductivity and ORR performance for alkaline fuel cell. Carbon, 174, 531–539. https://doi.org/10.1016/j.carbon.2020.12.070.)*

plots that demonstrate the synergy of Fe doping and CNTs significantly influences the enhancement of oxygen reduction reaction catalytic sites (Fig. 7.4A). Moreover, their results further reveal that while the onset potential remains nearly constant, the limiting current density increases markedly with higher rotation speeds (Fig. 7.4B), indicating the importance of the mass transport process. These results thus demonstrate LSV measurements as a powerful tool for probing catalytic performance and underlying mechanisms in the fuel cell system (Chai et al., 2021).

Collectively, these characterization techniques provide a comprehensive framework for understanding electrode behavior, allowing researchers and engineers to optimize electrode design rationally, thereby enabling advancements in performance and durability. The employment of these comprehensive approaches is thus essential for advancing the development of alkaline fuel cells and ensuring their viability for diverse industrial and consumer applications.

3 Membrane characterization techniques

The performance and long-term durability of alkaline fuel cells are critically dependent on the physicochemical properties of the polymer electrolyte membranes employed. The membranes serve as the central component needed to facilitate ion transport while maintaining structural integrity under harsh operating conditions. A comprehensive understanding of membrane characteristics, covering ionic conductivity, mechanical robustness, and other factors, thus requires a multifaceted analytical approach. This section summarizes the state-of-the-art characterization methods for polymer electrolyte membranes, emphasizing their applications in alkaline fuel cells.

The ionic conductivity of the polymer electrolyte membrane stands as a pivotal metric, directly influencing the performance of the alkaline fuel cell. For instance, in an AEM, the high ionic conductivity could ensure efficient hydroxide ion transport, thereby minimizing ohmic losses and enhancing overall cell performance. Electrochemical impedance spectroscopy (EIS) represents the most widely employed technique for evaluating membrane conductivity. In a typical EIS measurement, a membrane sample is sandwiched between two blocking electrodes, and a small sinusoidal voltage perturbation is applied across a range of frequencies. The resulting Nyquist plot yields the ohmic resistance (R) of the membrane, which is then used to calculate ionic conductivity (σ) through the following equation:

$$\sigma\left(S\,\mathrm{cm}^{-1}\right)=\frac{d}{RS} \tag{1}$$

where d denotes membrane thickness, R represents the measured ohmic resistance, and S corresponds to the effective contacting area. For accurate measurements, environmental conditions such as temperature and humidity must be strictly controlled in the process due to their significant impact on ion mobility. As shown in Fig. 7.5, high operating temperature and relative humidity typically assist with membrane conductivity (Li et al., 2013). Comparative studies are also often needed to provide a direct comparison with commercialized products such as the Nafion series membranes, thereby ensuring the reliability of the experimental data.

A major challenge in alkaline fuel cell operation arises from membrane degradation under highly alkaline conditions, which can lead to a reduced lifespan. To systematically evaluate membrane chemical stability, membranes are typically subjected to accelerated aging tests involving immersion in concentrated potassium hydroxide (KOH) solutions at elevated temperatures between 60 and 80°C for extended periods spanning days to weeks. Using this method, Han et al.

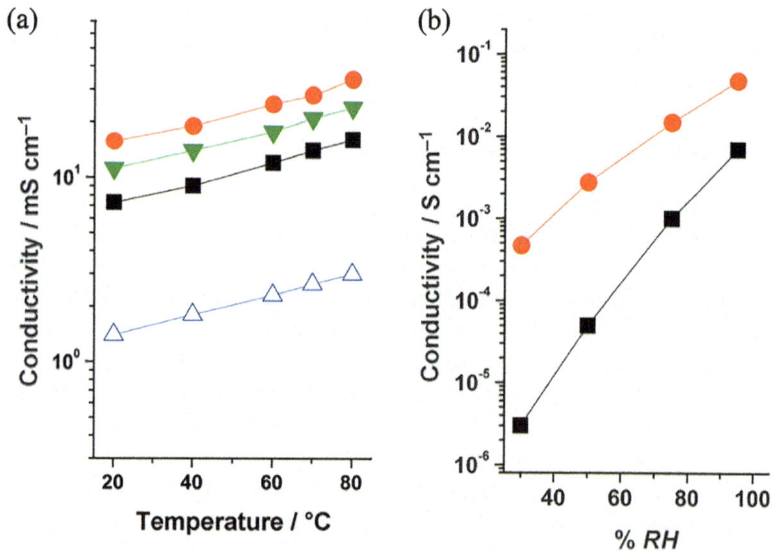

FIG. 7.5 Membrane conductivity. Membrane conductivity at different (A) operating temperatures and (B) humidity levels. *(Reproduced with permission from Li, N., Guiver, M.D., Binder, W.H., 2013. Towards high conductivity in anion-exchange membranes for alkaline fuel cells. ChemSusChem, 6 (8), 1376–1383. https://doi.org/10.1002/cssc.201300320.)*

evaluated the membrane's chemical stability. A 30-day exposure test was conducted, during which a substantial mass loss exceeding 45% was observed for one membrane sample (Fig. 7.6), indicating pronounced chemical degradation (Han et al., 2018). This approach thus underscores the significance of chemical stability testing in determining membrane durability under harsh operating conditions, which is pivotal before the widespread application of membranes.

In addition to the weight changes, postexposure analysis also encompasses other evaluation criteria. For instance, visual inspection could be performed to identify macroscopic failures, such as cracking or discoloration, in comparison to unexposed control samples (Merle et al., 2011). Meanwhile, tensile tests could be used to quantify the degradation of mechanical properties, including tensile strength and elasticity. Dimensional stability is another important property that could also be measured by precisely examining membrane swelling or shrinkage in thickness and planar area. Furthermore, electrochemical performance reassessment through repeated EIS tests could provide critical data on conductivity retention after chemical stability examination.

TGA serves as a complementary technique for evaluating the membrane's thermal stability, where controlled heating of membrane samples under inert atmospheres allows for the measurement of mass loss profiles. The resulting data, including degradation onset temperatures and residual mass percentages, offer valuable insights into the thermal degradation behavior of the membrane and its suitability for high-temperature alkaline fuel cell operations.

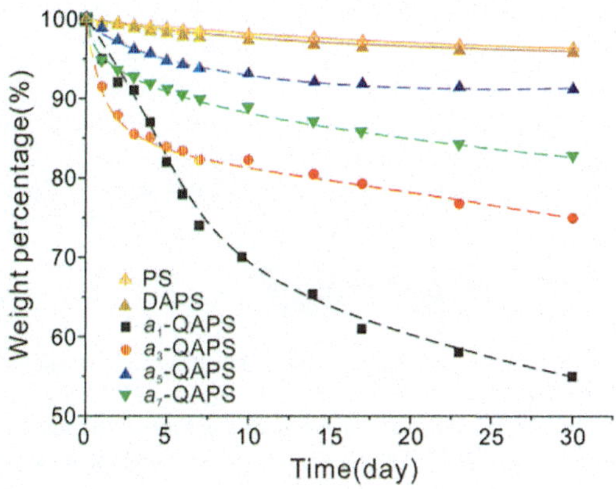

FIG. 7.6 Stability test. Stability test of the backbone in 1 mol L^{-1} KOH solution at 80°C. *(Reproduced with permission from Han, J., Pan, J., Chen, C., Wei, L., Wang, Y., Pan, Q., Zhao, N., Xie, B., Xiao, L., Lu, J., Zhuang, L., 2019. Effect of Micromorphology on Alkaline Polymer Electrolyte Stability. ACS Appl. Mater. Interfaces, 11 (1), 469–477. https://doi.org/10.1021/acsami.8b09481.)*

Beyond chemical and thermal properties, mechanical robustness represents another fundamental requirement for polymer electrolyte membranes, as the membrane is required to withstand the substantial compressive forces encountered during fuel cell assembly and operation while maintaining structural integrity over a prolonged operational life. The membrane tensile test is usually performed using a universal tensile testing machine (Fig. 7.7), which provides quantitative data on key mechanical parameters, including tensile strength, elongation at break, and Young's modulus (Su et al., 2017). Membranes demonstrating high tensile strength values coupled with moderate elongation capabilities are generally preferred for alkaline fuel cell applications, as they can ensure the membrane has good resistance to cracking during handling.

The water uptake and swelling behavior are also critical properties. Proper hydration is essential for maintaining adequate ionic conductivity, yet excessive water uptake can lead to detrimental swelling that compromises mechanical integrity and dimensional stability (Msomi et al., 2020). To measure the water uptake ratio, the weight of dried membrane samples is first measured, followed by immersion in deionized water for 24 hours. After carefully removing surface moisture, the wet weight of the membrane is then recorded, and the water uptake percentage is calculated according to:

$$\text{Water uptake}(\%) = \frac{W_w - W_d}{W_d} \times 100\%$$

where W_w and W_d represent the weight of the wet and dry membrane, respectively. The swelling ratio measurements, on the other hand, track dimensional

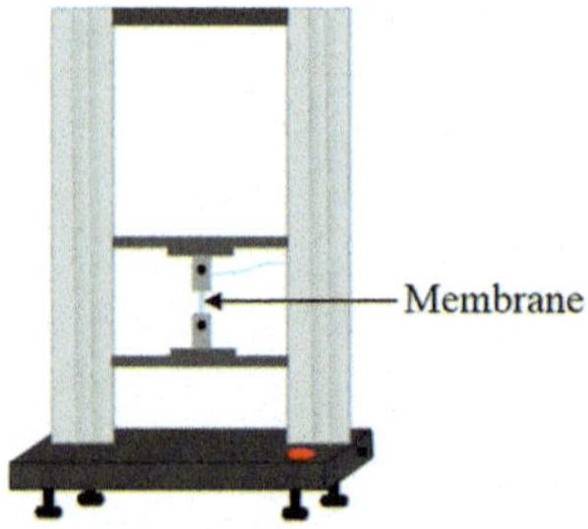

FIG. 7.7 Tensile test schematic. A schematic of a tensile test machine. *(Reproduced with permission from Su, L., Zhang, D., Peng, S., Wu, X., Luo, Y., He, G., 2017. Orientated graphene oxide/Nafion ultra-thin layer coated composite membranes for vanadium redox flow battery. Int. J. Hydrogen Energy, 42 (34), 21806–21,816. https://doi.org/10.1016/j.ijhydene.2017.07.049.)*

changes in length, width, and thickness using precision instruments, with results expressed as percentage changes relative to the dry state dimensions. These measurements provide crucial data for understanding the interplay between hydration and dimensional stability, enabling researchers to optimize membrane formulations for balanced performance. The integration of these characterization approaches provides a comprehensive understanding of membrane behavior under various operational conditions, facilitating the development of next-generation polymer electrolyte membranes with enhanced performance and durability characteristics tailored specifically for alkaline fuel cell applications.

4 Cell performance characterization techniques

In addition to the electrode- and membrane-level performance characterization techniques, the widespread adoption of alkaline fuel cells also requires rigorous characterization at the cell level to assess their performance, durability, and operational stability under realistic conditions. Unlike the aforementioned characterization techniques that focus on individual components, full-cell testing provides a comprehensive understanding at the system-level, enabling optimization of cell design parameters and identification of failure mechanisms.

The polarization curve serves as one of the foundational diagnostic tools for fuel cell performance evaluation, offering critical insights into the various loss mechanisms that govern system output power and efficiency. Generated by sweeping current density and monitoring cell voltage, the polarization curves reveal the contributions of activation, ohmic, and mass transport polarizations (Fig. 7.8). The open-circuit voltage represents the cell voltage when no current is applied. Deviations from the theoretical value often indicate issues such as mixed potentials, which can significantly impact overall performance. At

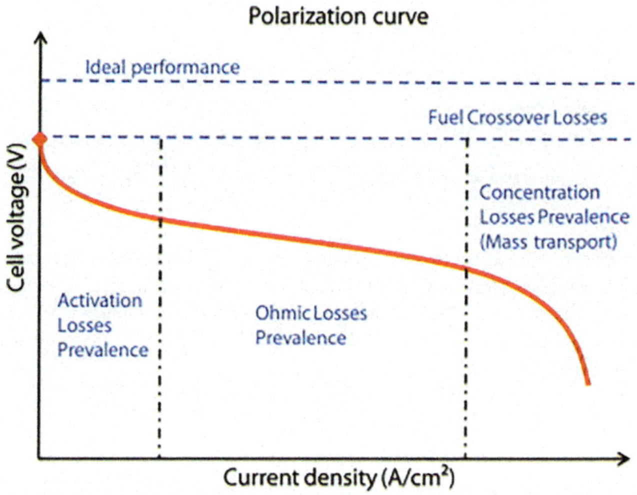

FIG. 7.8 Polarization curve. A typical fuel cell polarization curve. *(Reproduced with permission from Jouin, M., Gouriveau, R., Hissel, D., Péra, M.C., Zerhouni, N., 2013. Prognostics and Health Management of PEMFC—State of the art and remaining challenges. Int. J. Hydrogen Energy, 38 (35), 15,307–15,317. https://doi.org/10.1016/j.ijhydene.2013.09.051.)*

low current densities, activation polarization dominates due to sluggish reaction kinetics. This region of the polarization curve is highly sensitive to catalyst composition, electrode structure, and operating temperature, making it a key focus area for performance optimization. As the current density increases, ohmic polarization becomes the dominant loss, arising from contact resistances, membrane resistance, and ionic resistance. At even higher current densities, mass transport polarization emerges as the primary limiting factor, driven by insufficient reactant delivery to active sites and slow product removal from the electrode surface. This region of the polarization curve is strongly influenced by electrode porosity, gas diffusion layer design, and reactant concentration, highlighting the importance of structural optimization in mitigating mass transport losses.

In addition to the polarization curve test, full-cell EIS is also widely used to provide system-wide insights. By applying a small sinusoidal voltage perturbation across a range of frequencies and analyzing the resulting impedance response, this method provides critical insights into the resistive and capacitive processes within the electrochemical system. Specifically, EIS enables the decoupling of ohmic resistance, charge transfer resistance, and diffusion resistance, which are essential for understanding and optimizing cell performance. As illustrated in Fig. 7.9, increasing the KOH concentration from 1 to 3 M enhances catalytic activity, resulting in a reduction of charge transfer resistance (low-frequency region). However, concentrations beyond 3 M could

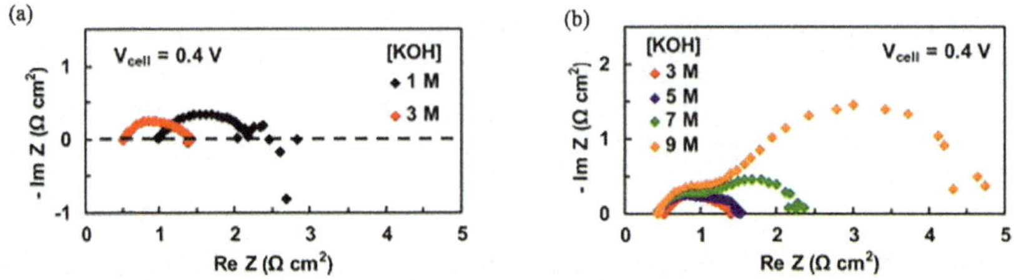

FIG. 7.9 EIS spectra. EIS spectra of an alkaline microfluidic H_2/O_2 fuel cell at 0.4 V with varying KOH concentrations. *(Reproduced with permission from Brushett, F.R., Naughton, M.S., Ng, J.W.D., Yin, L., Kenis, P.J.A., 2012. Analysis of Pt/C electrode performance in a flowing-electrolyte alkaline fuel cell. Int. J. Hydrogen Energy, 37 (3), 2559–2570. https://doi.org/10.1016/j.ijhydene.2011.10.078.)*

adversely induce anode flooding, which severely compromises hydrogen mass transport. This observation highlights that EIS can effectively decouple charge transfer and mass transport polarization (observed at low frequencies) from ohmic polarization (observed at the intersection of the x-axis at high frequency) (Bidault et al., 2010; Brushett et al., 2012).

Galvanostatic discharge testing represents another essential characterization method, providing a direct assessment of alkaline fuel cell performance under constant current conditions. This method provides critical insights into the cell voltage stability and efficiency over prolonged durations. During this test, the fuel cell is typically evaluated across a range of temperatures, pressures, and humidity levels to simulate diverse operating environments. The obtained discharge curves enable the determination of key performance metrics such as Faradaic efficiency, voltage efficiency, energy efficiency, and power output, all of which are critical for evaluating the practical viability of alkaline fuel cells. Moreover, by varying discharge current density, critical thresholds of fast performance decay can be identified, revealing the cell operation limits. Furthermore, long-term discharging tests could also help reveal gradual changes in voltage profiles, offering early indicators of degradation mechanisms such as electrode corrosion or membrane degradation. Posttest analysis, including microscopy and spectroscopy of cell components, helps correlate performance decay with physical and chemical changes in the electrodes and membranes, enabling a comprehensive understanding of failure modes. For instance, Sanetuntikul et al. reported that, through a 100-hour durability test, the Fe–N–C catalyst demonstrated better stability than the reference Pt/C catalyst (Fig. 7.10) (Sanetuntikul and Shanmugam, 2015).

By integrating these characterization techniques, a thorough alkaline fuel cell performance analysis can be achieved, which not only facilitates performance optimization but also accelerates the development of commercially viable alkaline fuel cell technologies.

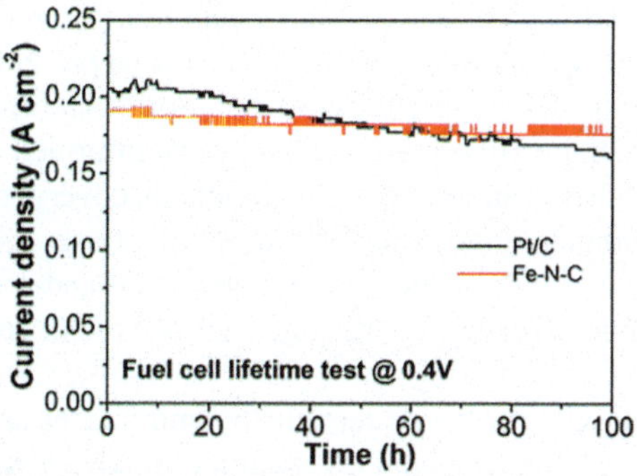

FIG. 7.10 Durability tests. Alkaline fuel cell performance durability tests. *(Reproduced with permission from Sanetuntikul, J., Shanmugam, S., 2015. High pressure pyrolyzed non-precious metal oxygen reduction catalysts for alkaline polymer electrolyte membrane fuel cells. Nanoscale, 7 (17), 7644–7650. https://doi.org/10.1039/C5NR00311C.)*

5 Summary

The performance and long-term durability of alkaline fuel cells are governed by a complex interplay of physicochemical properties among all internal components, including electrodes, electrolytes, catalysts, and membranes. Given this intricate interdependence, systematic and multifaceted characterization approaches are indispensable for understanding the mechanisms, optimizing material properties, and enhancing overall cell performance. This chapter provides a comprehensive overview of the common testing methodologies and advanced characterization techniques required to evaluate alkaline fuel cell performance under diverse operating conditions, including varied temperature, pressure, humidity, and reactant flow rates.

For instance, at the electrode level, techniques such as CV are critical for probing catalytic activity and charge transfer kinetics. At the membrane level, some of the important characterization techniques, including ionic conductivity, mechanical strength, and water uptake tests, are of pivotal importance. Meanwhile, cell-level assessments, including polarization curve analysis and galvanostatic discharge tests, provide insights at the cell-level. These methods, when combined with other material characterization techniques (e.g., SEM and X-ray photoelectron spectroscopy), allow for the understanding of structural and compositional changes that occur during operation.

Looking ahead, several transformative trends are poised to revolutionize alkaline fuel cell characterization. In situ and operando techniques, such as the development of specialized electrochemical cells compatible with XRD and Raman spectroscopy, are emerging as powerful tools for real-time observation of dynamic processes within operating fuel cells. These techniques enable researchers to observe changes during real-time operation, offering unprecedented insights into reaction mechanisms, phase transformations, and degradation pathways. Furthermore, the integration of multiscale computational modeling with advanced experimental data also presents a paradigm shift in fuel cell research. By coupling high-resolution spectroscopy, microscopy, and electrochemical measurements with physics-informed machine learning algorithms, researchers can accelerate materials discovery, optimize electrode structures, and predict long-term performance with greater accuracy. Such data-driven approaches not only reduce experimental trial-and-error but also facilitate the design of next-generation alkaline fuel cells with enhanced efficiency and durability.

In summary, this chapter has systematically outlined the critical performance testing and characterization techniques essential for advancing alkaline fuel cell technology. By applying both established methodologies and cutting-edge innovations, an in-depth understanding of alkaline fuel cells can be obtained. The insights gained from these approaches will be instrumental in overcoming existing challenges and unlocking the full potential of alkaline fuel cells for wide applications.

AI disclosure

During the preparation of this work, the author(s) used Deepseek and DeepAI to improve English writing and logic. After using these tools/services, the author(s) reviewed and edited the content as needed and take(s) full responsibility for the content of the publication.

Acknowledgement

The work described in this paper was supported by a grant from the National Natural Science Foundation of China (No. 52506278) and the Start-up Fund from City University of Hong Kong (Dongguan) (Grant No. B01040000082).

References

Bhandary, N., Basu, S., Ingole, P.P., 2016. Rudimentary simple, single step fabrication of nanoflakes like AgCd alloy electro-catalyst for oxygen reduction reaction in alkaline fuel cell. Electrochim. Acta. 00134686. 212, 122–129. Available from http://www.journals.elsevier.com/electrochimica-acta/. doi:10.1016/j.electacta.2016.06.143.

Bidault, F., Brett, D.J.L., Middleton, P.H., Abson, N., Brandon, N.P., 2010. An improved cathode for alkaline fuel cells. Int. J. Hydrog. Energy. 03603199. 35 (4), 1783–1788. doi:10.1016/j.ijhydene.2009.12.035.

Bidault, F., Brett, D.J.L., Middleton, P.H., Brandon, N.P., 2009. Review of gas diffusion cathodes for alkaline fuel cells. J. Power Sources. 03787753. 187 (1), 39–48. doi:10.1016/j.jpowsour.2008.10.106.

Brushett, F.R., Naughton, M.S., Ng, J.W.D., Yin, L., Kenis, P.J.A., 2012. Analysis of Pt/C electrode performance in a flowing-electrolyte alkaline fuel cell. Int. J. Hydrog. Energy. 03603199. 37 (3), 2559–2570. doi:10.1016/j.ijhydene.2011.10.078.

Chai, L., Hu, Z., Wang, X., Zhang, L., Li, T.T., Hu, Y., Pan, J., Qian, J., Huang, S., 2021. Fe7C3 nanoparticles with in situ grown CNT on nitrogen doped hollow carbon cube with greatly enhanced conductivity and ORR performance for alkaline fuel cell. Carbon. 00086223. 174, 531–539. Available from http://www.journals.elsevier.com/carbon/. doi:10.1016/j.carbon.2020.12.070.

Ergül, M.T., Türker, L., Eroğlu, I., 1997. An investigation on the performance optimization of an alkaline fuel cell. Int. J. Hydrog. Energy. 03603199. 22 (10-11), 1039–1045. Available from http://www.journals.elsevier.com/international-journal-of-hydrogen-energy/. doi:10.1016/s0360-3199(96)00244-3.

Fazil, A., Chetty, R., 2014. Synthesis and Evaluation of Carbon Nanotubes Supported Silver Catalyst for Alkaline Fuel Cell. Electroanalysis. 1040-0397. 26 (11), 2380–2387. doi:10.1002/elan.201400246.

Ferriday, T.B., Middleton, P.H., 2021. Alkaline fuel cell technology - A review. Int. J. Hydrog. Energy. 03603199. 46 (35), 18489–18510. Available from http://www.journals.elsevier.com/international-journal-of-hydrogen-energy/. doi:10.1016/j.ijhydene.2021.02.203.

Ganesh, P.A., Prakrthi, A., Chandrasekaran, S.S., Jeyakumar, D., 2021. Shape-tuned, surface-active and support-free silver oxygen reduction electrocatalyst enabled high performance fully non-PGM alkaline fuel cell. RSC Adv. 11, 24872–24882.

Han, J., Pan, J., Chen, C., Wei, L., Wang, Y., Pan, Q., Zhao, N., Xie, B., Xiao, L., Lu, J., 2018. Effect of micromorphology on alkaline polymer electrolyte stability. ACS Appl. Mater. Interfaces 11 (1), 469–477. doi:10.1021/acsami.8b0948110.1021/acsami.8b09481.s001.

Jo, J.H., Moon, S.K., Yi, S.C., 2000. Simulation of influences of layer thicknesses in an alkaline fuel cell. J. Appl. Electrochem. 0021891X. 30 (9), 1023–1031. doi:10.1023/A:1004046721157.

Kordesch, K., Hacker, V., Gsellmann, J., Cifrain, M., Faleschini, G., Enzinger, P., Fankhauser, R., Ortner, M., Muhr, M., Aronson, R.R., 2000. Alkaline fuel cells applications. J. Power Sources. 03787753. 86 (1), 162–165. doi:10.1016/S0378-7753(99)00429-2.

Li, N., Guiver, M.D., Binder, W.H., 2013. Towards high conductivity in anion-exchange membranes for alkaline fuel cells. ChemSusChem. 1864564X. 6 (8), 1376–1383. Available from http://www.interscience.wiley.com/jpages/1864-5631. doi:10.1002/cssc.201300320.

Maegawa, K., Hikima, K., Tan, W.K., Kawamura, G., Muto, H., Nagai, A., Jitianu, A., Matsuda, A., 2022. Enhancing water uptake and hydroxide ion conductivity of alkali fuel cell electrolyte membrane by layered double hydroxide. Solid State Ionics. 01672738. 385, 116021. doi:10.1016/j.ssi.2022.116021.

McLean, G.F., Niet, T., Prince-Richard, S., Djilali, N., 2002. An assessment of alkaline fuel cell technology. Int. J. Hydrog. Energy. 03603199. 27 (5), 507–526. doi:10.1016/S0360-3199(01)00181-1.

Merle, G., Wessling, M., Nijmeijer, K., 2011. Anion exchange membranes for alkaline fuel cells: a review. J. Membr. Sci. 03767388. 377 (1-2), 1–35. doi:10.1016/j.memsci.2011.04.043.

Msomi, P.F., Nonjola, P.T., Ndungu, P.G., Ramontja, J., 2020. Poly 2, 6-dimethyl-1, 4-phenylene)/polysulfone anion exchange membrane blended with TiO2 with improved water uptake for alkaline fuel cell application. Int. J. Hydrog. Energy. 03603199. 45 (53), 29465–29476. doi:10.1016/j.ijhydene.2020.08.012.

Sanetuntikul, J., Shanmugam, S., 2015. High pressure pyrolyzed non-precious metal oxygen reduction catalysts for alkaline polymer electrolyte membrane fuel cells. Nanoscale. 2040-3364. 7 (17), 7644–7650. doi:10.1039/C5NR00311C.

Su, L., Zhang, D., Peng, S., Xuemei, W., Luo, Y., Gaohong, H., 2017. Orientated graphene oxide/Nafion ultra-thin layer coated composite membranes for vanadium redox flow battery. Int. J. Hydrog. Energy. 03603199. 42 (34), 21806–21816. doi:10.1016/j.ijhydene.2017.07.049.

Tuan, C.M., Kim, D., 2016. Anion-exchange membranes based on poly(arylene ether ketone) with pendant quaternary ammonium groups for alkaline fuel cell application. J. Membr. Sci. 18733123. 511, 143–150. Available from www.elsevier.com/locate/memsci. doi:10.1016/j.memsci.2016.03.059.

Verhaert, I., De Paepe, M., Mulder, G., 2009. Thermodynamic model for an alkaline fuel cell. J. Power Sources. 03787753. 193 (1), 233–240. doi:10.1016/j.jpowsour.2009.02.095.

Verhaert, I., Verhelst, S., Huisseune, H., Poels, I., Janssen, G., Mulder, G., De Paepe, M., 2012. Thermal and electrical performance of an alkaline fuel cell. Appl. Therm. Eng. 13594311. 40, 227–235. doi:10.1016/j.applthermaleng.2012.01.042.

Verhaert, I., Verhelst, S., Janssen, G., Mulder, G., De Paepe, M., 2011. Water management in an alkaline fuel cell. Int. J. Hydrog. Energy. 03603199. 36 (17), 11011–11024. doi:10.1016/j.ijhydene.2011.05.172.

Verma, A., Basu, S., 2007. Experimental evaluation and mathematical modeling of a direct alkaline fuel cell. J. Power Sources. 03787753. 168 (1), 200–210. doi:10.1016/j.jpowsour.2007.02.069.

Yang, D., Hongmei, Y., Guangfu, L., Zhao, Y., Liu, Y., Zhang, C., Song, W., Shao, Z., 2014. Fine microstructure of high performance electrode in alkaline anion exchange membrane fuel cells. J. Power Sources. 03787753. 267, 39–47. doi:10.1016/j.jpowsour.2014.04.053.

Zhang, H., Lin, G., Chen, J., 2011. The performance analysis and multi-objective optimization of a typical alkaline fuel cell. Energy. 03605442. 36 (7), 4327–4332. doi:10.1016/j.energy.2011.04.009.

Chapter 8

Advanced theoretical modeling of alkaline fuel cells

Yutong Mu[a] and Zhiguo Qu[b]

[a]*School of Human Settlements and Civil Engineering, Xi'an Jiaotong University, Xi'an, Shaanxi, China*

[b]*School of Energy and Power Engineering, Xi'an Jiaotong University, Xi'an, Shaanxi, China*

Nomenclature

A	activation area, m^2
a	water activity, anode
$\mathbf{a}_{\beta\gamma}$	specific interface area between phases β and γ, m^{-1}
c	molar concentration, $mol\,m^{-3}$
Cp:	specific heat capacity, $J\,kg^{-1}K^{-1}$
D	mass diffusivity, $m^2\,s^{-1}$
EW	equivalent weight of membrane
F	Faraday's constant
h	latent heat, $J\,kg^{-1}$
I	current density, $A\,cm^{-2}$
j	reaction rate, $A\,m^{-3}$
K	permeability, m^2
M	molecular weight, $kg\,kmol^{-1}$
$\mathbf{n}_d$	electro-osmotic drag coefficient
p	pressure, Pa
R	universal gas constant, 8314 $J\,kmol^{-1}K^{-1}$
RH	relative humidity
$\mathbf{R}_u$	gas constant, $J\,kg^{-1}K^{-1}$
s	liquid water saturation
S	source term, entropy, $J\,kmol^{-1}K^{-1}$
T	temperature, K
u	velocity, $m\,s^{-1}$
Y	mass fraction

Greek symbols

α	Transfer coefficient
β	liquid phase

Alkaline Fuel Cells. https://doi.org/10.1016/B978-0-443-36638-3.00015-2

γ	water phase change rate, s^{-1}, gas phase, activity coefficient
ε	porosity
ζ	water transfer rate, s^{-1}
η	overpotential, V
θ	contact angle, °
ι	interfacial drag coefficient
κ	electrical conductivity, $S\,m^{-1}$
λ	water content, thermal conductivity, $W\,m^{-1}K^{-1}$
μ:	dynamic viscosity, $kg\,m^{-1}s^{-1}$, chemical potential, $J\,kmol^{-1}$
ξ	stoichiometry ratio
ρ	density, $kg\,m^{-3}$
σ	surface tension, $N\,m^{-1}$
φ	electronic potential, V
Φ	electric potential at the electrolyte phase, V

Superscripts and subscripts

0	standard state
a	anode
BP	bipolar plate
c	cathode, capillary
CL	catalyst layer
cond	condensation
E	energy
eff	effective
ele	electronic
EOD	electro-osmotic drag
equil	equilibrium
evap	evaporation
fl	fluid phase
g	gas phase
GDL	gas diffusion layer
$\mathbf{H_2}$	hydrogen
$\mathbf{H_2O}$	water
lh	latent heat
lq	liquid
m	mass
mem	membrane
m–l	membrane water to liquid
mw	membrane water
$\mathbf{O_2}$	oxygen
ref	reference state
rev	reversible
sat	saturation
sl	solid phase

u	momentum
v–l	vapor to liquid
vp	water vapor

Abbreviation

Alkaline anion exchange membrane fuel cell	AAEMFC
Alkaline anion exchange membrane	AAEM
Alkaline fuel cell	AFC
Bipolar plate	BP
Catalyst layer	CL
Gas diffusion layer	GDL

1 Introduction

Stationary fuel cells, particularly alkaline fuel cells (AFCs) initially deployed in Apollo missions of National Aeronautics and Space Administration (NASA), represent a promising renewable energy conversion technology characterized by high reliability, 60%–70% efficiency enabled by accelerated oxygen reduction kinetics in alkaline media, compatibility with cost-effective catalysts (e.g., Ni/Ag), and robust dynamic operation capabilities (Yang et al., 2014; Aremo et al., 2015; Verhaert et al., 2016). Despite these merits, AFCs confront critical bottlenecks, including CO_2-sensitive carbonate crystallization, corrosive liquid electrolyte management demanding precious metal components (Ni/Au), and constrained power density (0.1–0.3 W/cm^2) (Azzam et al., 2023). Emerging solid-state alkaline anion-exchange membranes demonstrate transformative potential by replacing volatile liquid potassium hydroxide (KOH) electrolytes with noncarbonating polymer membranes, thereby mitigating CO_2 poisoning risks while enabling simplified system integration (Saebea et al., 2019). However, persistent challenges in ionic conductivity optimization and power density enhancement remain unresolved (Raya et al., 2016), as fuel cell performance fundamentally relies on intricate multiphysics coupling: ohmic losses stem from electron transport in electrodes and hydroxide ion migration across catalyst layers (CL)/electrolytes; concentration polarization arises from gas diffusion through electrolyte-coated electrode films and liquid water flooding in porous media; activation losses depend on triple-phase boundary architectures in CLs (Cheng et al., 2022). Deciphering these interdependent charge-mass-heat transport phenomena provides critical guidance for engineering next-generation high-power-density AFCs through multiparameter optimization of electrode microstructures and interfacial dynamics.

Conventional experimental methods encounter inherent limitations in resolving the multiphysics interactions governing AFC performance. Critical degradation mechanisms—such as CO_2-triggered K_2CO_3 crystallization at catalytic interfaces (Bidault and Kucernak, 2011), HO_2^--mediated oxidative corrosion of carbon-polytetrafluoroethylene (PTFE) composites in concentrated

KOH electrolytes (Zadick et al., 2015), and electrolyte management failures caused by wettability-driven weeping/flooding phenomena (Burchardt, 2004)—demand advanced in situ diagnostics with prohibitive costs for nanoscale temporal-spatial resolution. These challenges intensify when replicating extreme operational conditions. Such experimental constraints amplify the critical role of multiscale computational frameworks in bridging knowledge gaps: atomic-scale molecular dynamics (MDs) simulations optimize hydroxide ion transport; macroscale multiphysics models integrate electrochemical kinetics with thermal-fluid transport to optimize system-level operational parameters and structural configurations. This hierarchical modeling paradigm enables cost-effective exploration of design-performance interdependencies across seven orders of magnitude in spatial-temporal scales, while reducing reliance on costly trial-and-error experimentation.

As shown in Fig. 8.1, AFC electrodes feature a multilayered PTFE-bonded carbon-black structure, with the three-layer configuration [backing material (BM)/gas diffusion layer (GDL)/CL] being the most widely adopted (Azzam et al., 2023). The BM requires high gas permeability, electronic conductivity, and corrosion resistance, additionally serving as the current collector in monopolar designs. The GDL facilitates reactant gas distribution to the CL while blocking electrolyte leakage, necessitating hydrophobic properties and electrical conductivity. The CL integrates carbon-black-supported catalysts bound by PTFE, where high carbon surface area is critical for maximizing power density.

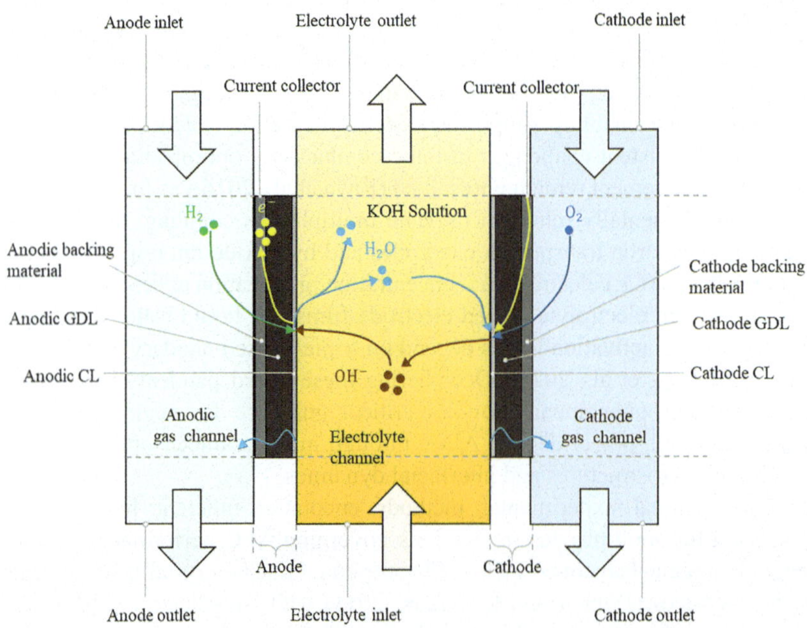

FIG. 8.1 Schematic illustrations of the transport process in a typical alkaline fuel cell.

During operation, hydrogen diffuses from the anode gas channel through the GDL to the CL, dissolving into the electrolyte film coating the catalyst surface before reacting electrochemically with hydroxyl ions. Similarly, oxygen diffuses through the cathode GDL, dissolves into the electrolyte film, and undergoes reduction with water-derived species at the catalyst interface. This structural hierarchy governs mass transport (gas diffusion through the GDL), charge transfer (ion migration through electrolyte films), and reaction kinetics (catalyst-electrolyte interactions) in AFC systems (Bidault and Middleton, 2012).

The anode and cathode two electron-transfer reactions are expressed as:

Anode:

$$H_2 + 2OH^- \rightarrow 2H_2O + 2e^- \tag{1}$$

Cathode:

$$O_2 + 2H_2O + 4e^- \rightarrow 4OH^- \tag{2}$$

The above cathode and anode reactions ultimately produce an overall reaction.

$$2H_2 + O_2 \rightarrow 2H_2O \tag{3}$$

This overall reaction produces water and heat as byproducts and generates four electrons per mole of oxygen in the process. This overall reaction has a theoretical cell potential, which is equivalent to +1.23 V.

AFC performance modeling primarily utilizes two methods: empirical correlations and mathematical frameworks. The empirical framework decomposes voltage losses into three constituent components through parameterized correlations, widely applied in AFC hybrid systems (Zhao et al., 2021; Chakma et al., 2022; Zeng et al., 2023; Zhang et al., 2024). In contrast, mathematical modeling has progressed through three evolutionary phases characterized by escalating spatial resolution and multiphysics integration. Initial efforts centered on 1D isothermal formulations to decode reaction-transport interactions in porous electrodes (Kimble and White, 1991). The seminal Kimble-White model established a five-layer architecture by implementing macrohomogeneous porous electrode theory, employing Stefan-Maxwell equations for gas-phase species transport and the Nernst-Planck formalism for liquid-phase ionic migration. Their analysis identified liquid-phase oxygen diffusion as the principal mass transport constraint at elevated current densities, with electronic conductivity dictating ohmic losses at lower currents. Subsequent refinements emerged through critical innovations: Björnbom and Yang (1993) contested these findings via concentrated electrolyte theory, revealing gas-phase diffusion bottlenecks and dynamic water flux reversals within CLs—a revelation that fueled sustained discourse on mass transport dominance. The Jo–Yi model enhanced kinetic fidelity by embedding water activity in Butler-Volmer kinetics and Nernst potentials, while exposing electrolyte concentration gradients as pivotal

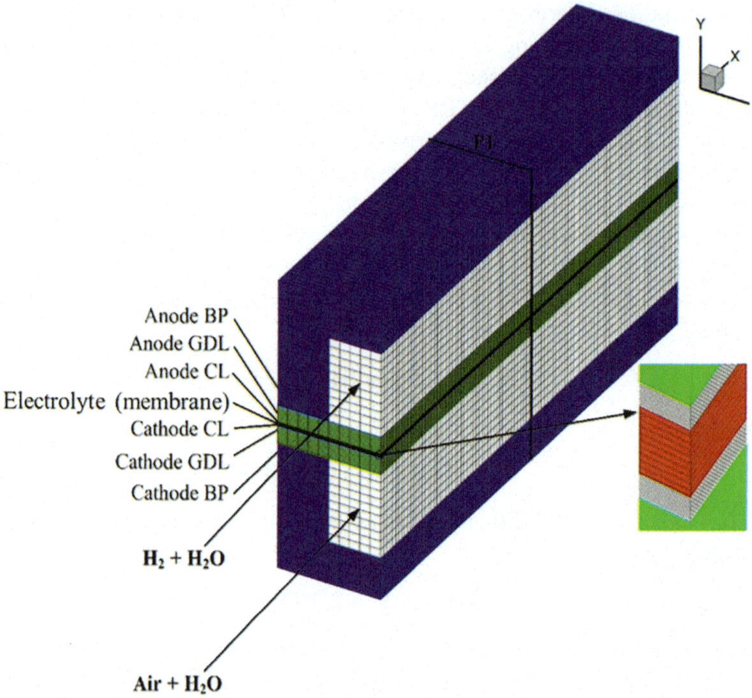

FIG. 8.2 Illustration of mathematical modeling of AFC and AEMFC (Jiao et al., 2014). *(From Jiao, K., He, P., Du, Q., Yin, Y., 2014. Three-dimensional multiphase modeling of alkaline anion exchange membrane fuel cell. Int. J. Hydrogen Energy 39 (11), 5981–5995, Elsevier Ltd. https://doi.org/10.1016/j.ijhydene.2014.01.180.)*

determinants of electrode efficacy (Jo and Yi, 1999). The third phase advanced multidimensional modeling, as depicted in Fig. 8.2. Zou et al. (2011a, 2011b) quantified a 50% underestimation in ideal-solution assumptions and 30% current gains under adiabatic conditions via coupled multiphysics simulations. Kunz et al. (Kunz, 2014, 2019) reconceptualized CLs as interfacial boundaries, integrating temperature-sensitive exchange currents and optimizing electrolyte concentration to 5 M KOH. Collectively, these resolved controversies by confirming liquid-phase oxygen diffusion as the current-limiting mechanism and establishing dry cathode and humidified anode gas configurations as paradoxically optimal for electrolyte management.

Regarding the AEM fuel cell, which evolved from conventional AFCs through solid electrolyte integration, it predominantly utilizes hydrogen for its high reactivity and energy density, though recent modeling efforts have expanded to liquid fuels like methanol and ethanol (Saebea et al., 2019; Pezzini et al., 2023). Inspired by proton exchange membrane fuel cells (PEMFCs), computational frameworks for AEM fuel cells span 3D multiphase computational fluid dynamics (CFD) models that dissect porous layer transport while addressing dynamic responses and pressure effects (Huo et al., 2012; An et al., 2013; Jiao et al., 2014). Water management challenges, including cathode dehydration, anode flooding, and interfacial saturation jumps (Jiao et al., 2015; Huo and Jiao, 2018), have been systematically addressed through nonisothermal simulations and asymmetric humidification strategies. For liquid fuel systems, methanol-focused studies quantify CO_2 bubble dynamics in anode CLs, porous media wettability impacts, and methanol utilization dependencies on anode water stabilization (Abdullah et al., 2016; Weinzierl and Krewer, 2016; Su et al., 2021). Meanwhile, ethanol, glycerol, and formate models integrate multiphysics transport to optimize electrochemical behavior and scalability, collectively emphasizing the critical balance between aqueous anode feeds and cathode dehydration to achieve operational durability and fuel efficiency (Han et al., 2014).

The hydroxide ion (OH^-) transport efficiency in anion exchange membranes (AEMs) is fundamentally governed by the interplay between their microstructural features—such as phase-separated hydrophilic/hydrophobic domains, continuity of hydrated channels, and spatial distribution of functional groups—and the dynamic coupling of vehicular (ion-water cluster diffusion) and Grotthuss mechanisms. To unravel this intricate relationship, multiscale computational approaches have been developed, each offering unique insights but constrained by inherent limitations (Dong et al., 2018). As depicted in Fig. 8.3, at the atomistic scale, MD simulations with polarizable force fields (e.g., APPLE&P) or reactive potentials (e.g., ReaxFF) enable precise characterization of local interactions, such as hydrogen-bond networks and cation-hydroxide electrostatic effects, revealing how Grotthuss mechanisms dominate OH^- migration through narrow channel bottlenecks. These methods also clarify the role of chemical modifications (e.g., hydrophobic alkyl chain elongation or hydrophilic side-chain grafting) in tuning water channel dimensions and connectivity. However, their computational cost restricts simulations to nanoscale systems and nanosecond timescales, failing to capture micrometer-scale phase separation or long-term degradation processes. In contrast, mesoscale coarse-grained (CG) models and dissipative particle dynamics efficiently predict equilibrium membrane morphologies (e.g., lamellar or bicontinuous structures in block copolymers) and correlate macroscopic properties like water uptake and ionic conductivity with parameters such as ion-exchange capacity (IEC) and crosslink density. Yet, these models oversimplify atomic-scale chemistry, neglecting critical details such as cation polarization effects or the relative contributions of Grotthuss

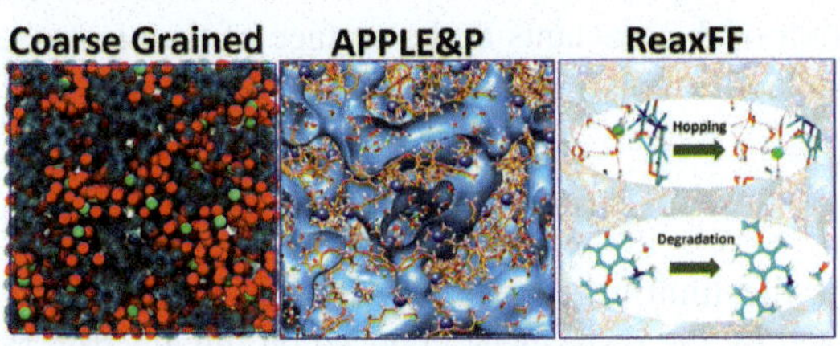

FIG. 8.3 Illustration of combining simulations at different scales (Dong et al., 2018). *With permission from—For content taken from a published or online source without alteration.*

versus vehicular transport. Bridging these scales, hierarchical multiscale frameworks integrate atomistic simulations (for local mechanisms), CG models (for morphology prediction), and continuum theories (for interfacial and concentration polarization effects), thereby coupling chemical degradation pathways with morphological evolution.

This chapter synthesizes empirical and mathematical modeling approaches for AFCs and anion exchange membrane fuel cells (AEMFCs), highlighting their role in deciphering multiphysics transport phenomena, electrochemical kinetics, and system performance. Advances in computational techniques have established numerical modeling as a cost-effective and versatile tool for simulating voltage-current characteristics, heat/mass transport dynamics, and operational responses to variables such as temperature, pressure, and fuel type (e.g., hydrogen, ammonia). Core frameworks integrate governing equations for charge conservation (ionic/electronic potentials), species transport, momentum, and energy balance, alongside empirical voltage-current correlations derived from polarization curves.

2 Theoretical modeling methods

2.1 Empirical model for AFC

The reversible thermodynamic cell voltage achieved for a typical AFC can be determined by the Nernst equation:

$$E_{\mathrm{rev}} = 1.229 - 0.846\times10^{-3}(T - 298.15) + \frac{RT}{2F}\ln\left(\frac{p_{\mathrm{H_2}}\left(p_{\mathrm{O_2}}\right)^{0.5}}{p_{\mathrm{H_2O}}}\right) \tag{4}$$

where T, R, and p are the operating temperature, the universal gas constant, and the partial gaseous pressure, respectively.

The irreversibilities taken into account within this analysis are: (a) Charge-transfer/activation overpotential losses that occur due to the slowness of the reaction taking place on the electrode surface and are represented by (Shen et al., 2021; Wang et al., 2021; Yu et al., 2021):

$$U_{\mathrm{act}} = \frac{RT}{\alpha n_{\mathrm{e}}F}\ln\left(\frac{j}{j_0}\right) \tag{5}$$

where α is the charge transfer coefficient, j is the electric current density, $j_0 = c_1\exp(-c_2/T)$ is the exchange current density, and c_1 and c_2 are two constants. (b) Concentration overpotential losses are losses that occur due to changes in the concentration of the reactants at the surface of the electrodes as the fuel is being consumed and are represented by:

$$U_{\mathrm{con}} = \frac{RT}{\alpha n_{\mathrm{e}}F}\ln\left(\frac{j_{\mathrm{L}}}{j_{\mathrm{L}} - j}\right) \tag{6}$$

where j_{L} is the limiting current density. (c) Ohmic overpotential losses are losses caused by the resistances imposed on the electron flow through the material of the electrodes, as well as the resistances imposed on the flow of ions through the electrolyte. This type of overpotential loss is represented by:

$$U_{ohm} = j\frac{t_{ele}}{k} \tag{7}$$

where t_{ele} is the electrolyte's thickness, and k is the specific conductivity of the KOH electrolyte. Gilliam et al. (GILLIAM et al., 2007) devised an empirical relation that can describe the relationship between the specific conductivity of aqueous KOH, temperature, and concentration, and it is represented by:

$$k = -2.041\times10^{-2}M - 2.8\times10^{-5}M^2 + 5.332\times10^{-5}MT + 2.072\frac{M}{T} + 1.043\times10^{-5}M^3 - 3\times10^{-9}M^2T^2 \tag{8}$$

This correlation is most accurate within temperature ranges of 0–100°C and KOH concentrations of 0–12 M, where M represents molarity.

The overall output voltage of the cell can now be determined by:

$$U_{cell} = E_{rev} - U_{act} - U_{con} - U_{ohm} \tag{9}$$

In the work of Hamada et al. (2023) the empirical model parameters are provided as listed in Table 8.1.

2.2 Mathematical model for AFC

2.2.1 Assumptions

To balance computational feasibility with predictive accuracy, critical simplifications are adopted in the mathematical model:

TABLE 8.1 AFC model parameters for the empirical method (Hamada et al., 2023).

Parameters	value
Partial pressure of H_2, p_{H2} (atm)	0.97
Partial pressure of O_2, p_{O2} (atm)	1
Partial pressure of H_2O, p_{H2O} (atm)	1
Constant, c_1 (A m^{-2})	174,512
Constant, c_2 (K)	5485
Temperature of environment, T_0 (K)	298
Operating temperature, T (K)	343
Transfer coefficient, α	0.1668
Limiting current density, j_L (A m^{-2})	2000

From Hamada, A.T., Orhan, M.F., Kannan, A.M., 2023. Alkaline fuel cells: Status and prospects. Energy Rep. 9, 6396–6418. https://doi.org/10.1016/j.egyr.2023.05.276.

1. Gas composition: Air (21% O_2, 79% N_2) is assumed for the cathode feed, with CO_2-induced electrolyte carbonate poisoning neglected due to rapid impurity removal via electrolyte circulation.
2. Electrolyte distribution: Hydrophobic GDLs confine the electrolyte to the channel and CLs, preventing pore flooding; liquid transport through electrodes is excluded.
3. Reaction kinetics: Single-step anode and cathode reactions are assumed, omitting intermediate mechanisms to maintain charge/mass balance consistency.
4. Gas transport: H_2 and O_2 crossover through the electrolyte film is presumed negligible, as their consumption at catalyst surfaces dominates over agglomerate-phase diffusion.
5. Water management: Convective water transport is prioritized over diffusion, with liquid product water from the anode reaction equally partitioned into anode/cathode gas streams and dynamically coupled to local current density.

These assumptions streamline multiphysics coupling while retaining fidelity for system-level performance prediction, avoiding computationally intensive simulations of secondary processes (e.g., carbonate formation, detailed reaction pathways, pore-scale flooding).

2.2.2 Governing equations

The conservation of continuity equations for the liquid and gas phases are:

$$\frac{\partial}{\partial t}\varepsilon_\beta \rho_\beta + \nabla \cdot \varepsilon_\beta \rho_\beta u_\beta = S^m_{\beta\gamma} \tag{10a}$$

$$\frac{\partial}{\partial t}\varepsilon_\gamma \rho_\gamma + \nabla \cdot \varepsilon_\gamma \rho_\gamma u_\gamma = -S^m_{\beta\gamma} \tag{10b}$$

where ε is the porosity, and the subscripts β, γ, σ denote the liquid, gas, and solid phases, respectively. The mass transfer rate at the liquid-gas interface of the CL for the anode and cathode layers is provided as follows.

Anode:

$$S^m_{\beta\gamma} = -a_{\beta\gamma}\frac{D_{\gamma,H_2O}}{\delta_g R_u T}\left(p_s - p_{H_2O}\right)M_{H_2O} + a_{\beta\gamma}D_{\beta,H_2}\frac{H_{H_2}p_{H_2} - C_{H_2}}{\delta_l}M_{H_2} \tag{11a}$$

Cathode:

$$S^m_{\beta\gamma} = -a_{\beta\gamma}\frac{D_{\gamma,H_2O}}{\delta_g R_u T}\left(p_s - p_{H_2O}\right)M_{H_2O} + a_{\beta\gamma}D_{\beta,O_2}\frac{H_{O_2}p_{O_2} - C_{O_2}}{\delta_l}M_{O_2} \tag{11b}$$

The model assumes that each of the three fluids—the anode gas, the cathode gas, and the electrolyte solution—enters the cell as laminar flow with moderate pressure and velocity. In the free regions, the pressure and velocity fields are calculated using the Navier-Stokes equation, whereas in the porous media, the Brinkman equation is used (Zhou et al., 2011ab).

$$\frac{\partial}{\partial t}(\varepsilon_\beta \rho_\beta u_\beta)+\nabla\cdot(\varepsilon_\beta \rho_\beta u_\beta u_\beta)=-\varepsilon_\beta \nabla p_\beta+\nabla\cdot(\mu_\beta^{eff}\nabla u_\beta)+\varepsilon_\beta \rho_\beta g+S_\beta^u \quad (12a)$$

$$\frac{\partial}{\partial t}(\varepsilon_\gamma \rho_\gamma u_\gamma)+\nabla\cdot(\varepsilon_\gamma \rho_\gamma u_\gamma u_\gamma)=-\varepsilon_\gamma \nabla p_\gamma+\nabla\cdot(\mu_\gamma^{eff}\nabla u_\gamma)+\varepsilon_\gamma \rho_\gamma g+S_\gamma^u \quad (12b)$$

where p is the pressure, μ is the viscosity, and **g** is the gravity constant. The fluid-phase viscous drag is given as follows:

$$\text{Catalyst layer: } S_\beta^u=-\frac{\varepsilon_\beta \mu_\beta u_\beta}{K_\beta} \quad (13a)$$

$$\text{Gas diffusion layer: } S_\gamma^u=-\frac{\varepsilon_\gamma \mu_\gamma u_\gamma}{K_\gamma} \quad (13b)$$

The species transport (H_2, O_2, $H_2O$and OH^-) in the liquid phase is given as:

$$\frac{\partial}{\partial t}(\varepsilon_\beta \rho_\beta Y_{\beta i})+\nabla\cdot(\varepsilon_\beta \rho_\beta u_\beta Y_{\beta i})=\nabla\cdot(\rho_\beta D_{\beta i}^{eff}\nabla Y_{\beta i})-\frac{M_{\beta i}}{z_{\beta i}F}i_\beta\cdot\nabla t_{\beta i}+S_{\beta,i}^m \quad (14)$$

where the source term for the hydroxide ions, hydrogen, and oxygen is provided by considering the reaction rate and the dissolution rate.

$$S_{\beta,OH^-}^m=\frac{M_{OH^-}}{z_{OH^-}F}(1-t_{OH^-})a_{\beta\sigma}i_n, [n=a \textit{ for anode, and } n=c \textit{ for cathode}] \quad (15a)$$

$$S_{\beta,H_2}^m=-\frac{M_{H_2}}{2F}a_{\beta\sigma}i_a+a_{\beta\sigma}D_{\beta,H_2}\frac{H_{H_2}p_{H_2}-C_{H_2}}{\delta_l}M_{H_2} \quad (15b)$$

$$S_{\beta,O_2}^m=-\frac{M_{H_2}}{4F}a_{\beta\sigma}i_c+a_{\beta\sigma}D_{\beta,O_2}\frac{H_{O_2}p_{O_2}-C_{O_2}}{\delta_l}M_{O_2} \quad (15c)$$

The gaseous transport is governed by

$$\frac{\partial}{\partial t}(\varepsilon_\gamma \rho_\gamma Y_{\gamma i})+\nabla\cdot(\varepsilon_\gamma \rho_\gamma u_\gamma Y_{\gamma i})=\nabla\cdot(\rho_\gamma D_{\gamma i}^{eff}\nabla Y_{\gamma i})+S_{\gamma,i}^m \quad (16)$$

where the source term for the hydrogen, oxygen, and hydrogen is provided as.

$$S_{\gamma,H_2}^m=-a_{\beta\sigma}D_{\beta,H_2}\frac{H_{H_2}p_{H_2}-C_{H_2}}{\delta_l}M_{H_2} \quad (17a)$$

$$S_{\gamma,O_2}^m=-a_{\beta\sigma}D_{\beta,O_2}\frac{H_{O_2}p_{O_2}-C_{O_2}}{\delta_l}M_{O_2} \quad (17b)$$

$$S_{\gamma,H_2O}^m=-a_{\beta\gamma}D_{\beta,H_2O}\frac{p_s-p_{H_2O}}{\delta_g R_u T}M_{H_2O} \quad (17c)$$

The energy transport equation can be written as:

$$\frac{\partial}{\partial t}(\rho c_p^{eff}T)+\nabla\cdot(\varepsilon_\beta c_{p\beta}^{eff}\rho_\beta u_\beta+\varepsilon_\gamma c_{p\gamma}^{eff}\rho_\gamma u_\gamma)T=\nabla\cdot(\lambda^{eff}\nabla T)+S_{react}^h+S_{phase}^h+S_{joule}^h \quad (18)$$

where the source terms are expressed as

$$S^h_{joule} = -i_\beta \cdot \nabla\Phi_\beta \tag{19a}$$

$$S^h_{react} = a_{\beta\sigma}\left(i_n\eta_n - \frac{i_n}{nF}T\Delta S\right) \tag{19b}$$

$$S^h_{phase} = -S^h_{\gamma,\mathrm{H_2O}} h_{\beta\gamma,\mathrm{H_2O}} \tag{19c}$$

The charge conservation equation is given as

$$\nabla\cdot\left(\kappa_\beta^{eff}\nabla\Phi_\beta\right) + \nabla\cdot\left(\kappa_\beta^{D,eff}\nabla\ln C_{\beta e}\right) = S^i_{\beta\sigma} \tag{20}$$

where the charge generation rate is expressed as

$$S^i_{\beta\sigma} = -a_{\beta\sigma} i_n,\ [n = a\ for\ anode,\ and\, n = c\ for\ cathode] \tag{21}$$

Consider the diluted electrolyte theory to describe the overall flux of an ion in the electrolyte solution. For the ideal solution, the flux of the charged species k is obtained.

$$j_k = -\rho D_k \nabla Y_k - \rho Y_k u_k z_k F\nabla\Phi + \rho Y_k u \tag{22}$$

The RHS of the equation is the Fickian diffusion, migration, and convective terms, respectively. For concentrated solutions, the generalized Stefan–Maxwell equation is used.

$$\sum_{i=1;i\neq k}^{N}\left(\frac{x_k x_i}{D_{ki}}\right)\left(\frac{j_i}{\rho_i} - \frac{j_k}{\rho_k}\right) = \frac{x_k}{R_u T}(\nabla\mu_k + z_k F\nabla\Phi) \tag{23}$$

where the chemical potential of species k is given by.

$$\mu_k = \mu_k^0 + R_u T\ln\alpha_k \tag{24}$$

Is the standard-state chemical potential, and is the activity of species k. The current density is calculated by summing the fluxes of all charged species:

$$\boldsymbol{i} = \sum_{k=1}^{N}\frac{z_k F\boldsymbol{j}_k}{M_k} \tag{25}$$

The electrolyte solution in AFCs consists of charged species (e.g., K^+, OH^-) and solvent (H_2O). The mass flux of species k is calculated by

$$\boldsymbol{j}_k = \left(\frac{t_k \rho Y_k \boldsymbol{i}}{z_k C_k F}\right) - \rho D_m \nabla Y_k + \rho Y_k \boldsymbol{u} \tag{26}$$

where the transference number t_k is given as:

$$t_k = \frac{z_k^2 u_k C_k}{\sum_{j=1}^{N} z_j^2 u_j C_j} \tag{27}$$

The current density is calculated by

$$i = -\kappa\nabla\Phi - \sum_{k=1}^{N} z_k F D_k \nabla C_k \tag{28}$$

where κ is the KOH solution's electrical conductivity.

$$\kappa = \sum_{k=1}^{N} z_k^2 F^2 u_k C_k \tag{29}$$

And D_m is the KOH diffusion coefficient.

$$D_m = \frac{2D_+ D_-}{D_+ + D_-} \tag{30}$$

where subscripts "+" and "–" denote the positive and negative ions, respectively. For an ideal solution, the Nernst–Einstein equation is used to calculate the mass diffusivity of species *k*.

$$D_k = u_k R_u T \tag{31}$$

where $u_k = t_k \kappa_e / z_k^2 u_k F^2 C_e$.

The specific gas–electrolyte interface area, α_g, and the thickness of the electrolyte film, δ_l, are specified using the reported values 10^5 m^{-1} and 10^{-1} μm, respectively. The thickness of the GDL characteristic length, δ_g, is set to 10 μm. The Butler–Volmer equation is used to calculate the HOR and ORR rates.

$$i_a = i_0^a \left[\left(\frac{C_{H_2}}{C_{H_2,ref}} \right)^{q_{H_2}^a} \left(\frac{C_{OH}}{C_{OH^-,ref}} \right)^{q_{OH^-}^a} \exp\left(\frac{\alpha_a^a n_a \eta_a}{R_u T} \right) - \left(\frac{C_{H_2O}}{C_{H_2O,ref}} \right)^{q_{H_2O}^a} \exp\left(-\frac{\alpha_c^a n_a \eta_a}{R_u T} \right) \right] \tag{32a}$$

$$i_c = i_0^c \left[\left(\frac{C_{OH}}{C_{OH^-,ref}} \right)^{q_{OH^-}^c} \exp\left(\frac{\alpha_a^c n_c \eta_c}{R_u T} \right) - \left(\frac{C_{O_2}}{C_{O_2,ref}} \right)^{q_{O_2}^c} \left(\frac{C_{H_2O}}{C_{H_2O,ref}} \right)^{q_{H_2O}^c} \exp\left(-\frac{\alpha_c^c n_c \eta_c}{R_u T} \right) \right] \tag{32b}$$

Typically, the catalyst surface is assumed to be covered by the liquid phase and not easily accessible from the gas phase. The flooded agglomerate model assumes the agglomerates, in which the catalyst particles are surrounded by the electrolyte. In analogy with the thin film model, the agglomerates are generally assumed to be covered by a thin liquid film, as portrayed in Fig. 8.4, where the transport processes of oxygen from the gas phase to the catalyst surface are exemplarily illustrated. Consequently, hydrogen and oxygen must dissolve in and diffuse through the electrolyte to the catalyst surface in order to react electrochemically. Correspondingly, the dissolution rate of gaseous species should be corrected to include the effect of the agglomerates.

The capillary pressure related to the liquid saturation in the porous electrode is defined as.

$$p_c = \frac{\sigma_s \cos\theta_c}{(K/\varepsilon)^{1/2}} \left(1.417S_r - 2.120S_r^2 + 1.263S_r^3\right),\ S_r = \frac{s - s_{im}}{1 - s_{im}} \tag{33}$$

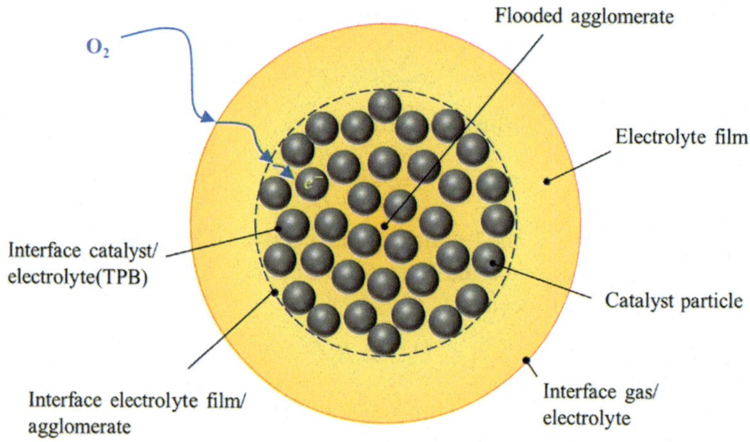

FIG. 8.4 Illustration of the agglomerate model in the catalyst layer of AFCs.

where K is the absolute permeability, σ_s is the surface tension, θ is the contact angle, and s_{im} is the immobile liquid saturation. The current flux is calculated.

$$i_\beta = -\kappa_\beta^{eff} \nabla \Phi_\beta - \kappa_\beta^{D,eff} \nabla \ln C_{\beta e} \tag{34}$$

The diluted theory only applies in cases of very low electrolyte concentrations and neglects multicomponent interactions. In cases of moderate to high concentrations, the solutes in the solution strongly affect one another, so their interactions with the solvent, taken singly, are not representative of their transport in the solution. The mass flux of species k is calculated by

$$j_k = \frac{\rho t_k Y_k i}{z_k C_k F} - \rho D_m \nabla Y_k + \rho Y_k u \tag{35}$$

where

$$D_m = \left(\frac{D C_T M_0}{\rho}\right)\left(1 + \frac{d\ln\gamma_{\pm,C}}{d\ln C_e}\right) \tag{36a}$$

$$D = \frac{D_{0+} D_{0-}(z_+ - z_-)}{z_+ D_{0+} - z_- D_{0-}} \tag{36b}$$

$$C_T = C_+ + C_- \tag{36c}$$

$$t_+ = \frac{\rho_0 t_+^0 + \rho_-}{\rho} \tag{36d}$$

where D_{0^+} and D_{0^-} are the binary diffusion coefficients, C_e is the KOH concentration, C_T is the total molar concentration, $\gamma_{\pm,C}$ is the mean molar activity coefficient, and t^+ is the transference number. The dependency of the transference number on electrolyte concentration is calculated by

$$\frac{t_i}{t_i^0} = \frac{\lambda_i^0}{\lambda_+^0 + \lambda_-^0} \tag{37}$$

where λ_i^0 is the limiting ionic equivalent conductance of ion i. The current density is calculated from

$$i = -\kappa \nabla \Phi - \kappa^D \nabla \ln C_e \tag{38}$$

where κ^D is the diffusion conductivity

$$\kappa^D = \left(\frac{2R_u T\kappa}{F}\right)\left(1 - t_-^0 + \frac{C_e}{2C_0}\right)\left(1 + \frac{d\ln\gamma_{\pm,C}}{d\ln C_e}\right) \tag{39}$$

The nonideal solution effects are calculated from the nonunity activity coefficients that appear in the species diffusion coefficients and diffusion conductivity. In the work of Zhou et al. (Zhou et al., 2011a, 2011b), the model parameters for the mathematical method are provided in Table 8.2.

TABLE 8.2 AFC model parameters for the empirical method (Zhou et al., 2011a, 2011b).

Anode		Cathode	
Parameter	Value	Parameter	Value
n_a	2	n_c	4
s_{H2}	1	s_{H2}	−1
s_{OH^-}	2	s_{OH^-}	4
s_{H2O}	−2	s_{H2O}	−2
q_{H2}	1	q_{H2}	1
q_{OH^-}	2	q_{OH^-}	4
q_{H2O}	2	q_{H2O}	2
v_{K+}	1	v_{K+}	1
v_{OH^-}	1	v_{OH^-}	1
z_{K+}	1	z_{K+}	1
z_{OH^-}	−1	z_{OH^-}	−1
$\alpha_a^a n_a$	1.5	$\alpha_a^c n_c$	2.5
$\alpha_c^a n_a$	0.5	$\alpha_c^c n_c$	1.5
i_0^a	5×10^{-4} A cm^{-2}	i_0^c	5×10^{-8} A cm^{-2}
a_l^a	1×10^7 m^{-1}	a_l^c	1×10^7 m^{-1}
a_g^a	6×10^5 m^{-1}	a_g^c	6×10^5 m^{-1}
δ_l^a	1×10^{-7} m	δ_l^c	1×10^{-7} m
δ_g^a	1×10^{-5} m	δ_g^c	1×10^{-5} m
$C_{H2,ref}$	0.60913 mol m^{-3}	$C_{O2,ref}$	0.34905 mol m^{-3}
ΔS_a	161.2 J mol^{-1}K^{-1}	ΔS_c	−648.0 J mol^{-1}K^{-1}

From Zhou, G., Chen, L.D., Seaba, J.P., 2011b. Effects of property variation and ideal solution assumption on the calculation of the limiting current density condition of alkaline fuel cells. J. Power Sour. 196 (11), 4923–4933. https://doi.org/10.1016/j.jpowsour.2011.02.026.

2.3 Mathematical model for anion exchange membrane fuel cell

2.3.1 Assumptions

Unlike the AFC, where the transport phenomena of charged ions in the electrolytes are driven by migration, convection, and diffusion, the transport of charged ions in the solid electrolyte of AEMs is comparatively simple. Besides, the water phases in AEMs and the gaseous transport are quite different from those in the liquid electrolyte. To explore the multiphysics coupling processes in AEMFCs, several assumptions are made (Cheng et al., 2022; Huo et al., 2012; Jiao et al., 2014):

1. Laminar flow regime governs all fluid dynamics.
2. Ideal gas behavior applies to all gaseous components.
3. Anode liquid water exclusion: The straight, short flow channels enable rapid liquid water removal in the anode, justifying a zero-liquid-water-content assumption. Cathode liquid water simplification: A constant liquid water volume fraction is prescribed in cathode channels due to minimal spatial variations.
4. Liquid-phase anode water: Liquid water formation at the anode is dictated by humidified inlet gases and low operating temperatures (≤80°C).
5. Cathode water phase selection: Liquid water consumption dominates under cathodic vapor supersaturation; gaseous water participation occurs only after complete liquid water depletion.

It is worth noting that the electrochemical reactions occur in the whole volume of the CL of AEMs, rather than in the thin film or agglomerates of the electrolyte in the vicinity of the CL.

2.3.2 Governing equations

The mass conservation equation is expressed as:

$$\frac{\partial}{\partial t}\left(\varepsilon\left(1-s_{lq}\right)\rho_g\right)+\nabla\cdot\left(\rho_g u_g\right)=S_m \tag{40}$$

The momentum equation is given as:

$$\frac{\partial}{\partial t}\left(\frac{\rho_g u_g}{\varepsilon\left(1-s_{iq}\right)}\right)+\nabla\cdot\left(\frac{\rho_g u_g u_g}{\varepsilon^2\left(1-s_{iq}\right)^2}\right)=-\nabla p_g+\mu_g\nabla\cdot\left(\nabla\left(\frac{u_g}{\varepsilon\left(1-s_{iq}\right)}\right)+\left(\frac{u_g^T}{\varepsilon\left(1-s_{iq}\right)}\right)\right)$$
$$-\frac{2}{3}\mu_g\nabla\left(\nabla\cdot\left(\frac{u_g}{\varepsilon\left(1-s_{lq}\right)}\right)\right)+S_u \tag{41}$$

The gaseous species transport in the gas channel and porous electrodes is given as:

$$\frac{\partial}{\partial t}\left(\varepsilon\left(1-s_{lq}\right)\rho_g Y_i\right)+\nabla\cdot\left(\rho_g u_g Y_i\right)=\nabla\cdot\left(\rho_g D_i^{eff}\nabla Y_i\right)+S_i \tag{42}$$

The volumetric electrochemical reaction rates occurring in the anode and cathode CLs are described by Butler-Volmer equations.

$$j_a = (1 - s_{lq}) j_{0,a}^{ref} \left(\frac{c_{H_2}}{c_{H_2}^{ref}} \right)^{0.5} \left[\exp\left(\frac{2\alpha_a F}{RT} \eta_{act} \right) - \exp\left(-\frac{2\alpha_c F}{RT} \eta_{act} \right) \right] \quad (43)$$

$$j_c = (1 - s_{lq}) j_{0,c}^{ref} \left(\frac{c_{O_2} c_{H_2O}}{c_{O_2}^{ref} c_{H_2O}^{ref}} \right) \left[-\exp\left(\frac{4\alpha_a F}{RT} \eta_{act} \right) + \exp\left(-\frac{4\alpha_c F}{RT} \eta_{act} \right) \right] \quad (44)$$

During AEMFC operation, all three water phases participate in electrochemical reactions. Therefore, the effective water molar concentration in the Butler-Volmer equation describing the cathode half-reaction is calculated using the following modified expression (Jiao et al., 2014):

$$c_{H_2O} = \frac{\varepsilon \rho_{lq} S_{lq}}{M_{H_2O}} + \frac{\omega \rho_{mem} \lambda_{mw}}{EW} + \frac{\varepsilon (1 - s_{lq}) Y_{H_2O} \rho_g}{M_{H_2O}} \quad (45)$$

where ρ_{mem} is the density of the dry AEM, EW is the effective equivalent mass, λ is the membrane water content, and ρliq is the liquid water density.

The conservation of liquid water is given as.

$$\frac{\partial (\varepsilon s_{lq} \rho_{lq})}{\partial t} + \nabla \cdot (\iota \rho_{lq} u_g) = \nabla \cdot (\rho_{lq} D_{lq} \nabla s_{lq}) + S_{lq} \quad (46)$$

where the interfacial drag function is defined as $\iota = K_{lq} \mu_g / K_g \mu_{lq}$. Due to the impact of

Capillary forces, the capillary diffusivity of liquid water in GDL and CL, is defined as

$$D_{lq} = -\frac{K_{lq}}{\mu_{lq}} \frac{dp_c}{ds_{lq}} \quad (47)$$

Note that the above treatment of the liquid water results in the continuity of the liquid water saturation at different interfaces of the porous component. A more comprehensive treatment is to solve the liquid pressure equation as follows:

$$\frac{\partial (\varepsilon s_{lq} \rho_{lq})}{\partial t} + \nabla \cdot \left(\rho_{\mathbf{lq}} \frac{K k_{\mathbf{lq}}}{\mu_{\mathbf{lq}}} \nabla P_{\mathbf{lq}} \right) + S_{\mathbf{lq}} = 0 \quad (48)$$

The source term of phase change between liquid water and water vapor can be calculated as:

$$S_{v-l} = \begin{cases} \gamma_{cond} \varepsilon (1 - s_{lq}) \dfrac{(p_{H_2O} - p_s) M_{H_2O}}{RT} & if \ p_{H_2O} \geq p_s \\ \gamma_{evap} \varepsilon s_{lq} \dfrac{(p_{H_2O} - p_s) M_{H_2O}}{RT} & if \ p_{H_2O} \leq p_s \end{cases} \quad (49)$$

The conservation of membrane water is given as.

$$\frac{\rho_{mem}}{EW} \frac{\partial (\omega \lambda_{nf})}{\partial t} = \frac{\rho_{max}}{EW} \nabla \cdot (D_{max}^{eff} \nabla \lambda_{max}) + S_{max} \quad (50)$$

The transfer between liquid water and membrane water in CL can be defined as.

$$S_{\text{m-l}} = \zeta_{\text{m-l}} \frac{\rho_{\text{mem}}}{EW} (\lambda_{\text{mw}} - \lambda_{\text{sat}}) \tag{51}$$

where λ_{mw} and λ_{sat} are the membrane water content and saturated membrane water content, respectively. Considering the impact of EOD, water can be dragged from the cathode to the anode by the transport of hydroxyl ions; therefore, the source term of EOD is expressed as:

$$\begin{cases} S_{\text{EOD}}^{\text{a}} = \dfrac{n_{\text{d}} j_{\text{a}} M_{\text{H}_2\text{O}}}{F} & \text{(in anode CL)} \\ S_{\text{EOD}}^{\text{c}} = -\dfrac{n_{\text{d}} j_{\text{c}} M_{\text{H}_2\text{O}}}{F} & \text{(in cathode CL)} \end{cases} \tag{52}$$

where the EOD coefficient, n_{d}, is calculated as:

$$n_{\text{d}} = 0.183\lambda_{\text{mw}} + 1.3 \tag{53}$$

In the conservation equation of membrane water, the relationship between membrane water content and water concentration is defined as

$$\lambda_{\text{mw}} = \frac{EW}{\rho_{\text{mem}}} c_{\text{mw}} \tag{54}$$

The diffusivity of membrane water is fitted as (Huo et al., 2012)

$$D_{\text{mw}} = \begin{cases} (0.0051\lambda_{\text{mw}} \times T_0 - 1.44\lambda_{\text{mw}}) \times 10^{-10} \quad 0.0 \le \lambda_{\text{mw}} \le 14.0 \\ \begin{bmatrix} (-23.2404 + 4.513\lambda_{\text{mw}} - 0.28926\lambda_{\text{mw}}^2 + 0.006131\lambda_{\text{mw}}^3) \\ \times (T_0 - 303.15) - 79.826 + 17.928\lambda_{\text{mw}} - 1.3329\lambda_{\text{mw}}^2 + 0.03337\lambda_{\text{mw}}^3 \end{bmatrix} \\ \times 10^{-10} \quad 14.0 < \lambda_{\text{mw}} \le 19.0 \\ [(-41.916 + 0.00613\lambda_{\text{mw}}^3) \times (T_0 - 303.15) + 8.5139] \times 10^{-10} \quad \lambda_{\text{mw}} > 19.0 \end{cases} \tag{55}$$

The equilibrium water content is defined as (Huo et al., 2012)

$$\lambda_{\text{equil}} = \begin{cases} (-0.605a^3 + 0.85a^2 - 0.205a + 0.153) \times (T_0 - 313.15) \\ \quad + 39.0a^3 - 47.7a^2 + 23.4a + 0.117 \quad 0 \le a \le 1.0 \\ (-0.00265a + 0.05795) \times (T_0 - 313.15) + 1.5915(a-1) \\ \quad + 14.817 \quad 1.0 < a \le 3.0 \end{cases} \tag{56}$$

In the solid AEMs, the conservation of charge is expressed as

$$0 = \nabla \cdot \left(\kappa_{\text{ele}}^{\text{eff}} \nabla \phi_{\text{ele}} \right) + S_{\text{ele}} \tag{57a}$$

$$0 = \nabla \cdot \left(\kappa_{\text{ion}}^{\text{eff}} \nabla \phi_{\text{ion}} \right) + S_{\text{ion}} \tag{57b}$$

The ion conductivity and the water vapor activity are fitted to the experimental data as.

$$\begin{aligned} \kappa_{\text{ion}} = {} & 0.1334 - 3.882 \times 10^{-4} T_0 + (0.01148 T_0 - 3.909) a_{\text{w}} - (0.06690 T_0 - 23.01) a_{\text{w}}^2 \\ & + (0.1227 T_0 - 42.61) a_{\text{w}}^3 - (0.06021 T_0 - 21.80) a_{\text{w}}^4 \end{aligned} \tag{58a}$$

$$\begin{aligned} a_{\text{w}} = {} & 0.8118 - 2.296 \times 10^{-3} T_0 + \left(5.815 \times 10^{-3} T_0 - 2.005\right) \lambda_{\text{mw}} \\ & - \left(2.977 \times 10^{-3} T_0 - 1.046\right) \lambda_{\text{mw}}^2 + \left(4.825 \times 10^{-4} T_0 - 0.1676\right) \lambda_{\text{mw}}^3 \\ & - \left(3.179 \times 10^{-5} T_0 - 0.01094\right) \lambda_{\text{mw}}^4 + \left(7.427 \times 10^{-7} T_0 - 2.539 \times 10^{-4}\right) \lambda_{\text{mw}}^5 \end{aligned} \tag{58b}$$

The energy conservation equation is expressed as:

$$\frac{\partial}{\partial t}\left((\rho C_p)_{fl,sl}^{eff} T \right) + \nabla \cdot \left((\rho C_p \vec{u})_{\text{fl}}^{\text{eff}} T \right) = \nabla \cdot \left(\kappa_{\text{fl,sl}}^{\text{eff}} \nabla T \right) + S_{\text{E}} \tag{59}$$

All the source terms mentioned above are listed in Table 8.3.

TABLE 8.3 Source terms of the governing equations of AEMFCs (Jiao et al., 2014, 2015).

Source term	Units
$S_m = S_{H_2} + S_{O_2} + S_{vp}$	kg m^{-3} s^{-1}
$S_u = \begin{cases} -\frac{\mu_g}{K_g} u_g \text{(in CL and GDL)} \\ 0 \text{(in other zones)} \end{cases}$	kg m^{-2} s^{-2}
$S_{H_2} = \begin{cases} -\frac{j_a}{2F} M_{H_2} \text{(in anode CL)} \\ 0 \text{(in other zones)} \end{cases}$	kg m^{-3} s^{-1}
$S_{O_2} = \begin{cases} -\frac{j_c}{4F} M_{O_2} \text{(in cathode CL)} \\ 0 \text{(in other zones)} \end{cases}$	kg m^{-3} s^{-1}

Continued

TABLE 8.3 Source terms of the governing equations of AEMFCs—cont'd

Source term	Units
$S_{vp} = \begin{cases} -S_{v-1} (\text{in CL and GDL}) \\ 0 (\text{in other zones}) \end{cases}$	kg m^{-3} s^{-1}
$S_{lq} = \begin{cases} \frac{j_a}{F} M_{H_2O} + S_{v-1} + S_{m-1} M_{H_2O} + S_{EOD}^{a} (\text{in anode CL}) \\ -\frac{j_c}{2F} M_{H_2O} + S_{v-1} + S_{m-1} M_{H_2O} + S_{EOD}^{c} (\text{in cathode CL}) \\ S_{v-1} (inGDL) \\ 0 (\text{in other zones}) \end{cases}$	kg m^{-3} s^{-1}
$S_{mw} = \begin{cases} -S_{m-1} (inCL) \\ -\frac{S_{EOD}^{a} V_{CL}^{a} + S_{EOD}^{c} V_{CL}^{c}}{V_{mem} M_{H_20}} (\text{in membrane}) \end{cases}$	Kmol m^{-3} s^{-1}
$S_{ele} = \begin{cases} -j_a (\text{in anode CL}) \\ j_c (\text{in cathode CL}) \\ 0 (\text{in other zones}) \end{cases}$	A m^{-3}
$S_{ion} = \begin{cases} j_a (\text{in anode CL}) \\ -j_c (\text{in cathode CL}) \\ 0 (\text{in other zones}) \end{cases}$	A m^{-3}
$S_E = \begin{cases} \frac{j_a T \Delta S}{2F} + j_a \lvert \eta_{act} \rvert + \nabla\phi_{ele}{}^2 \kappa_{ele}^{eff} + \nabla\phi_{icm}{}^2 \kappa_{icm}^{eff} + S_{ln} (\text{in anode CL}) \\ j_c \lvert \eta_{act} \rvert + \nabla\phi_{ele}{}^2 \kappa_{ele}^{eff} + \nabla\phi_{em}{}^2 \kappa_{icm}^{eff} + S_{ln} (\text{in cathode CL}) \\ \nabla\phi_{ele}{}^2 \kappa_{ele}^{eff} + S_{ln} (inGDL) \\ \nabla\phi_{ele}{}^2 \kappa_{ele}^{eff} (inBP) \\ \nabla\phi_{icm}{}^2 \kappa_{icm}^{eff} (\text{in membrane}) \\ 0 (\text{in other zones}) \end{cases}$	W m^{-3}

2.4 Multiscale modeling approach for the hydroxide transport in AEM

The strategy to explore anion transport in AEM using the multiscale modeling method is to sequentially integrate CG, Atomistic Polarizable Potential for Liquids, Electrolytes, and Polymers (APPLE&P), and ReaxFF simulations to investigate hydrated anion-exchange membranes (Dong et al., 2018). Initial

CG simulations establish equilibrium membrane morphology by capturing thermodynamically driven microphase segregation and enabling extended timescales for polymer relaxation and domain maturation. This mesoscale architecture—featuring interconnected hydrophilic channels within a glassy polymer matrix—is preserved when mapped to atomistic resolution, where subsequent simulations refine local hydration structures and water distribution profiles within nanochannels. Finally, ReaxFF simulations probe hydroxide transport mechanisms (combining Grotthuss hopping and vehicular diffusion). The hierarchical framework maintains structural fidelity across scales while systematically resolving phenomena from morphological evolution to bond reorganization, with detailed methodological parameters and multilevel property comparisons discussed in subsequent sections.

The CG model of the hydrated PPO/TMA membrane was established, where water molecules were represented using the mW potential. Tetramethylammonium (TMA) cations and chloride (Cl^-) anions were CG into neutral particles, while the polymer backbone employed a united atom representation. Interparticle interactions integrated the two-body Stillinger-Weber potential and Yukawa-type screened Coulomb interactions.

The APPLE&P polarizable force field, parameterized to reproduce dynamical and thermodynamic properties of alkyl-ammonium aqueous systems, employs multitime-step integration alongside constraint algorithms for bond geometries. Electrostatic interactions are calculated using Ewald summation and reciprocal-space polarization modeling. To transition from CG to atomistic resolution, hydrogen atoms were reconstructed on the polymer backbone using predefined bond parameters, TMA functional groups were atomistically positioned at CG cationic sites, and hydroxide ions replaced chloride anions with randomized O–H orientations.

ReaxFF simulations employed the CHON-2017_weak force field, optimized for modeling weak interactions in functionalized hydrocarbon/water systems. Seamless atomistic mapping from the APPLE&P framework required minimal structural relaxation due to shared resolution fidelity. Trajectory analysis focused on bond rupture events and reactive intermediates, leveraging ReaxFF's explicit bond-order formalism to quantify hydroxide transport mechanisms.

3 Results and discussions

3.1 Operational performance based on the empirical model for AFC

3.1.1 Temperature impact on the performance of alkaline fuel cells

The performance of AFCs is critically influenced by operating conditions, including fuel composition, temperature, and pressure. Temperature plays a dominant role by modulating hydroxyl ion conductivity, electrochemical reaction kinetics, and mass transport efficiency (Hamada et al., 2023). As illustrated

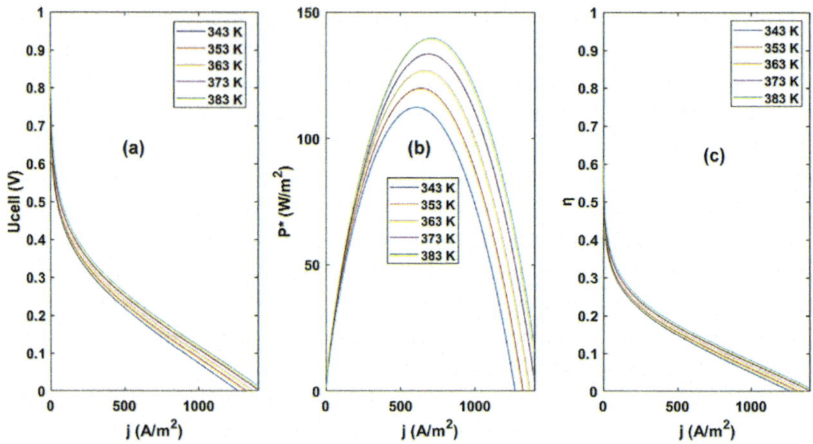

FIG. 8.5 Effect of cell operating temperatures on the (a) output cell voltage, (b) power density, and (c) efficiency of an AFC (Hamada et al., 2023). *(From Hamada, A.T., Orhan, M.F., Kannan, A.M., 2023. Alkaline fuel cells: Status and prospects. Energy Rep. 9, 6396–6418. https://doi.org/10.1016/j.egyr.2023.05.276.)*

in Fig. 8.5, increasing the operational temperature enhances cell performance through faster ion mobility, reduced cell resistance, and higher exchange current densities, leading to elevated power densities and output voltages. For KOH-based electrolytes, optimal ionic conductivity requires temperature-dependent concentration adjustments, rising from ~30 wt% at 0°C to 34 wt% at 80°C. However, excessively high temperatures risk membrane dehydration and humidity loss, which paradoxically degrade performance despite initial kinetic improvements. This further demonstrates that AFC efficiency peaks within a narrow thermal range, emphasizing the necessity of balancing enhanced electrochemical activity with sustained hydration to avoid irreversible performance decline.

3.1.2 KOH concentration impact on the performance of AFCs

Fig. 8.6 demonstrates the performance of AFCs influenced by the electrolyte concentration (Hamada et al., 2023). Increasing KOH concentration enhances output voltage and power density by improving ionic conductivity and reducing ohmic losses, thereby elevating overall efficiency. It can also be observed that higher concentrations correlate with lower cell resistance, though performance gains diminish beyond 41 mol/L due to excessive hydroxyl ion coverage on the anode CL, which limits fuel adsorption sites. Consequently, while moderate KOH concentration adjustments optimize conductivity and electrochemical output, excessively high concentrations yield minimal improvements, necessitating a balance between ionic transport efficiency and catalytic activity preservation.

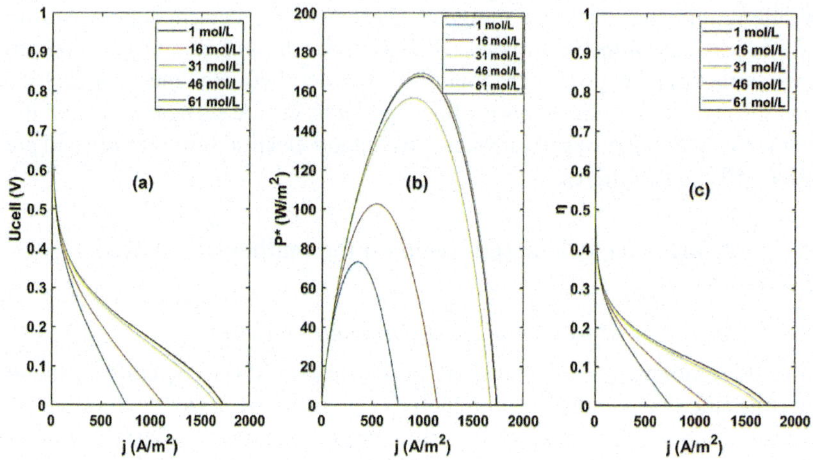

FIG. 8.6 Effect of KOH concentrations on the (A) output cell voltage, (B) power density, and (C) efficiency of an AFC (Hamada et al., 2023). *(From Hamada, A.T., Orhan, M.F., Kannan, A.M., 2023. Alkaline fuel cells: Status and prospects. Energy Rep. 9, 6396–6418. https://doi.org/10.1016/j.egyr.2023.05.276.)*

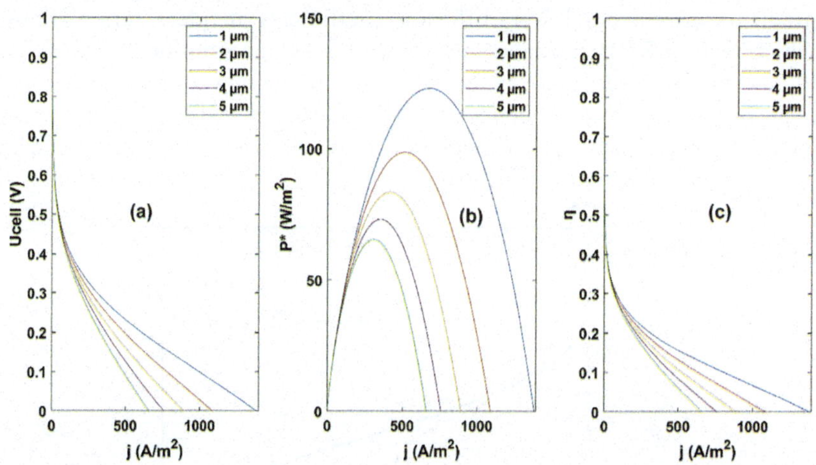

FIG. 8.7 Effect of electrolyte thickness on the (A) output cell voltage, (B) power density, and (C) efficiency of an AFC (Hamada et al., 2023). *(From Hamada, A.T., Orhan, M.F., Kannan, A.M., 2023. Alkaline fuel cells: Status and prospects. Energy Rep. 9, 6396–6418. https://doi.org/10.1016/j.egyr.2023.05.276.)*

3.1.3 Electrolyte thickness impact on the performance of AFCs

Electrolyte thickness critically governs fuel cell performance, as shown in Fig. 8.7, which correlates increased thickness (1–5 μm) with reduced output voltage, power density, and efficiency due to elevated ohmic losses from higher internal resistance to hydroxyl ion transport (Hamada et al., 2023).

While minimizing thickness mitigates resistance, it underscores practical limits: excessively thin electrolytes risk electrode short-circuiting and compromise structural integrity, necessitating a balance between ionic conductivity and mechanical stability. Optimal thickness must ensure sufficient ion mobility while maintaining physical robustness to support electrode architecture and prevent operational failures.

3.2 Performance calculated based on the mathematical model for AFC

3.2.1 *Impacts of the ideal solution assumption*

The thermodynamic and transport properties of AFCs critically influence performance, as electrolyte conductivity governs IR losses, while reactant dissolution rates dictate limiting current density (Zhou et al., 2011b). Fig. 8.8 demonstrates that relaxing the ideal solution assumption extends the calculated limiting current density from 1.0 to 1.8 A cm^{-2} due to enhanced KOH diffusion, whereas Fig. 8.9 reveals significant concentration gradients under ideal assumptions (cathode: 5.3–9.5 M; anode: 5.5–7.0 M at 0.8 A cm^{-2}) compared to the nonideal model's uniform distribution (6.4–7.7 M). These gradients arise from lower diffusion coefficients in ideal formulations, which restrict KOH transport, elevate cathode concentration, suppress O_2 dissolution, and reduce current density. Since maximum conductivity occurs near 7 M KOH, uniform concentrations (as shown in Fig. 8.9) minimize IR losses and improve cell voltage in the 0.4–0.8 A cm^{-2} regime. The ideal

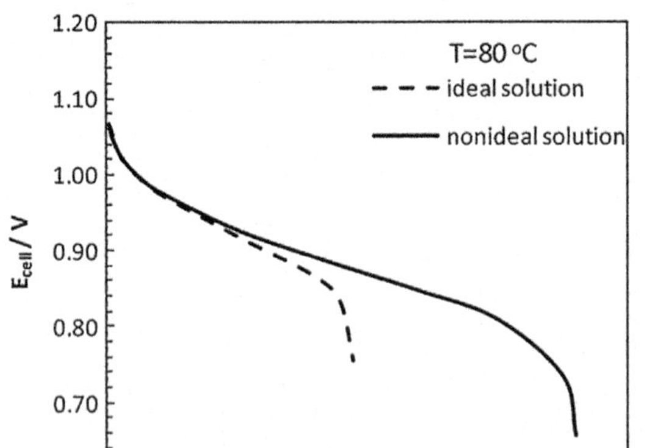

FIG. 8.8 Comparison of nonideal solution versus ideal solution results—polarization curve under isothermal boundary conditions (Zhou et al., 2011b). *(From Zhou, G., Chen, L.D., Seaba, J.P., 2011b. Effects of property variation and ideal solution assumption on the calculation of the limiting current density condition of alkaline fuel cells. J. Power Sour. 196 (11), 4923–4933. https://doi.org/10.1016/j.jpowsour.2011.02.026.)*

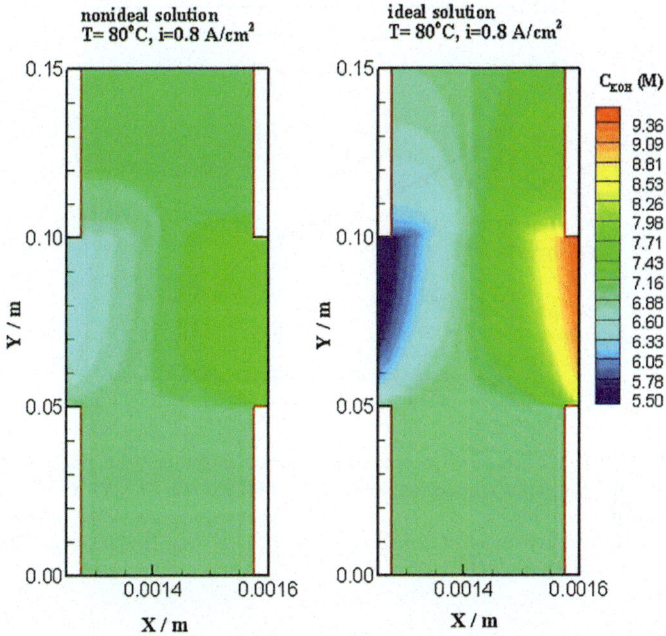

FIG. 8.9 Comparison of nonideal solution versus ideal solution results: isopleths of KOH concentration in the separator and extended channel; $i = 0.8$ A cm^{-2}; anode and cathode located at $x = 0$ and 0.0016 m (Zhou et al., 2011b). *(From Zhou, G., Chen, L.D., Seaba, J.P., 2011b. Effects of property variation and ideal solution assumption on the calculation of the limiting current density condition of alkaline fuel cells. J. Power Sour. 196 (11), 4923–4933. https://doi.org/10.1016/j.jpowsour.2011.02.026.)*

solution model underestimates limiting current density by ~50%, highlighting the necessity of nonideal assumptions in AFC simulations to avoid significant errors in predicting performance under realistic concentration gradient conditions.

3.2.2 *Impacts of the thermal boundary conditions*

Fig. 8.10 illustrates the impact of thermal conditions on polarization curves, showing that replacing the isothermal boundary condition (80°C) with an adiabatic one increases the cathode temperature from 82°C at the inlet to over 100°C at the exit (Zhou et al., 2011ab). While oxygen solubility remains stable in this temperature range, dissolved O_2 levels rise significantly at the exit due to a temperature-driven order-of-magnitude increase in KOH diffusion coefficients (80°C–100°C), enhancing KOH transport from the cathode to the anode and reducing cathode-side concentration. This effect extends the limiting current density from 1.8 to 2.3 A cm^{-2} under adiabatic conditions, as shown in Figure 8.10. Despite similar cell voltages for currents below 1.6 A cm^{-2}, the

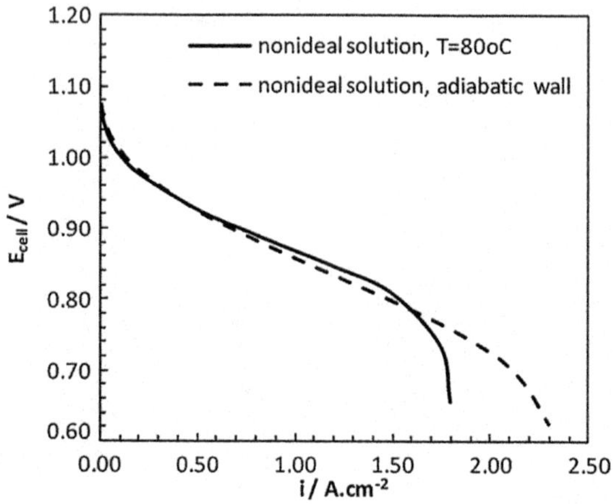

FIG. 8.10 Effects of adiabatic versus isothermal boundary conditions on the calculated polarization curve; nonideal solution formulation (Zhou et al., 2011b). *(From Zhou, G., Chen, L.D., Seaba, J.P., 2011b. Effects of property variation and ideal solution assumption on the calculation of the limiting current density condition of alkaline fuel cells. J. Power Sour. 196 (11), 4923–4933. https://doi.org/10.1016/j.jpowsour.2011.02.026.)*

adiabatic model predicts markedly higher limiting current densities, emphasizing the critical role of temperature-dependent transport properties in optimizing AFC performance under high-current operation.

3.2.3 *Impacts of the operating pressure and temperature*

AFCs with mobile electrolytes require balanced gas-electrolyte pressure to prevent gas intrusion into electrolyte channels or electrolyte leakage into gas spaces, thereby preserving three-phase contact areas. Elevated pressure shifts polarization curves upward (due to increased equilibrium potential via the Nernst equation) and enhances limiting current density (via pressure-driven solubility gains per Henry's law). Higher pressures improve cell performance by raising both voltage and current density limits. However, excessive or unstable pressure increases the risk of gas breakthrough across the electrolyte, electrolyte penetration into electrodes, or disrupted gas-liquid interfaces, all of which must be rigorously mitigated during system operation.

Increasing the operating temperature from 20°C to 80°C improves AFC performance by enhancing ohmic and mass transport regions while minimally affecting the activation region. Although rising temperature thermodynamically reduces reversible cell voltage, this effect is offset by accelerated kinetics: cathode exchange current density grows approximately 27-fold at 80°C versus 20°C, drastically lowering activation losses (Kunz, 2019). Ohmic losses decrease as KOH electrolyte conductivity more than doubles with temperature, while mass transport improves due to increased oxygen diffusion coefficients, which rise fivefold in 5 M KOH at 80°C. This reduces oxygen concentration gradients, lowering mass transport losses and raising limiting current density. Earlier models predicted voltage declines in activation and ohmic regions with temperature but agreed on limiting current gains. Kunz (2019) incorporated temperature-dependent kinetics in their models and found that the voltage reduction is predicted, and the cell performance trends can be better captured. This advancement highlights the critical role of temperature in optimizing ionic conductivity, diffusion rates, and reaction kinetics for improved high-temperature AFC operation.

3.3 Operational performance based on the mathematical model for AEMFC

3.3.1 Impacts of anode inlet relative humidity

In AEMFC, water is generated at the anode during operation and migrates from the cathode to the anode under electro-osmotic drag. Anode inlet humidification may lead to excessive liquid water accumulation; however, reverse diffusion of liquid water from the anode to the cathode occurs due to concentration gradients, necessitating systematic investigation of anode humidification. AEMFC performance under varying anode inlet relative humidity (RH) levels (50%–100%) is shown in Fig. 8.11 (Jiao et al., 2014). AEMFC performance improves significantly with increased anode humidification, though the enhancement diminishes progressively at higher RH. Fig. 8.11 indicates that liquid water forms only at 100% RH, altering water removal mechanisms: partial humidification promotes water evaporation and vapor-phase removal, while saturation (100% RH) shifts transport to slower liquid-phase pathways. The liquid water volume fraction in anode porous media surges from 0 to 0.11 as RH increases from 90% to 100%, exacerbating flooding in gas diffusion and CLs. Notably, water production from anode reactions and cathode-to-anode electro-osmotic

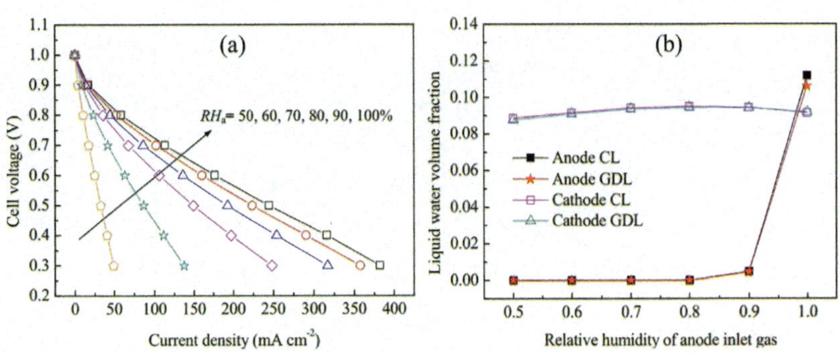

FIG. 8.11 Figure 11. Effect of anode inlet relative humidity on the polarization curve and liquid water volume fraction (0.5 V) (Jiao et al., 2014). *(From Jiao, K., He, P., Du, Q., Yin, Y., 2014. Three-dimensional multiphase modeling of alkaline anion exchange membrane fuel cell. Int. J. Hydrogen Energy 39 (11), 5981–5995, Elsevier Ltd. https://doi.org/10.1016/j.ijhydene.2014.01.180.)*

migration collectively elevate anode water content. Reduced polarization curve slopes suggest lower ohmic losses with higher anode humidification. In addition, while simulations assume optimal cathode humidification with liquid water supply in flow channels, anode humidification becomes critical under cathode drying conditions.

3.3.2 Impacts of AEM thickness

Membrane thickness is a critical factor influencing fuel cell performance, primarily due to its direct impact on ohmic losses during ion conduction and water diffusion rates between the anode and cathode. As shown in Fig. 8.12, AEMFC polarization curves demonstrate significant performance improvements with reduced membrane thickness (Jiao et al., 2014). This enhancement stems from two mechanisms: thinner membranes lower ohmic losses by shortening ion conduction pathways and mitigate mass transport losses at high current densities by balancing water content across the electrodes. However, thinner membranes exhibit reduced durability, necessitating a compromise between performance gains and operational longevity in membrane design.

3.3.3 Impacts of metal porous bipolar plate

Bipolar plates (BPs), critical components of alkaline AEMFCs (AAEMFCs), distribute reactants, conduct electricity and heat, and remove water. Metal foam (MF) flow fields offer advantages such as high electrical and thermal conductivity, a compressible porous structure, and tunable porosity. As shown in Fig. 8.13, the MF case exhibits a lower maximum temperature and a more uniform thermal distribution at 0.1 A cm^{-2} compared to serpentine

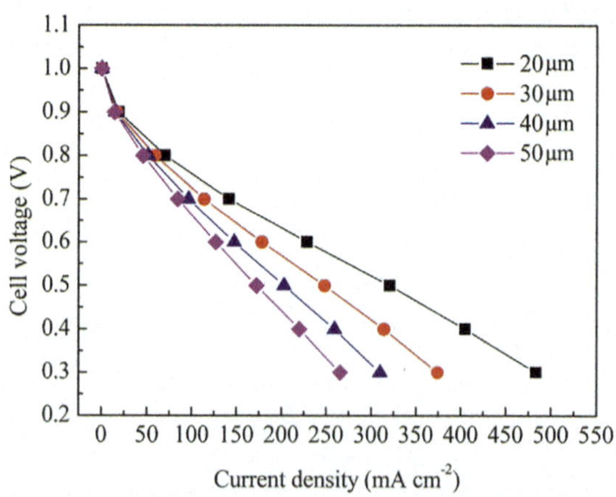

FIG. 8.12 Polarization curves at different membrane thicknesses (Jiao et al., 2014). *(From Jiao, K., He, P., Du, Q., Yin, Y., 2014. Three-dimensional multiphase modeling of alkaline anion exchange membrane fuel cell. Int. J. Hydrogen Energy 39 (11), 5981–5995, Elsevier Ltd. https://doi.org/10.1016/j.ijhydene.2014.01.180.)*

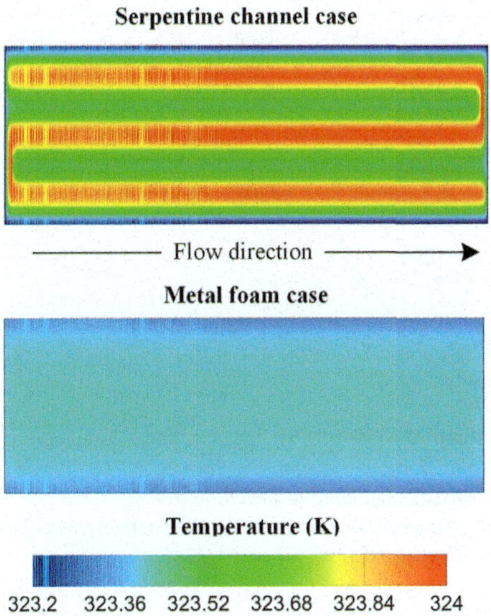

FIG. 8.13 Distributions of temperature at the middle plane of the ACL for the AEMFCs with serpentine channels and metal foam flow fields at a current density of 0.1 A cm^{-2} (Cheng et al., 2022). *(From Dong, D., Zhang, W., Barnett, A., Lu, J., van Duin, A.C.T., Molinero, V., Bedrov, D., 2018. Multiscale modeling of structure, transport and reactivity in alkaline fuel cell membranes: Combined coarse-grained, atomistic and reactive molecular dynamics simulations. Polymers 10 (11). https://doi.org/10.3390/polym10111289.)*

channels, which is attributed to MF's homogeneous heat transfer. In contrast, serpentine channels display localized temperature variations due to channel-rib structural heterogeneity, demonstrating MF's thermal superiority (Cheng et al., 2022).

3.4 AEM physical properties with multiscale strategy

The diffusion coefficient of molecules was calculated based on the Einstein relation:

$$D = \frac{1}{6Nt}\lim_{t\to\infty}\frac{d}{dt} < \sum_{m=1}^{M}[r_m(t) - r_0(t)]^2 \tag{60}$$

In which the MSD is the mean square displacement, and M is the total number of molecules of the species of interest. Notably, only ReaxFF simulations explicitly incorporated the Grotthuss mechanism, enabling concurrent contributions from vehicular motion and proton hopping. By contrast, APPLE&P and CG force fields restricted anion and H_2O diffusivity to vehicular transport, though the CG model aimed to approximate total diffusivity, while APPLE&P prioritized resolving vehicular dynamics. In hydrated membranes, the interconnected subnanometer water channels govern membrane morphology.

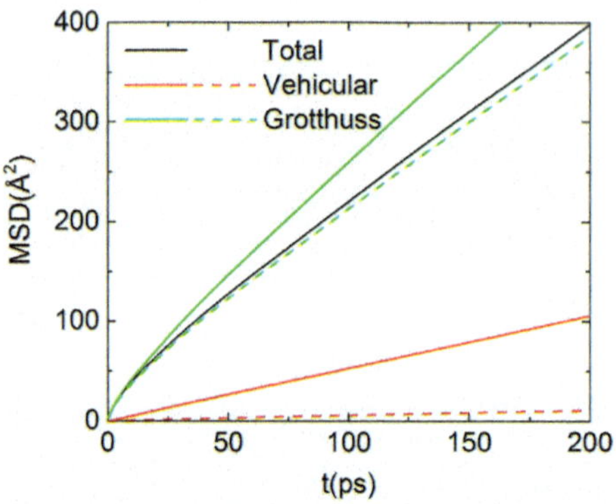

FIG. 8.14 Decomposition of MSD of OH^- into Grotthuss hopping and vehicular contribution (Dong et al., 2018). *(From Dong, D., Zhang, W., Barnett, A., Lu, J., van Duin, A.C.T., Molinero, V., Bedrov, D., 2018. Multiscale modeling of structure, transport and reactivity in alkaline fuel cell membranes: Combined coarse-grained, atomistic and reactive molecular dynamics simulations. Polymers 10 (11). https://doi.org/10.3390/polym10111289.)*

ReaxFF's inclusion of hopping mechanisms induces localized perturbations to hydrogen-bond networks within these nanochannels during proton transfer events.

To understand the OH^- transport mechanism, the MSD of OH^- from ReaxFF simulation can be decomposed into contributions from the vehicular and Grotthuss motions. The decomposed mean squared displacement (MSD) has been obtained from ReaxFF simulations and is shown in Fig. 8.14 (Dong et al., 2018). Grotthuss hopping dominates OH^- diffusion in hydrated membranes, with vehicular contributions (including coupling effects) constituting less than 5% of total mobility.

4 Summary

This chapter systematically reviews mainstream modeling methodologies for AFCs and AAEMFCs, with emphasis on multiscale modeling approaches that optimize anion transport in membranes to achieve high ion conductivity and cell performance. Critical insights from representative studies reveal the operational-structural interdependencies: (1) empirical models efficiently identify AFC design parameters but require experimental calibration for validity; (2) ideal solution assumptions underestimate limiting current density and demand rigorous validation; (3) thermal management extends current thresholds, while pressure elevation enhances gas dissolution and temperature increases improve gas diffusivity/conductivity synergistically. Membrane-electrode optimization strategies are further highlighted: water management via controlled anode humidity balances membrane hydration and flooding risks, while MF BPs enable superior reactant/temperature uniformity compared to conventional designs—a promising direction for AEMFC development. Notably, multiscale modeling confirms Grotthuss-dominated hydroxide conduction in hydrated membranes, providing atomic-level guidance for AEM molecular engineering. Collectively, this analysis establishes a framework for correlating operating conditions (pressure, temperature, humidity), structural designs (BPs, membrane architecture), and performance metrics, offering foundational insights for next-generation AFC optimization.

Acknowledgements

This research was supported by the National Natural Science Foundation of China (52576090) and the Joint Fund of the National Natural Science Foundation of China (U2441281).

References

Abdullah, S., Kamarudin, S.K., Hasran, U.A., Masdar, M.S., Daud, W.R.W., 2016. Electrochemical kinetic and mass transfer model for direct ethanol alkaline fuel cell (DEAFC). J. Power Sources. 03787753. 320, 111–119. doi:10.1016/j.jpowsour.2016.04.003.

An, L., Chai, Z.H., Zeng, L., Tan, P., Zhao, T.S., 2013. Mathematical modeling of alkaline direct ethanol fuel cells. Int. J. Hydrog. Energy. 03603199. 38 (32), 14067–14075. doi:10.1016/j.ijhydene.2013.08.080.

Aremo, B., Adeoye, M.O., Obioh, I.B., 2015. A simplified test station for alkaline fuel cell. J. Fuel Cell Sci. Technol. 1550-624X. 12 (2). doi:10.1115/1.4029421.

Azzam, M., Qaq, Z., Orhan, M., 2023. Design and analysis of an alkaline fuel cell. J. Therm. Eng. 21487847. 9 (1), 138–160. Available from: http://eds.yildiz.edu.tr/journal-of-thermal-engineering/Content. doi:10.18186/thermal.1243498.

Bidault, F., Kucernak, A., 2011. Cathode development for alkaline fuel cells based on a porous silver membrane. J. Power Sources. 03787753. 196 (11), 4950–4956. doi:10.1016/j.jpowsour.2011.02.008.

Bidault, F., Middleton, P.H., 2012. 4.07 - Alkaline Fuel Cells: Theory and Application Comprehensive Renewable Energy. Elsevier, United Kingdom, pp. 179–202. Available from: https://www.sciencedirect.com/book/9780080878737. doi:10.1016/B978-0-08-087872-0.00405-4.

Björnbom, P., Yang, S.C., 1993. Modeling and simulation of the operation of a whole alkaline fuel cell by using concentrated electrolyte solution transport theory. Electrochim. Acta. 00134686. 38 (17), 2599–2609. doi:10.1016/0013-4686(93)80158-V.

Burchardt, T., 2004. An evaluation of electrocatalytic activity and stability for air electrodes. J. Power Sources. 03787753. 135 (1-2), 192–197. doi:10.1016/j.jpowsour.2004.03.072.

Chakma, R., Karmaker, A.K., Roy, N.K., 2022. Performance evaluation of 2.4 kW alkaline fuel cell stack for emergency applications. 4th International Conference on Sustainable Technologies for Industry 4.0 (STI).

Cheng, C., Wang, B., Liu, Z., Zhang, G., Xie, B., Tongsh, C., Xi, F., Chen, W., Jiao, K., 2022. Numerical investigation of design and operating parameter effects on permeability-differentiated alkaline fuel cell with metal foam flow field. Appl. Therm. Eng. 13594311. 207. Available from: http://www.journals.elsevier.com/applied-thermal-engineering/. doi:10.1016/j.applthermaleng.2022.118183.

Dong, D., Zhang, W., Barnett, A., Lu, J., van Duin, A.C.T., Molinero, V., Bedrov, D., 2018. Multiscale modeling of structure, transport and reactivity in alkaline fuel cell membranes: combined coarse-grained, atomistic and reactive molecular dynamics simulations. Polymers

20734360. 10 (11). Available from: https://www.mdpi.com/2073-4360/10/11/1289/pdf. doi:10.3390/polym10111289.

GILLIAM, R., GRAYDON, J., KIRK, D., THORPE, S., 2007. A review of specific conductivities of potassium hydroxide solutions for various concentrations and temperatures. Int. J. Hydrog. Energy. 03603199. 32 (3), 359–364. doi:10.1016/j.ijhydene.2006.10.062.

Hamada, A.T., Orhan, M.F., Kannan, A.M., 2023. Alkaline fuel cells: status and prospects. Energy Rep. 23524847. 9, 6396–6418. Available from: http://www.journals.elsevier.com/energy-reports/. doi:10.1016/j.egyr.2023.05.276.

Han, X., Chadderdon, D.J., Qi, J., Xin, L., Li, W., Zhou, W., 2014. Numerical analysis of anion-exchange membrane direct glycerol fuel cells under steady state and dynamic operations. Int. J. Hydrog. Energy 03603199. 39 (34), 19767–19779. Available from: http://www.journals.elsevier.com/international-journal-of-hydrogen-energy/. doi:10.1016/j.ijhydene.2014.08.144.

Huo, S., Deng, H., Chang, Y., Jiao, K., 2012. Water management in alkaline anion exchange membrane fuel cell anode. Int. J. Hydrog. Energy. 03603199. 37 (23), 18389–18402. doi:10.1016/j.ijhydene.2012.09.074.

Huo, S., Jiao, K., 2018. Mathematical modeling of alkaline anion exchange membrane fuel cells. In: An, L., Zhao, T.S. (Eds.), Anion Exchange Membrane Fuel Cells, vol. 63. Springer Cham, Switzerland, pp. 169–215. Available from: https://www.springer.com/series/8874. doi:10.1007/978-3-319-71371-7_6.

Jiao, K., He, P., Du, Q., Yin, Y., 2014. Three-dimensional multiphase modeling of alkaline anion exchange membrane fuel cell. Int. J. Hydrog. Energy. 03603199. 39 (11), 5981–5995. doi:10.1016/j.ijhydene.2014.01.180.

Jiao, K., Huo, S., Zu, M., Jiao, D., Chen, J., Du, Q., 2015. An analytical model for hydrogen alkaline anion exchange membrane fuel cell. Int. J. Hydrog. Energy 03603199. 40 (8), 3300–3312. Available from: http://www.journals.elsevier.com/international-journal-of-hydrogen-energy/. doi:10.1016/j.ijhydene.2014.12.091.

Jo, J.H., Yi, S.C., 1999. Computational simulation of an alkaline fuel cell. J. Power Sources. 03787753. 84 (1), 87–106. doi:10.1016/S0378-7753(99)00309-2.

Kimble, M.C., White, R.E., 1991. A mathematical model of a hydrogen/oxygen alkaline fuel cell. J. Electrochem. Soc. 19457111. 138 (11), 3370–3382. doi:10.1149/1.2085416.

Kunz, F., 2014. Verification and Validation of a Simulation Model of an Alkaline Fuel Cell (AFC) (Master thesis). Universität Duisburg-Essen, Duisburg, Germany.

Kunz, F., 2019. Modeling and Simulation of an Alkaline Fuel Cell (Ph.D. thesis, Dissertation). Universität Duisburg-Essen, Duisburg, Germany. doi:10.17185/duepublico/70291.

Pezzini, A., de Castro, U.J., de Oliveira, D.S.B.L., Tremiliosi-Filho, G., de Sousa Júnior, R., 2023. Mathematical modeling of alkaline direct glycerol fuel cells. Energies 19961073. 16 (19). Available from: http://www.mdpi.com/journal/energies/. doi:10.3390/en16196762.

Raya, I.P., Ellis, M.W., Hernandez-Guerrero, A., Elizalde-Blancas, F., 2016. Modeling the effect of membrane conductivity on the performance of alkaline fuel cells. J. Power Sources. 03787753. 307, 898–906. doi:10.1016/j.jpowsour.2015.12.141.

Saebea, D., Chaiburi, C., Authayanun, S., 2019. Model based evaluation of alkaline anion exchange membrane fuel cells with water management. Chem. Eng. J. 13858947. 374, 721–729. Available from: https://www.elsevier.com/inca/publications/store/6/0/1/2/7/3/index.htt. doi:10.1016/j.cej.2019.05.200.

Shen, Y.K., Li, X.Q., Wang, N., Li, L., Hoseyni, A., 2021. Introducing and investigation of a pumped hydro-compressed air storage based on wind turbine and alkaline fuel cell and electrolyzer. Sustain. Energy Technol. Assess. 22131388. 47. Available from: http://www.journals.elsevier.com/sustainable-energy-technologies-and-assessments. doi:10.1016/j.seta.2021.101378.

Su, X., Pan, Z., an, L., Yu, Y., 2021. Mathematical modeling of direct formate fuel cells incorporating the effect of ion migration. Int. J. Heat Mass Transf. 164, 120629. doi:10.1016/j.ijheatmasstransfer.2020.120629.

Verhaert, I., Mulder, G., De Paepe, M., 2016. Evaluation of an alkaline fuel cell system as a micro-CHP. Energy Convers. Manag. 01968904. 126, 434–445. doi:10.1016/j.enconman.2016.07.058.

Wang, Z., Zhang, X., Rezazadeh, A., 2021. Hydrogen fuel and electricity generation from a new hybrid energy system based on wind and solar energies and alkaline fuel cell. Energy Rep. 23524847. 7, 2594–2604. Available from: http://www.journals.elsevier.com/energy-reports/. doi:10.1016/j.egyr.2021.04.060.

Weinzierl, C., Krewer, U., 2016. Model-based analysis of water management at anode of alkaline direct methanol fuel cells. Chem. Eng. Sci. 00092509. 143, 181–193. doi:10.1016/j.ces.2015.12.006.

Yang, P., Zhu, Y., Zhang, P., Zhang, H., Hu, Z., Zhang, J., 2014. Performance evaluation of an alkaline fuel cell/thermoelectric generator hybrid system. Int. J. Hydrog. Energy 03603199. 39 (22), 11756–11762. Available from: http://www.journals.elsevier.com/international-journal-of-hydrogen-energy/. doi:10.1016/j.ijhydene.2014.05.166.

Yu, X., She, C., Gholizadeh, F., Xu, Y.P., 2021. Numerical investigation of a new combined energy cycle based on Miller cycle, Organic Rankine cycle, Stirling engine and alkaline fuel cell. Energy Rep. 23524847. 7, 5406–5419. Available from: http://www.journals.elsevier.com/energy-reports/. doi:10.1016/j.egyr.2021.08.111.

Zadick, A., Dubau, L., Sergent, N., Berthomé, G., Chatenet, M., 2015. Huge instability of Pt/C catalysts in alkaline medium. ACS Catal. 21555435. 5 (8), 4819–4824. Available from: http://pubs.acs.org/page/accacs/about.html. doi:10.1021/acscatal.5b01037.

Zeng, R., Tang, X., Deng, Y., Zhang, X., Li, H., Yin, W., Zhang, G., 2023. Design and optimization of solar energy system with hydrogen energy storage and alkaline fuel cell. Energy Convers. Manag. 01968904. 295. Available from: https://www.journals.elsevier.com/energy-conversion-and-management. doi:10.1016/j.enconman.2023.117628.

Zhang, C., Li, H., Zhang, X., Shen, M., Jin, X., 2024. The performance evaluation of a hybrid system combining an alkaline fuel cell with an inhomogeneous thermoelectric generator. Energies 19961073. 17 (9). Available from: http://www.mdpi.com/journal/energies/. doi:10.3390/en17092066.

Zhao, Q., Zhang, H., Hu, Z., Li, Y., 2021. An alkaline fuel cell/direct contact membrane distillation hybrid system for cogenerating electricity and freshwater. Energy 03605442. 225. Available from: https://www.journals.elsevier.com/energy. doi:10.1016/j.energy.2021.120303.

Zhou, G., Chen, L.D., Seaba, J.P., 2011a. CFD prediction of shunt currents present in alkaline fuel cells. J. Power Sources. 03787753. 196 (20), 8180–8187. doi:10.1016/j.jpowsour.2011.04.029.

Zhou, G., Chen, L.D., Seaba, J.P., 2011b. Effects of property variation and ideal solution assumption on the calculation of the limiting current density condition of alkaline fuel cells. J. Power Sources. 03787753. 196 (11), 4923–4933. doi:10.1016/j.jpowsour.2011.02.026.

Chapter 9

Alkaline fuel cells fueled by liquid fuels: Fuels, mechanisms, materials and perspectives

Yun Liu, Yao Zhu, and Liang An
Department of Mechanical Engineering, The Hong Kong Polytechnic University, Hung Hom, Hong Kong SAR, China

1 Introduction

Alkaline liquid fuel cells (ALFCs) represent an emerging class of electrochemical energy conversion technologies (Ong et al., 2017; Shaari et al., 2021). Unlike traditional hydrogen-oxygen fuel cell technologies, ALFCs primarily utilize liquid fuels, such as direct alcohol fuels like methanol and ethanol, to convert chemical energy into electrical energy at low or ambient temperatures. Examples include direct methanol fuel cells (DMFCs) and direct ammonia fuel cells (DAFCs) (Demirci, 2007). ALFCs are based on alkaline membranes, which facilitate electrochemical energy conversion through the conduction of OH^- ions. Compared to conventional liquid fuel cells, ALFCs offer several advantages (Tran et al., 2014): 1) Enhanced reaction kinetics: Under alkaline conditions, the kinetics of fuel oxidation and oxygen reduction reactions (ORRs) are accelerated compared to acidic conditions, reducing the overpotential and thereby increasing the overall voltage of the fuel cell (Lin et al., 2010). 2) Use of nonprecious metal catalysts: In alkaline environments, nonprecious metal catalysts exhibit significant electrochemical activity. Transition metal catalysts such as Fe, Co, and Ni can catalyze the oxidation of anode fuels or the reduction of oxygen at the cathode, significantly reducing catalyst costs (Pan et al., 2018). 3) Mitigated fuel crossover: ALFCs rely on OH^- conduction, where the direction of OH^- transfer within the membrane is from the cathode to the anode, opposite to the direction of liquid fuel transfer (anode to cathode). This theoretically reduces the rate of fuel diffusion (Li et al., 2025). 4) Versatile liquid fuel application: Unlike traditional fuel cells that primarily use hydrogen, ALFCs can utilize a variety of fuels, including direct alcohol fuels like methanol, ethanol, ethylene glycol, as well as emerging energy carriers such

Alkaline Fuel Cells. https://doi.org/10.1016/B978-0-443-36638-3.00017-6

as ammonia, hydrazine, and urea. Compared to gaseous hydrogen, liquid fuels are easier to store and transport, have higher volumetric energy density, and are widely applicable (Liu et al., 2022).

In recent years, significant advancements have been made in catalyst development, membrane engineering, and system integration for ALFCs. For instance, Pd/C catalysts have emerged as cost-effective alternatives to platinum, achieving high current densities (e.g., 450 mA/cm^2 for ethanol oxidation) and tolerance to intermediates like CO (Mabhulusa et al., 2024). Bimetallic Pd–Pd and Pd–Ag alloys have further enhanced activity through synergistic electronic effects (Pan et al., 2023). Beyond catalyst development, Tokuyama Corporation has developed anion exchange membranes (AEMs) that prevent carbonate formation, enabling prolonged ALFC operation by promoting OH^- transport and minimizing CO_2 poisoning (McLean et al., 2002). Recent advancements in quaternary ammonium-functionalized polymers, such as polyarylene ether sulfone, have improved ion conductivity (>100 mS/cm) and chemical stability (Wang et al., 2019). Research has expanded from methanol to ethanol, glycerol, formate esters, and ammonia to leverage their high energy density and renewability. For example, ALFCs using glycerol as fuel can achieve power densities of approximately 120 mW/cm^2, making them suitable for wastewater-to-energy applications (Fan et al., 2021). Additionally, Li et al. developed a hybrid ammonia fuel cell with a power density of approximately 600 mW/cm^{-2} (Wenzhi et al., 2024). Advances in stack design, including planar and tubular configurations, have also improved power density and scalability.

Despite significant progress, direct liquid fuel cells (DLFCs) still face challenges such as limited power density, relatively low energy conversion efficiency, and insufficient long-term stability. To enhance the understanding of liquid fuel cells and clarify future development directions, this section will use DMFCs and DAFCs as examples to provide a detailed overview of the working principles, key advancements, and internal mass and charge transfer mechanisms of ALFCs. Additionally, it will discuss future applications, thereby paving the way for the future development of ALFCs.

2 Liquid fuels

Liquid fuels, such as methanol, ethanol, ammonia, and sodium borohydride, offer distinct advantages over gaseous hydrogen, including higher volumetric energy density, safer and easier handling, and compatibility with existing fuel infrastructure (Speight, 2011). Liquid fuels also simplify system design by eliminating the need for external reformers. For example, DMFCs oxidize methanol directly at the anode, producing water and carbon dioxide as byproducts, while ammonia-fed AFCs generate nitrogen and water, minimizing environmental impact (Fukuzumi, 2017; Viswanathan and Scibioh, 2007). Furthermore, the alkaline environment enhances ORR kinetics at the cathode, enabling efficient operation at lower temperatures (20–90°C) compared to high-temperature solid oxide fuel cells.

A detailed comparison of different liquid fuels has been presented in Table 9.1, which includes the fuel, anode oxidation products, transferred electrons, theoretical potential, and energy density. It can be seen that the energy density of direct alcohol fuels, as well as ammonia, is higher than that of other fuels. For example, methanol (CH_3OH) is widely used due to its high energy density and ease of storage (Zhao et al., 2011). It is primarily synthesized from natural gas or biomass-derived syngas, though "green methanol" produced via CO_2 hydrogenation using renewable energy is gaining traction. While methanol oxidation in AFCs is efficient, challenges include catalyst poisoning by intermediates like formaldehyde and membrane degradation due to methanol

TABLE 9.1 Comparison of liquid fuels (Soloveichik, 2014).

Fuel	Anode products	Number of electrons	$E°$, V	Energy density (Wh/L, neat)	Energy density (Wh/L, solution)
Liquid H_2	H_2O*	2	1.23	2350	–
H_2 gas (70 MPa)	H_2O*	2	1.23	1300	–
Formic acid	$CO_2 + H_2O$	2	1.45	2103	1190 (10 M)
Formate	$CO_2 + H_2O$	2	1.45	–	145 (1 M)
Methanol	$CO_2 + H_2O$	8	1.17	5897	305 (2 M)
Ethanol	$CO_2 + H_2O$	12	1.14	6307	915 (3 M)
Ethanol	C_2H_4O	2	0.95	872	109 (3 M)
Ethylene glycol	$C_2H_2O_2$	4	0.87	1652	168 (1 M)
Ethylene glycol	$(COOH)_2$	8	1.09	4180	546 (2 M)
Ethylene glycol	$CO_2 + H_2O$	10	0.87	5800	168 (1 M)
2-Propanol	C_2H_5CHO	2	1.02	695	105 (2 M)
2-Propanol	$(CH_3)_2CO$	2	1.07	750	114 (2 M)
Glycerol	$CO_2 + H_2O$	14	1.21	5965	–
1,4-Butanediol	$C_4H_6O_2$	4	1.13	1361	–
2,4-Pentane-diol	$C_5H_8O_2$	4	1.27	1105	–
Cyclohexane	C_6H_6	6	1.06	1578	–
Decalin	$C_{10}H_8$	10	1.09	1893	–
Dodecahydro-N-ethylcarbazole	$C_{14}H_{13}N$	12	1.18	1715	–

Continued

TABLE 9.1 Comparison of liquid fuels—cont'd

Fuel	Anode products	Number of electrons	$E°$, V	Energy density (Wh/L, neat)	Energy density (Wh/L, solution)
Ammonia	N_2	3	1.17	–	1704 (35 wt %)
Ammonia borane	NH_4BO_2	6	1.62	–	655 (2 M)
Hydrazine hydrate	N_2	4	1.56	4269	873 (4 M)
Sodium borohydride	$NaBO_2$	8	1.64	–	2940 (30 wt %)

Data from Soloveichik, G.L., 2014. Liquid fuel cells. Beilstein Journal of Nanotechnology, 5(1), 1399–1418. https://doi.org/10.3762/bjnano.5.153.

crossover. Ethanol (C_2H_5OH) offers a safer, less toxic alternative with a higher energy density than methanol. It is predominantly produced from biomass (e.g., sugarcane, corn) or waste lignocellulose, aligning with circular economy principles. However, ethanol's incomplete oxidation to acetic acid reduces efficiency, necessitating advanced catalysts like palladium-nickel composites to promote complete conversion. Ammonia (NH_3) is a carbon-free fuel with high hydrogen content (17.6 wt%) (Kojima and Yamaguchi, 2022). It can be synthesized via the Haber-Bosch process using renewable hydrogen or through electrochemical pathways powered by solar/wind energy. Sodium borohydride ($NaBH_4$) provides exceptional energy density and rapid hydrogen release via hydrolysis. However, its high cost and byproduct management (e.g., sodium metaborate) limit scalability (Ma et al., 2010). Advances in recyclable catalysts and membrane-electrode assemblies have improved its viability for portable applications.

Sustainable fuel production is critical to promoting the future application of liquid fuels. Currently, the two most popular liquid fuels, i.e., methanol and ammonia, which are considered promising hydrogen carriers, can both be produced via a green process. Taking methanol as an example, it can be produced by combining CO_2 and water, powered by renewable electricity. The methanol obtained in this way is also known as "liquid sunshine." Regarding ammonia, it can also be produced via the improved Haber-Bosch process, which directly couples nitrogen and green hydrogen. The green hydrogen is derived directly from water electrolysis technology. Another perspective for promoting the application of liquid fuels is to reduce operational costs and improve storage and transportation technology. With these measures, liquid fuels could be a potential candidate for a future zero-carbon society.

3 Working principle

The working principle of ALFCs is similar to that of direct alkaline fuel cells, which convert the chemical energy stored in the liquid fuel into electrical energy for end-users. As shown in Fig. 9.1, taking ammonia as an example, the fuel cells consist of various key components, which can generally be divided into three parts. These mainly include the anode, cathode, and the anion exchange membrane. While the structure of the anode and cathode is the same, it features a dual-layer structure consisting of a catalyst layer (CL) and a diffusion layer. Regarding the CL, it mainly consists of catalyst powders for catalyzing chemical reactions, carbon support for enhancing the efficiency of electron transfer and increasing exposure of the catalyst active sites, and binder for creating the layer structure as well as providing the transfer path for ions such OH^-. The component next to the CL is the diffusion layer, whose main function is to deliver reactants and products in the fuel cells. To meet this function, the structure of the diffusion layer is mainly porous, designed with a specified size and porosity. In this case, fuel and oxidants can be delivered more efficiently, resulting in an even distribution at the CL. In addition to the above function, an

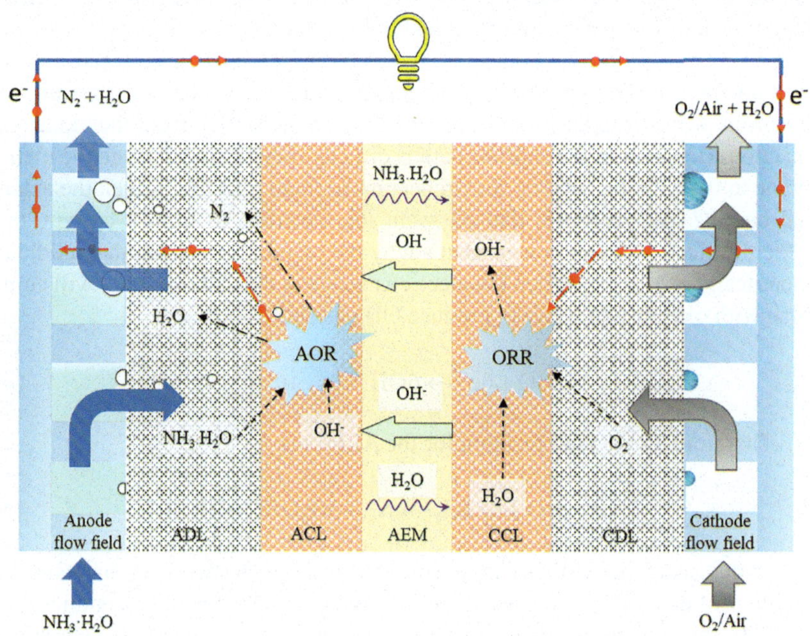

FIG. 9.1 The working principle of ALFCs (taking ammonia as the fuel, for example) (Liu et al., 2022). *(Reproduced with permission, Copyright © 2022, American Chemical Society. Data from Liu, Y., Pan, Z., Esan, O.C., Xu, X., An, L., 2022. Performance Characteristics of a Direct Ammonia Fuel Cell with an Anion Exchange Membrane. Energy Fuels, 36 (21), 13,203–13,211. https://doi.org/10.1021/acs.energyfuels.2c02951.)*

important role of the diffusion layer is to provide support for the CL. In terms of the peripheral components, which are named the flow field, these consist of ribs and flow channels. The main function of this component is to supply the reactants to the fuel cells and to remove products from the fuel cells in a timely manner. That's a brief introduction to the key components of the ALFCs. Next, to gain a better understanding of how the chemical energy stored in the fuel is transferred to electrical energy, a detailed description of the mass and charge transfer mechanism will be introduced (Zhao et al., 2009).

As demonstrated, the chemical energy transferred to electrical energy was realized through two combined electrochemical reactions that occurred at the anode and the cathode. At the anode, the fuel is oxidized; taking ammonia as an example, an electrochemical reaction occurs as follows:

$$2NH_3{\cdot}H_2O + 6OH^- \rightarrow N_2 + 8H_2O + 6e^- E_a^0 = -0.77V \text{ vs. SHE} \quad (1)$$

As demonstrated, the fuel (ammonia) will be oxidized to produce nitrogen, water, and release electrons. The produced electrons, of course, will transfer from the anode to the cathode through the external circuit. You may have two questions about this movement. First, why will the electrons move from the anode to the cathode? Second, why will the electrons move via the external circuit rather than being directly transferred to the cathode through the membrane? Answering the first question, the electrons want to move because there is a potential difference between the anode and the cathode. As a result, the electrons will be dragged to move when they are generated from the oxidation reaction. However, if the shortest path is directly transferring through the membrane to the cathode, why do researchers make them transfer through the external circuit? The reason is that we want the electrons to form a current and generate electricity and power. That's why the membrane used in the middle is electrically insulating. When the electrons arrive at the cathode, they will combine with oxygen and water to produce OH^-, as shown below:

$$3/2O_2 + 3H_2O + 6e^- \rightarrow 6OH^- E_c^0 = +0.40V \text{ vs. SHE} \quad (2)$$

Therefore, the overall reaction for the ALFCs is:

$$2NH_3{\cdot}H_2O + 3/2O_2 \rightarrow N_2 + 5H_2O E^0 = 1.17V \quad (3)$$

In this case, the ALFCs can generate electricity. However, to maintain the stability of electricity generation, the fuel and oxidants should be continuously fed into the fuel cell system to ensure the supply of reactants as well as the removal of products. The key parameters used to describe the performance of ALFCs include power densities, energy efficiency, and stability, which will be mentioned in the following section when the specified case study is introduced. Similar to traditional gas-fed fuel cells, fuel crossover will also occur.

This means that the fuel at the anode will be transferred to the cathode through the membrane. This crossover behavior leads to several effects, for example, cathode catalyst poisoning as well as cathode flooding. Meanwhile, two-phase flow behavior can also be observed at the anode since the fuel is liquid while the products could be gaseous, such as CO_2 and N_2. In addition, the wettability of the membrane and the effect of temperature on cell performance will change, which will be discussed in the following section.

4 Alkaline liquid fuel cells

4.1 Typical ALFCs

DLFCs utilize liquid fuels such as methanol, ethanol, urea, formate, and aqueous ammonia as anode reactants (see Table 9.2). These fuels share common advantages, including ease of storage and the ability to convert chemical

TABLE 9.2 An overview of direct liquid fuel cells (Ong et al., 2017).

Type of DLFC	Anode reaction	Cathode reaction	Overall reaction
DBFC	$BH_4^- + 8OH^- \rightarrow 8BO_2^- + H_2O + 8e^-$	$2O_2 + 4H_2O + 8e^- \rightarrow 8OH^-$	$BH_4^- + 2O_2 \rightarrow BO_2^- + 2H_2O$
DMFC	$(CH_3O)_2CH_2 + 4H_2O \rightarrow 3CO_2 + 16H^+ + 16e^-$	$4O_2 + 16H^+ + 16e^- \rightarrow 8H_2O$	$(CH_3O)_2CH_2 + 4O_2 \rightarrow 3CO_2 + 4H_2O$
DEFC	$(CH_3)_2O + 3H_2O \rightarrow 2CO_2 + 12H^+ + 12e^-$	$3O_2 + 12H^+ + 12e^- \rightarrow 6H_2O$	$(CH_3)_2O + 3O_2 \rightarrow 2CO_2 + 3H_2O$
DEFC	$C_2H_5OH + 3H_2O \rightarrow 2CO_2 + 12H^+ + 12e^-$	$3O_2 + 12H^+ + 12e^- \rightarrow 6H_2O$	$C_2H_5OH + 3O_2 \rightarrow 2CO_2 + 3H_2O$
DEGFC	$C_2H_6O_2 + 2H_2O \rightarrow 2CO_2 + 10H^+ + 10e^-$	$5/2O_2 + 10H^+ + 10e^- \rightarrow 5H_2O$	$C_2H_6O_2 + 5/2O_2 \rightarrow 2CO_2 + 3H_2O$
DFAFC	$HCOOH \rightarrow CO_2 + 2H^+ + 2e^-$	$1/2O_2 + 2H^+ + 2e^- \rightarrow H_2O$	$HCOOH + 1/2O_2 \rightarrow CO_2 + H_2O$
DHFC	$N_2H_4 \rightarrow N_2 + 4H^+ + 4e^-$	$O_2 + 4H^+ + 4e^- \rightarrow 2H_2O$	$N_2H_4 + O_2 \rightarrow N_2 + 2H_2O$
DAFC	$2NH_3 \cdot H_2O + 6OH^- \rightarrow N_2 + 8H_2O + 6e$	$3/2O_2 + 3H_2O + 6e^- \rightarrow 6OH^-$	$2NH_3 \cdot H_2O + 3/2O_2 \rightarrow N_2 + 5H_2O$
DMFC	$CH_3OH + H_2O \rightarrow CO_2 + 6H^+ + 6e^-$	$3/2O_2 + 6H^+ + 6e^- \rightarrow 3H_2O$	$CH_3OH + 3/2O_2 \rightarrow CO_2 + 2H_2O$
DGFC	$C_3H_8O_3 + 3H_2O \rightarrow 3CO_2 + 14H^+ + 14e^-$	$3/2O_2 + 6H^+ + 6e^- \rightarrow 3H_2O$	$C_3H_8O_3 + 7/2O_2 \rightarrow 3CO_2 + 4H_2O$

Data from Ong, B.C., Kamarudin, S.K., Basri, S., 2017. Direct liquid fuel cells: A review. Int. J. Hydrogen Energy, 42 (15), 10,142–10,157. https://doi.org/10.1016/j.ijhydene.2017.01.117.

energy into electrical energy under ambient temperature and pressure conditions. This section summarizes the fundamental operating principles of these fuel cell types. Among existing research and practical applications, the DMFCs remain the most extensively studied and applied variant. Additionally, recent advances in hydrogen storage carriers have driven growing interest in DAFCs. Consequently, this section will focus on DMFCs and DAFCs, encompassing their operating principles, historical development, innovations in materials and cell architectures, investigations into their mass and charge transport mechanisms, and discussions of their potential applications.

4.2 Direct alkaline alcohol fuel cells

4.2.1 Working principle

Direct alkaline alcohol fuel cells (DAAFCs) represent a subclass of low-temperature fuel cells that directly oxidize small organic alcohols in an alkaline environment to generate electricity. Unlike their acidic counterparts, such as proton exchange membrane fuel cells (PEMFCs), DAAFCs benefit from faster electrochemical kinetics and the ability to use nonprecious metal catalysts, significantly reducing overall costs.

Alcohols such as methanol, ethanol, and propanol are attractive fuels due to their high energy density, low toxicity, and ease of storage and transport. In an alkaline medium, both the alcohol oxidation at the anode and the ORR at the cathode proceed more readily. However, DAAFCs still face challenges, including alcohol crossover through the membrane, catalyst poisoning, and limited long-term membrane stability. Among the various DAAFCs, direct alkaline methanol fuel cells (DAMFCs) have received particular attention due to the favorable properties of methanol (Yaqoob et al., 2021; Alias et al., 2020).

DAMFCs operate based on the electrochemical oxidation of methanol in an alkaline medium, coupled with the ORR at the cathode (see Fig. 9.2). The typical cell configuration comprises an anode, a cathode, and an electrolyte—either a liquid alkaline solution (e.g., KOH) or a solid-state AEM—which facilitates the transport of hydroxide ions (OH^-) from the cathode to the anode (Yu et al., 2010).

At the anode, methanol undergoes catalytic oxidation in the presence of OH^- to generate carbon dioxide, water, and electrons. The corresponding half-reaction is:

$$CH_3OH + 6OH^- \rightarrow CO_2 + 5H_2O + 6e^- \tag{4}$$

At the cathode, oxygen is electrochemically reduced, consuming the electrons from the external circuit and producing hydroxide ions:

$$\frac{3}{2}O_2 + 3H_2O + 6e^- \rightarrow 6OH^- \tag{5}$$

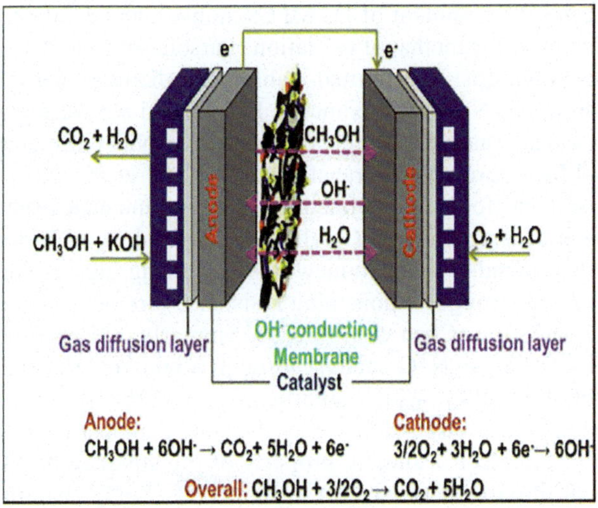

FIG. 9.2 Working principle of a DMFC single cell (Jamil et al., 2022). *(Reproduced with permission, © 2022 John Wiley & Sons, Ltd. Data from Jamil, Y.M.S., Awad, M.A.H., Al-Maydama, H.M.A., EL-Ghoul, Y., Al-Hakimi, A.N., 2022. Synthesis and study of enhanced electrochemical properties of NiO nanoparticles deposited on TiO2 nanotubes. Appl. Organometall. Chem., 36(9). https://doi.org/10.1002/aoc.6795.)*

The overall net cell reaction is:

$$CH_3OH + \frac{3}{2}O_2 \rightarrow CO_2 + 2H_2O \tag{6}$$

This process results in the direct conversion of chemical energy stored in methanol into electrical energy, with water and carbon dioxide as the primary byproducts.

Compared with acidic DMFCs, DAMFCs benefit from more favorable reaction kinetics in alkaline media, particularly for the ORR. Furthermore, the alkaline environment enables the use of nonprecious metal catalysts (e.g., Ni, Co, Ag)(Zhang et al., 2007), thereby enhancing the cost-effectiveness and sustainability of the technology. The selection of electrolyte and catalyst materials, as well as the optimization of operating conditions, critically influences the performance and long-term stability of DAMFC systems.

4.2.2 Catalyst material

The development of effective electrode and electrolyte materials has been pivotal to the advancement of DAMFCs. In the early stages, noble metal catalysts such as Pd and Pt were commonly employed due to their high catalytic activity toward methanol oxidation. However, the high cost and limited stability of these materials in alkaline environments necessitated the exploration of alternative, nonprecious metal-based options.

In the early development of DAMFCs, noble metal catalysts such as Pt were widely used for methanol oxidation (Parsons and VanderNoot, 1988), despite their high cost and limited stability in alkaline media, as demonstrated in Fig. 9.3. Nonprecious metals like Ni and Ag were also explored as lower-cost alternatives. Yu et al. reported a Ni@PdPt electrocatalyst synthesized by a three-phase-transfer method (Yu et al., 2015). In recent years, research has focused on bimetallic catalysts and nanostructured materials. Chen et al. prepared a NiCo alloy electrocatalyst on N-doped carbon nanofibers via metal-organic frameworks, achieving high methanol oxidation activity and strong CO poisoning resistance through surface electronic structure modulation (Chen et al., 2023). Similarly, Shi et al. developed a dual-metal site Fe/Co–N–C catalyst through a combined chemical-doping and adsorption strategy, which significantly improved the ORR catalytic activity (Shi et al., 2020). Additionally, Su et al. demonstrated the effective regulation of the electrocatalytic property of the commercial Pt/C catalyst by ultrasonic mixing with oxygen vacancy-rich CeO_2 quantum dots (Su et al., 2023).

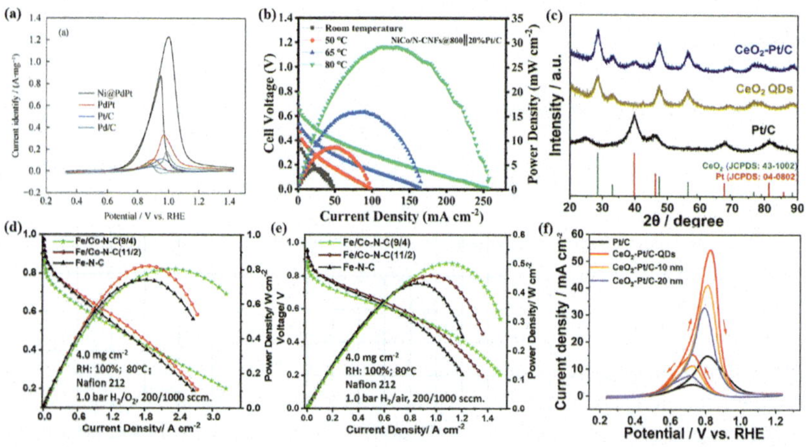

FIG. 9.3 (A) Cyclic voltammograms of methanol electrooxidation on Ni@PdPt and other commercial catalysts (Yu et al., 2015). (B) Cell voltage and power density polarization plots for NiCo/N-CNFs@800 (Chen et al., 2023). (C) XRD patterns of CeO_2-Pt/C, CeO_2, and Pt/C catalysts (Su et al., 2023). (D) H_2–O_2 and (E) H_2–air cells using two Fe/Co–N–C and the Fe–N–C catalysts (Shi et al., 2020). (F) CV curves for methanol oxidation by CeO_2-Pt/C catalysts with CeO_2 QDs, CeO_2-10 nm, and CeO_2-20 nm NPs (Su et al., 2023). *(Panel (A) Reproduced with permission, Copyright © 2015, University of Science and Technology Beijing and Springer-Verlag Berlin Heidelberg. Data from Yu, P.S., Liu, C.T., Feng, B., Wan, J.F., Li, L., Du, C.Y., 2015. Highly efficient anode catalyst with a Ni@PdPt core–shell nanostructure for methanol electrooxidation in alkaline media. Int. J. Miner. Metall. Mater., 22 (10), 1101–1107. https://doi.org/10.1007/s12613-015-1173-0. Panel (B) Reproduced with permission, © 2023 Elsevier Inc. Panel (C) Reproduced with permission, © 2025 Copyright Clearance Center, Inc. Panel (D, E and F) Reproduced with permission, Copyright © 2023, American Chemical Society.)*

4.2.3 Electrode

The electrode structure of DAMFCs plays a pivotal role in determining their electrochemical performance. Early configurations, borrowed from acidic fuel cell designs, typically consisted of CLs applied to gas diffusion layers, such as carbon paper, compressed into membrane electrode assemblies (MEAs). However, in alkaline environments, these designs often suffer from high interfacial resistance and inefficient three-phase boundary formation, as presented in Fig. 9.4.

Fan et al. reported a tailored synthesis of Pt nanorings decorated with surface-enriched single Bi atoms (SE-Bi_1/Pt NRs) for methanol oxidation, using PtBi intermetallic nanoplates, followed by etching and annealing (Fan et al., 2024). Liu et al. reported that 3D porous PdNi@Pt core-shell nanostructures (CSNSs) were facilely synthesized by first preparing the Pd–Ni alloy networks (Pd–Ni ANWs) core via a cyanogel-reduction method, followed by a galvanic displacement reaction to generate the Pt-rich shell (Liu et al., 2015).

4.2.4 Cell

The performance of DAMFCs at the cell level was initially constrained by issues such as fuel crossover, ion transport resistance, and nonoptimized structural

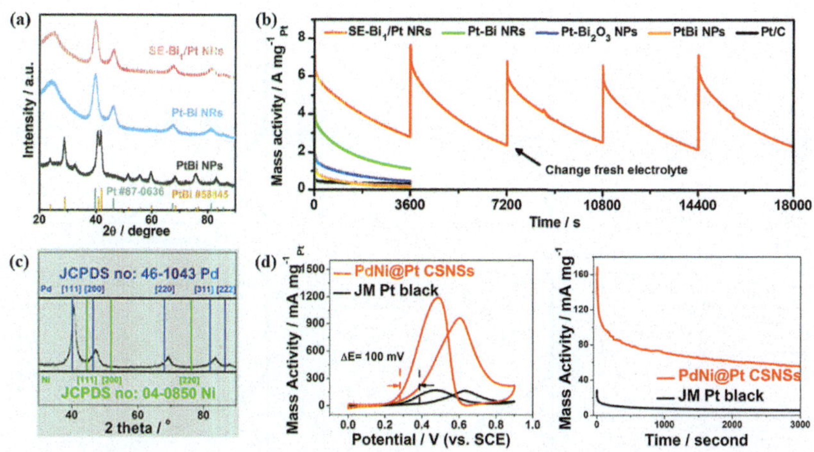

FIG. 9.4 (A) XRD patterns of the intermetallic PtBi NPs, Pt–Bi NRs, and SE-Bi1/Pt NRs. (B) Chronoamperometry of different electrocatalysts at − 0.4 V versus SCE. (C) XRD pattern of Pd–Ni ANWs. (D) The Pt metal mass-normalized cyclic voltammograms for PdNi@Pt CSNSs and Pt black, along with chronoamperometry curves for PdNi@Pt CSNSs and Pt black. *(Panel (B) (Fan, X., et al.) Reproduced with permission, © 2024 Wiley. Panel (D) (Liu, X., et al) Reproduced with permission, open access. Data from Fan, X., Chen, W., Xie, L., Liu, X., Ding, Y., Zhang, L., Tang, M., Liao, Y., Yang, Q., Fu, X.Z., Luo, S. & Luo, J.-L., 2024). Surface-Enriched Single-Bi-Atoms Tailoring of Pt Nanorings for Direct Methanol Fuel Cells with Ultralow-Pt-Loading. Adv. Mater., 36 (21). https://doi.org/10.1002/adma.202313179.)*

assembly. In early studies, liquid alkaline electrolytes were typically used, and no standardized MEA configuration had been established. For example, Scott et al. examined the effects of methanol concentrations and flow rates, finding that higher methanol levels increased reaction rates but also aggravated crossover losses, resulting in voltage drops (Scott et al., 2008).

With a better understanding of interfacial resistance and mass transport phenomena, recent work has focused on optimizing operational parameters and structural configurations to enhance cell performance (see Fig. 9.5). Okech et al. designed and fabricated three new cathode flow fields to improve the performance of the cell in terms of current density, power density, and byproduct water removal from the cathode (Okech et al., 2024). Galvan et al. showed that the cation structure of anion exchange ionomers (AEIs) significantly affects methanol

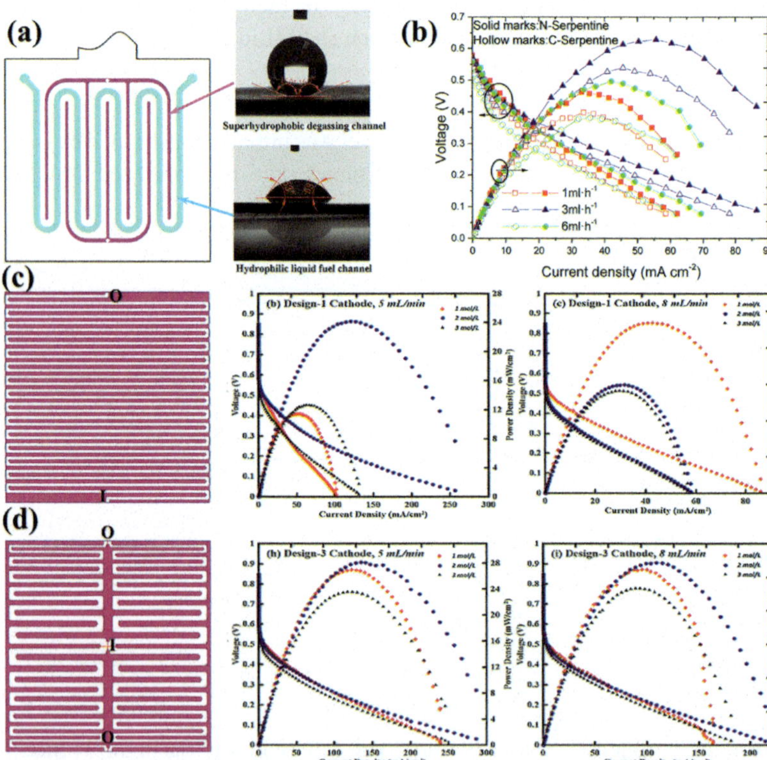

FIG. 9.5 (A) AFFs with N-serpentine (Liang et al., 2017). (B) Performance comparison for μDMFCs equipped with N-serpentine and C-serpentine AFFs (Liang et al., 2017), AFFs (Speight, 2011). (C) Performance of the fuel cell at different flow rates for Design-1 (C) and Design-3 (D) (Okech et al., 2024). *(Panel (C) Reproduced with permission, © 2017 Elsevier. Data from Liang, J,, Luo, Y,, Zheng, S., Wang, D., 2017. Enhance performance of micro direct methanol fuel cell by in situ CO 2 removal using novel anode flow field with superhydrophobic degassing channels. J. Power Sources, 351, 86–95. https://doi.org/10.1016/j.jpowsour.2017.03.099.)*

oxidation reaction (MOR) activity on PtRu/C, using a poly(terphenylene) (TPN) membrane in alkaline direct methanol fuel cell (ADMFC) half-cell tests (Galvan et al., 2021). Liang et al. proposed an innovative (Anode flow field) AFF design featuring a nested arrangement of hydrophilic fuel channels and superhydrophobic degassing channels, effectively removing most CO_2 before it enters the fuel channels (Liang et al., 2017).

4.2.5 System integration

Despite substantial progress in materials and electrode architecture, the practical deployment of DAMFCs still faces considerable engineering challenges at the system level. In early studies (see Fig. 9.6), DAMFCs were typically constructed as lab-scale single cells using open-loop fuel supply and recirculated liquid KOH electrolyte, which limited their stability, safety, and portability. In 2010, Zenich et al. developed a dynamic model of a DMFC system, highlighting that autonomous operation depends on environmental conditions and air excess ratio. However, water retention in the degasser and condenser distorts methanol concentration, impacting performance (Zenith and Krewer, 2010). Based on this finding, Na et al. proposed that a highly integrated system can minimize methanol losses by optimizing the concentration control to improve cell efficiency (Na et al., 2017). Wang et al. developed a bipolar passive DMFC stack for portable applications, showing that vertical cathode air channels provide more stable high-current performance. The stack powered a fan for 100 days, though some performance degradation was observed (Wang et al., 2018).

4.2.6 Application and perspectives

As a low-temperature, liquid-fueled power source, DAMFCs have long been considered promising candidates for portable energy systems. Early application scenarios were envisioned in military equipment, handheld communication devices, and emergency power supplies. More recently, with advances in

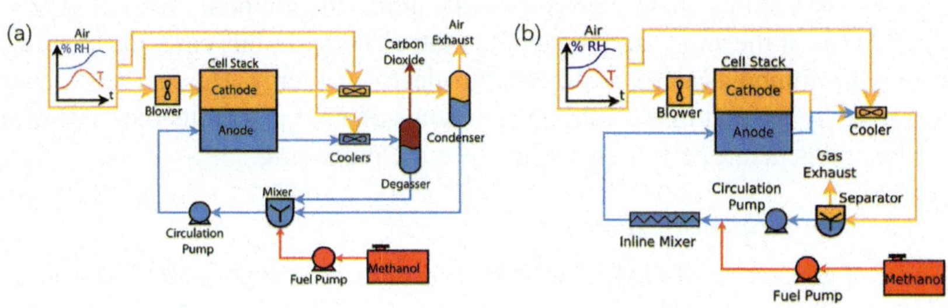

FIG. 9.6 The process scheme of (a) the reference system (Zenith and Krewer, 2010) and (b) the mingled-outlet system (Na et al., 2017). *(Part (A) Reproduced with permission, Copyright © 2010 Elsevier. Data from Zenith, F., Krewer, U., 2010. Modelling, dynamics and control of a portable DMFC system. J. Process Control, 20 (5), 630–642. https://doi.org/10.1016/j.jprocont.2010.02.014. Part (B) Reproduced with permission, © 2017 Elsevier.)*

miniaturization, system integration, and thermal management, DAMFCs have been demonstrated in real-world platforms. As a liquid-fueled, low-temperature fuel cell system, DMFCs offer great potential in portable electronics, electric mobility, and backup power systems due to their high energy density, simple operation, and compact form. Early industrial deployments of DMFCs focused on portable electronic devices. The German company Smart Fuel Cell AG launched the C25 system, which delivers 25 W of power for over 7 hours and uses lightweight methanol cartridges for rapid refueling (Osman et al., 2025). Commercial trials by Toshiba, Samsung, and Panasonic included DMFCs in notebooks, mobile players, and handheld chargers. In the transportation sector, DMFCs have been integrated into light electric vehicles, forklifts, and military equipment. The Jülich Forschungszentrum developed a 1.3 kW DMFC module for hybrid scooters, achieving significant platinum reduction and lightweight design via graphite stack components. Moreover, the advancement of micro-DMFCs (μ-DMFCs) has enabled their application in wearables and medical electronics. A system developed by the Danish Technological Institute for hearing aids operated continuously for 24 hours using just 200 mL of methanol, while achieving weight reduction compared to traditional batteries. In 2014, Kimiaie et al. conducted a 20,000-hour lifetime test of a 7 kW hybrid DMFC system for forklift applications. The system, combining a 1 kW DMFC stack and a 45 Ah lithium-ion battery, showed strong durability under dynamic load conditions, with a stack degradation rate of 8.5 mV h^{-1} and battery state of health above 93% after 5000 cycles, demonstrating its commercial viability (Kimiaie et al., 2014).

4.3 Direct alkaline ammonia fuel cells

4.3.1 Working principle

DAFCs fueled by liquid ammonia have recently attracted the attention of researchers because they can directly convert the chemical energy in ammonia into electrical energy at low temperatures (Liu et al., 2022). The operating principle of the DAFCs mainly consists of coupling the ammonia oxidation reaction (AOR) at the anode and the ORR at the cathode. In this case, the chemical energy in ammonia can be converted into electrical energy. The fuel cell is characterized by the fact that combining ammonia and oxygen emits only nitrogen, water, and generates electricity, which is quite promising.

4.3.2 Development history

The first generation of DAFCs was developed in 1968 by researchers from the General Electric Company, in which ammonia was directly used as the fuel with aqueous potassium hydroxide as the electrolyte (Gabor, 1968). Following that, a series of research works have been carried out, mainly focusing on the application of precious metal-based catalysts for the efficient oxidation of ammonia, especially platinum-based catalysts. For example, Silva et al. (Silva et al.,

2015) applied a PtAu/C catalyst as the ammonia oxidation catalyst at the anode, achieving a peak power density of 2.64 MW cm^{-2} and an open-circuit voltage of 0.588 V. Meanwhile, the PtRh/C catalyst was also tested by Assumpção et al. (Assumpção et al., 2015), which achieved a slight improvement, demonstrating a peak power density of 5.37 MW cm^{-2} and an open-circuit voltage of 0.68 V.

Despite the mentioned progress, the achieved performance for the DAFCs remains quite limited. Luckily, a performance breakthrough for the DAFCs was realized in 2019, carried out by the research group led by Professor Yushan Yan at the University of Delaware (Zhao et al., 2019). They achieved a higher power density of 135 mW cm^{-2} with a high-conductivity ion exchange membrane (see Fig. 9.7). Following that, another performance breakthrough was achieved by the same group, which attained a performance of 314 mW cm^{-2} by using an advanced catalyst (PtIrZn/CeO_2-ZIF-8) to promote the AOR (Li et al., 2021). In addition, by further optimizing the operating conditions, the fuel cell performance was improved to 410 mW cm^{-2} (Wang et al., 2022). Recently, another performance record for the DAFCs was achieved by our research group. Through the application of H_2O_2 to replace oxygen reduction at the cathode, the highest performance was achieved with a power density of 600 mW cm^{-2} (Wenzhi et al., 2024).

Regarding the nonprecious catalyst development, Professor Tao from the University of Warwick has conducted a series of works (Zhang et al., 2022a, 2021; Jeerh et al., 2022a, 2022b). For example, they developed a bifunctional catalyst ($Ni_4Cu_5Fe_x$/C) for both the AOR and the ORR toward DAFCs, which demonstrated a power density of 8.9 mW cm^{-2} and an open circuit voltage of 0.62 V (Zhang et al., 2022b). Besides, they designed and fabricated a cathode CL using perovskite oxide as the catalyst, which achieved performance improvement with a peak power density of 30.1 mW cm^{-2} and a current density of 317 mA cm^{-2} (Jeerh et al., 2023).

4.3.3 Mass and transport behaviours

As previously discussed, the performance of fuel cells is primarily determined by the reaction rate at individual active sites within the catalytic layer and the number of triple-phase boundaries, under the precondition of fixed material compositions and structural parameters (Wenzhi et al., 2024). The quantity of triple-phase reaction sites depends on the physical structure of the CL, while the reaction rate at individual sites is governed by the transport kinetics of the triple-phases—namely, reactants, ions, and electrons—at the reaction interface. For electrochemical reactions, all three phases must simultaneously reach the reaction site. Consequently, the overall reaction rate at a three-phase boundary is limited by the slowest transport process among the three phases, analogous to the "bucket theory," where the water-holding capacity of a barrel is dictated by its shortest stave. For DAFCs, the anode three-phase boundaries involve solid-phase electron transport, liquid-phase reactant (ammonia) and ion (OH^-) transport, and gas-phase nitrogen transport. Sustaining continuous reactions at these

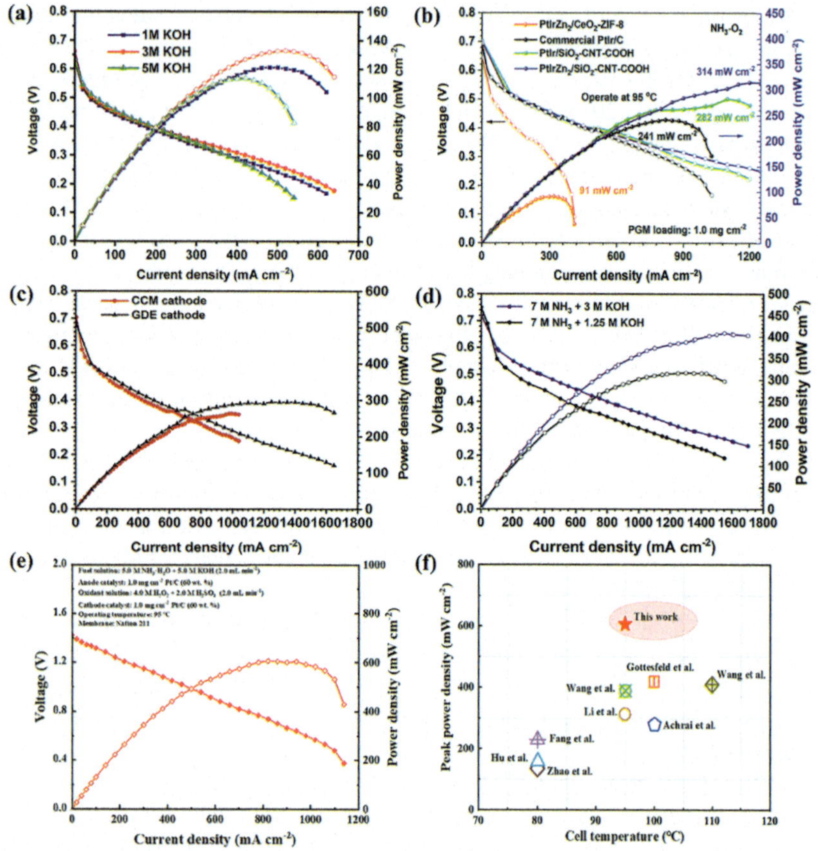

FIG. 9.7 Development trends of DAFCs: (A) membrane innovation (Zhao et al., 2019), (B) catalyst innovation (Li et al., 2021), (C, D) operating condition optimization, and (E, F) system innovation (Li et al., 2021; Wang et al., 2022; Wenzhi et al., 2024). *(Panel (A) Reproduced with permission, © 2019 Elsevier Inc. Panel (B) Reproduced with permission, Royal Society of Chemistry (RSC). Panel (C,D) Reproduced with permission, © 2022 The Authors1. Panel (E,F) Reproduced with permission, © 2023 Elsevier B.V. Data from Zhao, Y., Setzler, B.P., Wang, J., Nash, J., Wang, T., Xu, B., Yan, Y., 2019. An Efficient Direct Ammonia Fuel Cell for Affordable Carbon-Neutral Transportation. Joule, 3 (10), 2472–2484. https://doi.org/10.1016/j.joule.2019.07.005.)*

interfaces requires maintaining dynamic equilibrium in the transport rates of all three phases. Therefore, systematic investigation of mass and charge transport mechanisms across fuel cell components is critical for optimizing performance. Guided by this theoretical framework, our research group has focused on the following aspects: (1) *In situ* observation of two-phase flow behavior in DAFC anodes and its impact on cell performance; and (2) characterization of ammonia and water transport properties in AEMs for DAFCs.

In DAFC anode flow fields, a two-phase flow system theoretically exists, comprising a liquid-phase ammonia-KOH solution and gas-phase nitrogen (see Fig. 9.8). Experimental observations, however, revealed that the gas phase contains not only product nitrogen but also reactant ammonia due to its phase transition from liquid to gas at elevated operating temperatures. Furthermore, periodic bubble nucleation and removal were observed in the anode flow field. This phenomenon arises from progressive bubble accumulation and coalescence into elongated gas slugs. Delayed removal of these slugs induces reactant starvation, leading to concentration polarization. Our experimental results demonstrate that maintaining a fuel supply rate exceeding 1.0 mL min^{-1} effectively mitigates this issue by ensuring timely bubble expulsion (Liu et al., 2024).

Similar to PEMFCs, cathode water flooding remains a challenge in DAFCs due to the ORR. Specifically, oxygen mass transport is impeded by liquid-phase species (ammonia and water) crossing over from the anode through the AEM. To address this, we developed a novel experimental setup and theoretical model for quantifying trace species crossover. Our findings indicate that ammonia crossover reduces the ORR potential by approximately 150 mV. Additionally, operating conditions significantly influence membrane transport properties: increasing the temperature from 80°C to 90°C enhanced fuel efficiency by 33% while reducing ammonia crossover by 43% (Wenzhi et al., 2025). Membrane thickness also critically affects ammonia and water permeability. For instance, increasing AEM thickness from 13 μm to 60 μm decreased ammonia and water crossover rates by 52% and 19%, respectively.

4.3.4 The applications

The primary application of ammonia fuel cells lies in converting the chemical energy of ammonia into electrical energy. To enhance fuel cell output power and current density, Zhao et al. developed a compact DAFC stack with a designed power density of 75 W and an active area of 50 cm^2 per cell (see Fig. 9.9). Experimental results demonstrated that the developed stack operated stably for 10 hours under a total current of 15 A at 90°C (Wang et al., 2022). Beyond power generation, ammonia fuel cells exhibit potential for wastewater treatment by leveraging the AOR at the anode to degrade ammonia contaminants. Tao et al. implemented an ammonia electrolyzer employing a $La_{0.9}Ni_{0.6}Cu_{0.35}Fe_{0.05}O_{3-\delta}$ anode catalyst, achieving approximately 95.4% ammonia removal from simulated wastewater after 80 hours of continuous operation, a significant advancement for water quality improvement (Jeerh et al., 2022a). Furthermore, our research team proposed an open-cathode DAFC design (Liu et al., 2023). This configuration enables direct utilization of atmospheric oxygen and employs single fuel injection, eliminating the need for complex reactant supply systems, which demonstrates promising applicability for powering portable electronic devices.

4.3.5 Challenges

While significant progress has been made in DAFCs, their power density and current density remain inferior to those of alkaline hydrogen-oxygen fuel cells.

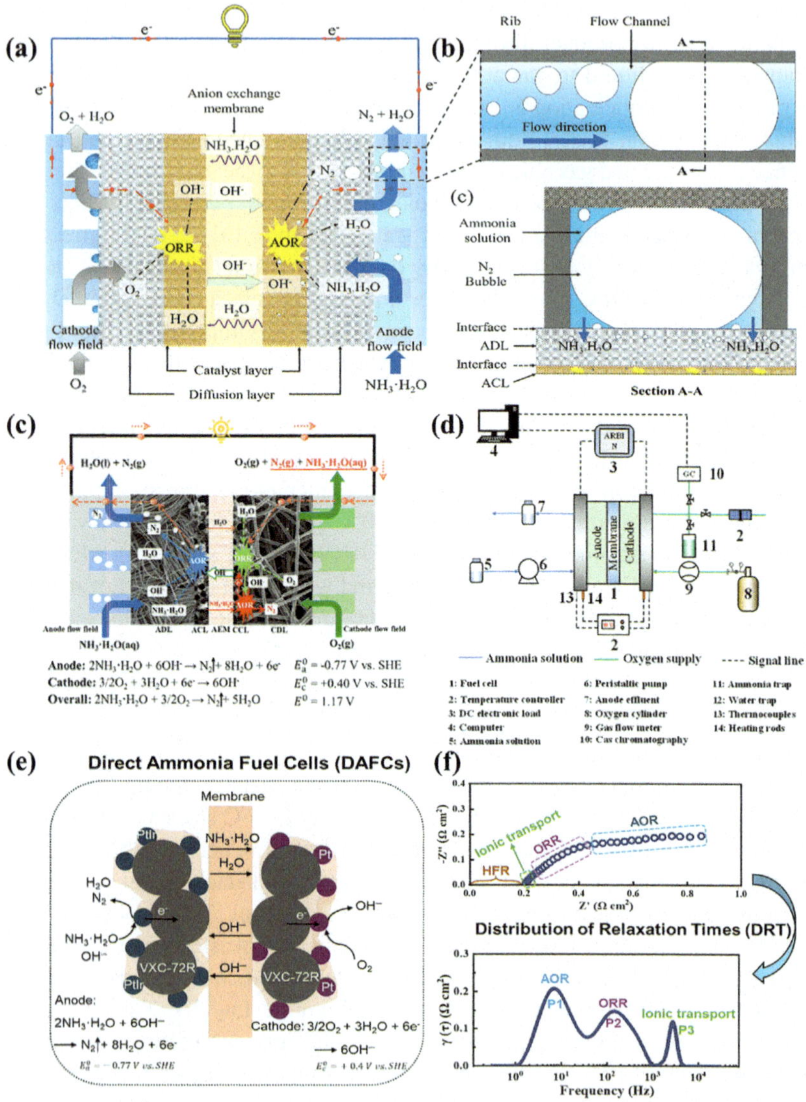

FIG. 9.8 Mass and charge transport in DAFCs: (A, B) two-phase flow (Liu et al., 2024); (C, D) ammonia and water transport behavior in the membrane (Wenzhi et al., 2025); (E, F) degradation analysis (Li et al., 2025). *(Panel (A,B) Reproduced with permission, © 2024 Hydrogen Energy Publications LLC. Published by Elsevier Ltd Panel (C,D) Reproduced with permission, © 2024 Elsevier B.V. Panel (E,F) Reproduced with permission, © 2025 The Author(s). Data from Liu, Y., Pan, Z., Huo, X., Li, W., Shi, X., Chen, R., An, L., 2024. Two-phase flow visualization in direct ammonia fuel cells. Int. J. Hydrogen Energy, 70, 159–169. https://doi.org/10.1016/j.ijhydene.2024.05.158.)*

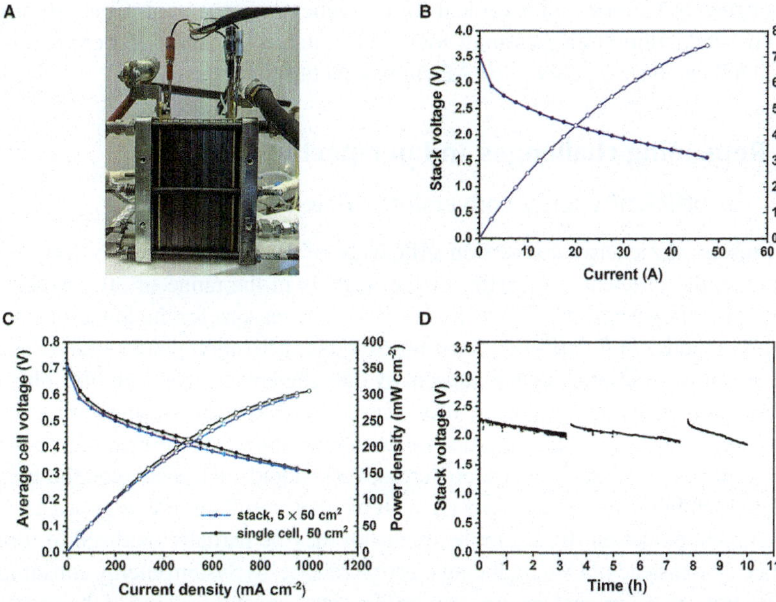

FIG. 9.9 DAFC stack (Wang et al., 2022). *(Reproduced with permission, © 2022 The Authors. From Wang, T., Zhao, Y., Setzler, B.P., Abbasi, R., Gottesfeld, S., Yan, Y., 2022. A high-performance 75 W direct ammonia fuel cell stack. Cell Rep. Phys. Sci., 3 (4). https://doi.org/10.1016/j.xcrp.2022.100829.)*

Current limitations and future challenges for DAFC performance enhancement primarily include:

1. Catalyst poisoning in the anodic AOR: Nitrogen-containing intermediates generated during AOR exhibit strong adsorption on catalyst surfaces, leading to irreversible catalyst deactivation and subsequent performance degradation. Mitigation strategies involve developing poisoning-resistant catalysts and implementing advanced antipoisoning protocols, such as pulsed voltage application, to facilitate intermediate desorption.
2. Operational instability of DAFCs: Catalyst poisoning not only reduces power output but also destabilizes long-term operation. Furthermore, reactant phase transitions (e.g., ammonia vaporization) induced by elevated temperatures exacerbate system instability. Addressing these issues requires optimization of reaction conditions and the development of high-temperature-tolerant AEMs to suppress two-phase flow formation.
3. Low energy conversion efficiency and high costs: Current DAFCs exhibit energy conversion efficiencies below 30%. Improving reaction kinetics through material structural optimization (e.g., nanostructured catalysts,

engineered interfaces) is critical for efficiency enhancement. Concurrently, cost-reduction strategies may focus on scalable synthesis of nonprecious metal catalysts and low-cost AEM manufacturing processes.

5 Remaining challenges and perspectives

5.1 Insufficient energy conversion efficiency

At present, the energy conversion efficiency of DLFCs is generally low. For example, the efficiency of methanol fuel cells is in the range of 30% to 45%, while the efficiency of DAFCs is below 30%. The reasons for the low efficiency mainly include: 1) Polarization loss: for example, activation polarization occurs due to slow reaction kinetics caused by insufficient catalyst activity, ohmic polarization arises due to voltage drop caused by the electrolyte membrane and contact resistance, and concentration polarization results from limited fuel or oxidant transport. 2) Fuel penetration (crossover): liquid fuel penetrates the electrolyte membrane and reacts directly with oxygen, resulting in energy waste. 3) Incomplete oxidation: for example, methanol may be partially oxidized to form formic acid instead of CO_2, reducing electron release. 4) System energy consumption: auxiliary equipment (pumps, controllers, etc.) consumes some of the energy.

5.2 Insufficient system stability

To improve the system stability of DLFCs, it is necessary to optimize from the multidimensional aspects of material durability, fuel management, thermodynamic control, and system integration. (1) Material level: Enhance catalyst durability through antipoisoning catalyst design and structural stability optimization. Examples include the use of Pt–Ru alloy catalysts to inhibit poisoning by CO intermediates and the use of core-shell catalysts (e.g., Pt shells over nonprecious metal cores) to reduce precious metal agglomeration and loss. (2) Chemical and mechanical stability of membrane materials: Develop anion exchange membranes that are resistant to swelling and aging, as well as employ plasma treatment or fluorination modification of membrane surfaces to enhance oxidation resistance. (3) Anticorrosion of bipolar plates and flow field structure: adopt titanium alloy plating or graphene composite bipolar plates to replace traditional graphite or metal, thereby reducing the corrosion rate. (4) Optimization of fuel and reactant management: Optimize the material management of the battery system, inhibit fuel infiltration, and balance hydrothermal management. (5) System control and operation strategy: apply artificial intelligence algorithms to optimize the reaction system and provide early warning of failures.

5.3 High cost of fuel cell system

Key strategies for cost reduction include: 1. Reduce precious metal catalyst loading: e.g., improve precious metal utilization through nano-structural design (e.g., nanowires, porous structure) and reduce platinum loading per unit of power (target: <0.1 mg/cm^2). 2. Develop low-cost membrane materials: develop polyaromatic hydrocarbon-based membranes (e.g. sulphonated polysulphone) to replace Nafion membranes, and the cost can be reduced by >50%). Optimize the component manufacturing process: e.g., use roll-to-roll manufacturing processes to optimize fuel cell system performance. Component manufacturing processes include, for example, continuous coating of CLs and membrane electrodes using the roll-to-roll process to reduce labor and energy costs.4. Simplification of system modules: e.g., integration of fuel storage, battery stack, and control system to reduce the complexity of piping and interfaces.5. Fuel selection and supply chain optimization: e.g., low-cost fuel adaptation and high fuel concentration.

5.4 Special application scenarios

With its high energy density, convenient fuel storage and transportation, and environmental adaptability, DLFCs show unique advantages in the following special application scenarios: (1) Military and field combat Advantages include strong concealment, long endurance, and fast fuel replenishment. For example, the U.S. Army has developed portable DMFC power supplies to provide power to soldiers' communication equipment and night vision devices. (2) Long-endurance UAVs and underwater equipment: dvantages include high energy density and fuel flexibility. For example, Israel Aeronautics' Orbiter 4 UAV is equipped with DMFC to achieve over 24 hours of endurance. (3) Off-grid power supply in remote areas: Advantages include convenient fuel storage and transportation, as well as modular expansion. For example, methanol fuel cell microgrids are used in rural areas of Africa to replace diesel generators, reducing carbon emissions and fuel costs. (4) Marine and ship power: Advantages include fuel compatibility and high space utilization. For example, the Viking Energy, a supply vessel of the Norwegian company Eidesvik Offshore, is testing an ammonia fuel cell hybrid power system. (5) Emergency disaster relief and mobile healthcare: Advantages include rapid deployment and stable power supply. For example, methanol fuel cell mobile charging stations were used to power cell phones and first aid equipment in earthquake-stricken areas of Japan. (6) Low temperature and extreme environments: Advantages include low-temperature adaptability and no risk of icing. For example, methanol fuel cell heating systems are used in Arctic research stations, supplying power and heat simultaneously. They are also used to power high-altitude border posts, adapting to environments with large temperature differences between day and night and oxygen scarcity. 7. Special transportation: Advantages include silent operation and no exhaust pollution. For example, DMFC-driven sightseeing cars in Disneyland achieve zero emissions and low noise. 8. Space exploration and satellites: Strengths include high specific energy and fuel reuse. NASA's research on methanol fuel cell lunar rovers utilizes in situ resources on the lunar surface to prepare methanol (CO_2 + hydrogen).

6 Concluding remarks

ALFCs have undergone significant advancements over the past few decades. Compared to conventional gas-fed fuel cells, ALFCs demonstrate operational advantages under ambient temperature and pressure conditions. The utilization of nonprecious metal catalysts in alkaline environments has further reduced their application costs. This chapter initially examines the characteristic properties of various liquid fuels, followed by a comprehensive analysis of the operational principles and mass/charge transport mechanisms in liquid fuel cells. Furthermore, using DMFCs and DAFCs as representative examples, it elaborates on the developmental trajectory of ALFCs, encompassing material development, device design, and system innovation. Critical technological breakthroughs are systematically reviewed, along with an in-depth discussion of existing challenges and potential applications in future carbon-neutral societies. The analysis concludes with a perspective on the pivotal role of DLFCs in enabling sustainable energy development, particularly through their integration into decarbonized energy systems. It is substantiated that these systems will serve as essential components in achieving carbon neutrality objectives, offering scalable solutions for renewable energy storage and conversion.

References

Alias, M.S., Kamarudin, S.K., Zainoodin, A.M., Masdar, M.S., 2020. Active direct methanol fuel cell: an overview. Int. J. Hydrogen Energy. 03603199. 45 (38), 19620–19641. doi:10.1016/j.ijhydene.2020.04.202.

Assumpção, M.H.M.T., Piasentin, R.M., Hammer, P., De Souza, R.F.B., Buzzo, G.S., Santos, M.C., Spinacé, E.V., Neto, A.O., Silva, J.C.M., 2015. Oxidation of ammonia using PtRh/C electrocatalysts: fuel cell and electrochemical evaluation. Appl. Catal. B: Environ. 09263373. 174-175, 136–144. Available from: www.elsevier.com/inca/publications/store/5/2/3/0/6/6/index.htt. doi:10.1016/j.apcatb.2015.02.021.

Chen, F., Guo, S., Shuyan, Y., Zhang, C., Guo, M., Congju, L., 2023. Hierarchical N-doped carbon nanofiber-loaded NiCo alloy nanocrystals with enhanced methanol electrooxidation for alkaline direct methanol fuel cells. J. Colloid Interface Sci. 00219797. 646, 43–53. doi:10.1016/j.jcis.2023.05.030.

Demirci, U.B., 2007. Direct liquid-feed fuel cells: thermodynamic and environmental concerns. J. Power Sources. 03787753. 169 (2), 239–246. doi:10.1016/j.jpowsour.2007.03.050.

Fan, L., Liu, B., Liu, X., Senthilkumar, N., Wang, G., Wen, Z., 2021. Recent Progress in Electrocatalytic Glycerol Oxidation. Energy Technol. 2194-4288. 9 (2), doi:10.1002/ente.202000804.

Fan, X., Chen, W., Xie, L., Liu, X., Ding, Y., Zhang, L., Tang, M., Liao, Y., Yang, Q., Xian-Zhu, F., Luo, S., Luo, J., 2024. Surface-Enriched Single-Bi-Atoms Tailoring of Pt Nanorings for Direct Methanol Fuel Cells with Ultralow-Pt-Loading. Adv. Mater. 0935-9648. 36 (21), doi:10.1002/adma.202313179.

Fukuzumi, S., 2017. Production of Liquid Solar Fuels and Their Use in Fuel Cells. Joule. 25424351. 1 (4), 689–738. doi:10.1016/j.joule.2017.07.007.

Gabor, D., 1968. Holographic model of temporal recall. Nature. 00280836. 217 (5128), 584. doi:10.1038/217584a0.

Galvan, V., Shrimant, B., Bae, C., Prakash, G.K.S., 2021. Ionomer Significance in Alkaline Direct Methanol Fuel Cell to Achieve High Power with a Quarternized Poly(terphenylene) Membrane. ACS Appl. Energy Mater. 25740962. Available from: https://pubs.acs.org/journal/aaemcq. doi:10.1021/acsaem.1c00681.

Jamil, Y.M.S., Awad, M.A.H., Al-Maydama, H.M.A., EL-Ghoul, Y., Al-Hakimi, A.N., 2022. Synthesis and study of enhanced electrochemical properties of NiO nanoparticles deposited on TiO2 nanotubes. Appl. Organomet. Chem. 10990739. 36 (9). Available from: http://onlinelibrary.wiley.com/journal/10.1002/(ISSN)1099-0739. doi:10.1002/aoc.6795.

Jeerh, G., Zou, P., Zhang, M., Chen, S., Humphreys, J., Tao, S., 2022a. Electrooxidation of ammonia on A-site deficient perovskite oxide La0.9Ni0.6Cu0.35Fe0.05O3-δ for wastewater treatment. Sep. Purif. Technol. 13835866. 297, 121451. doi:10.1016/j.seppur.2022.121451.

Jeerh, G., Zou, P., Zhang, M., Tao, S., 2022b. Perovskite oxide LaCr0.25Fe0.25Co0.5O3-δ as an efficient non-noble cathode for direct ammonia fuel cells. Appl. Catal. B: Environ. 09263373. 319, 121919. doi:10.1016/j.apcatb.2022.121919.

Jeerh, G., Zou, P., Zhang, M., Tao, S., 2023. Optimization of a Perovskite Oxide-Based Cathode Catalyst Layer on Performance of Direct Ammonia Fuel Cells. ACS Appl. Mater. Interfaces. 1944-8244. 15 (1), 1029–1041. doi:10.1021/acsami.2c17253.

Kimiaie, N., Wedlich, K., Hehemann, M., Lambertz, R., Müller, M., Korte, C., Stolten, D., 2014. Results of a 20 000 h lifetime test of a 7 kW direct methanol fuel cell (DMFC) hybrid system-degradation of the DMFC stack and the energy storage. Energy Environ. Sci. 17545706. 7 (9), 3013–3025. Available from: http://www.rsc.org/Publishing/Journals/EE/About.asp. doi:10.1039/c4ee00749b.

Kojima, Y., Yamaguchi, M., 2022. Ammonia as a hydrogen energy carrier. Int. J. Hydrogen Energy. 03603199. 47 (54), 22832–22839. doi:10.1016/j.ijhydene.2022.05.096.

Li, W., Wang, Q., Liu, Y., Lizhen, W., Pan, Z., Chen, R., Liang, A., 2025. Unravelling degradation mechanisms of anion exchange membrane direct ammonia fuel cells via distribution of relaxation times. Chem. Eng. J. 13858947. 512, 162204. doi:10.1016/j.cej.2025.162204.

Li, Y., Pillai, H.S., Wang, T., Hwang, S., Zhao, Y., Qiao, Z., Mu, Q., Karakalos, S., Chen, M., Yang, J., Su, D., Xin, H., Yan, Y., Wu, G., 2021. High-performance ammonia oxidation catalysts for anion-exchange membrane direct ammonia fuel cells. Energy Environ. Sci. 17545706. 14 (3), 1449–1460. Available from http://pubs.rsc.org/en/journals/journal/ee. doi:10.1039/d0ee03351k.

Liang, J., Luo, Y., Zheng, S., Wang, D., 2017. Enhance performance of micro direct methanol fuel cell by in situ CO 2 removal using novel anode flow field with superhydrophobic degassing channels. J. Power Sources. 03787753. 351, 86–95. doi:10.1016/j.jpowsour.2017.03.099.

Lin, B., Qiu, L., Jianmei, L., Yan, F., 2010. Cross-Linked Alkaline Ionic Liquid-Based Polymer Electrolytes for Alkaline Fuel Cell Applications. Chem. Mater. 0897-4756. 22 (24), 6718–6725. doi:10.1021/cm102957g.

Liu, X., Xu, G., Chen, Y., Lu, T., Tang, Y., Xing, W., 2015. A Strategy for Fabricating Porous PdNi@Pt Core-shell Nanostructures and Their Enhanced Activity and Durability for the Methanol Electrooxidation. Sci. Rep. 20452322. 5. Available from: www.nature.com/srep/index.html. doi:10.1038/srep07619.

Liu, Y., Pan, Z., Esan, O.C., Huo, X., Shi, X., An, L., 2023. Development and performance evaluation of a passive direct ammonia fuel cell. J. Power Sources. 03787753. 570, 233057. doi:10.1016/j.jpowsour.2023.233057.

Liu, Y., Pan, Z., Esan, O.C., Xinhai, X., Liang, A., 2022. Performance Characteristics of a Direct Ammonia Fuel Cell with an Anion Exchange Membrane. Energy & Fuels. 0887-0624. 36 (21), 13203–13211. doi:10.1021/acs.energyfuels.2c02951.

Liu, Y., Pan, Z., Huo, X., Wenzhi, L., Shi, X., Chen, R., Liang, A., 2024. Two-phase flow visualization in direct ammonia fuel cells. Int. J. Hydrogen Energy. 03603199. 70, 159–169. doi:10.1016/j.ijhydene.2024.05.158.

Ma, J., Choudhury, N.A., Sahai, Y., 2010. A comprehensive review of direct borohydride fuel cells. Renew. Sustain. Energy Rev. 13640321. 14 (1), 183–199. doi:10.1016/j.rser.2009.08.002.

Mabhulusa, W., Sekhosana, K.E., Fuku, X., 2024. The impact and performance of carbon-supported platinum group metal electrocatalysts for fuel cells. Int. J. Electrochem. Sci. 14523981. 19 (4), 100524. doi:10.1016/j.ijoes.2024.100524.

McLean, G.F., Niet, T., Prince-Richard, S., D, N., 2002. An assessment of AFC technology. Int. J. Hydrogen Energy 27 (5), 507–526.

Na, Y., Zenith, F., Krewer, U., 2017. Highly integrated direct methanol fuel cell systems minimizing fuel loss with dynamic concentration control for portable applications. J. Process Control. 09591524. 57, 140–147. doi:10.1016/j.jprocont.2017.06.014.

Okech, G., Emam, M., Mori, S., Ahmed, M., 2024. Enhancing the performance of direct methanol fuel cells using new cathode flow field designs: an experimental investigation. Int. J. Hydrogen Energy. 03603199. 57, 161–175. doi:10.1016/j.ijhydene.2024.01.015.

Ong, B.C., Kamarudin, S.K., Basri, S., 2017. Direct liquid fuel cells: a review. Int. J. Hydrogen Energy. 03603199. 42 (15), 10142–10157. doi:10.1016/j.ijhydene.2017.01.117.

Osman, S.H., Kamarudin, S.K., Shaari, N., Hanapi, I.H., Elham, O.S.J., Aminudin, M.A., 2025. Review on Direct Methanol Fuel Cells: bridging the Gap between Theory and Application for Sustainable Energy Solutions. Energy Fuels. 15205029. 39 (12), 5651–5671. Available from http://pubs.acs.org/journal/enfuem. doi:10.1021/acs.energyfuels.4c05357.

Pan, Z.F., An, L., Zhao, T.S., Tang, Z.K., 2018. Advances and challenges in alkaline anion exchange membrane fuel cells. Prog. Energy Combust. Sci. 03601285. 66, 141–175. doi:10.1016/j.pecs.2018.01.001.

Pan, Z., Zhang, Z., Wenzhi, L., Huo, X., Liu, Y., Esan, O.C., Qixing, W., Liang, A., 2023. Development of a High-Performance Ammonium Formate Fuel Cell. ACS Energy Lett. 2380-8195. 8 (9), 3742–3749. doi:10.1021/acsenergylett.3c01165.

Parsons, R., VanderNoot, T., 1988. The oxidation of small organic molecules. A survey of recent fuel cell related research. J. Electroanal. Chem. 00220728. 257 (1-2), 9–45. doi:10.1016/0022-0728(88)87028-1.

Scott, K., Yu, E., Vlachogiannopoulos, G., Shivare, M., Duteanu, N., 2008. Performance of a direct methanol alkaline membrane fuel cell. J. Power Sources. 03787753. 175 (1), 452–457. doi:10.1016/j.jpowsour.2007.09.027.

Shaari, N., Kamarudin, S.K., Bahru, R., Osman, S.H., Md Ishak, N.A.I., 2021. Progress and challenges: review for direct liquid fuel cell. Int. J. Energy Res. 1099114X. 45 (5), 6644–6688. Available from: http://onlinelibrary.wiley.com/journal/10.1002/(ISSN)1099-114X. doi:10.1002/er.6353.

Shi, Q., He, Y., Bai, X., Wang, M., Cullen, D.A., Lucero, M., Zhao, X., More, K.L., Zhou, H., Feng, Z., Liu, Y., Wu, G., 2020. Methanol tolerance of atomically dispersed single metal site catalysts: mechanistic understanding and high-performance direct methanol fuel cells. Energy Environ. Sci. 17545706. 13 (10), 3544–3555. Available from http://pubs.rsc.org/en/journals/journal/ee. doi:10.1039/d0ee01968b.

Silva, J.C.M., Silva, S.G.D., De Souza, R.F.B., Buzzo, G.S., Spinacé, E.V., Neto, A.O., Assumpção, M.H.M.T., 2015. PtAu/C electrocatalysts as anodes for direct ammonia fuel cell. Appl. Catal. A: Gen. 0926860X. 490, 133–138. Available from http://www.sciencedirect.com/science/journal/0926860X. doi:10.1016/j.apcata.2014.11.015.

Soloveichik, G.L., 2014. Liquid fuel cells . Beilstein-Institut Zur Forderung der Chemischen Wissenschaften. Beilstein J. Nanotechnol. 21904286. 5 (1), 1399–1418. Available from http://www.beilstein-journals.org/bjnano/content/pdf/2190-4286-5-153.pdf. doi:10.3762/bjnano.5.153.

Speight, J.G., 2011. Fuels for Fuel Cells Fuel Cells: Technologies for Fuel Processing. Elsevier, United States, pp. 29–48. Available from: https://www.sciencedirect.com/book/9780444535634. doi:10.1016/B978-0-444-53563-4.10003-3.

Su, H., Gao, W., Lanqing, L., Yuan, H., Wen, D., 2023. Oxygen Vacancy-Rich CeO 2 Quantum Dots Boost the Activity and Durability of Pt/C for Methanol Oxidation and Oxygen Reduction Reactions. ACS Sustain. Chem. Eng. 2168-0485. 11 (33), 12317–12325. doi:10.1021/acssuschemeng.3c02126.

Tran, K., Nguyen, T.Q., Bartrom, A.M., Sadiki, A., Haan, J.L., 2014. A fuel-flexible alkaline direct liquid fuel cell. Fuel Cells. 16156854. 14 (6), 834–841. Available from www.interscience.wiley.com/jpages/1615-6846/info.html. doi:10.1002/fuce.201300291.

Viswanathan, B., Scibioh, M.A., 2007. Fuel Cells: Principles and Applications. CRC Press. doi:10.1002/1439-7641(20001215)1.

Wang, J., Zhao, Y., Setzler, B.P., Rojas-Carbonell, S., Ben Yehuda, C., Amel, A., Page, M., Wang, L., Hu, K., Shi, L., Gottesfeld, S., Xu, B., Yan, Y., 2019. Poly(aryl piperidinium) membranes and ionomers for hydroxide exchange membrane fuel cells. Nat. Energy. 20587546. 4 (5), 392–398. Available from: www.nature.com/nenergy/. doi:10.1038/s41560-019-0372-8.

Wang, L., Yuan, Z., Wen, F., Cheng, Y., Zhang, Y., Wang, G., 2018. A bipolar passive DMFC stack for portable applications. Energy. 03605442. 144, 587–593. doi:10.1016/j.energy.2017.12.039.

Wang, T., Zhao, Y., Setzler, B.P., Abbasi, R., Gottesfeld, S., Yan, Y., 2022. A high-performance 75 W direct ammonia fuel cell stack. Cell Rep. Phys. Sci. 26663864. 3 (4), 100829. doi:10.1016/j.xcrp.2022.100829.

Wenzhi, L., Liu, Y., Luo, M., Shi, X., Pan, Z., Chen, R., Liang, A., 2025. Monitoring ammonia and water transport through anion exchange membranes in direct ammonia fuel cells. J. Power Sources. 03787753. 628, 235900. doi:10.1016/j.jpowsour.2024.235900.

Wenzhi, L., Liu, Y., Zhang, Z., Pan, Z., Chen, R., Liang, A., 2024. Performance of a hybrid direct ammonia fuel cell with hydrogen peroxide reduction. J. Power Sources. 03787753. 593, 233985. doi:10.1016/j.jpowsour.2023.233985.

Yaqoob, L., Noor, T., Iqbal, N., 2021. Recent progress in development of efficient electrocatalyst for methanol oxidation reaction in direct methanol fuel cell. Int. J. Energy Res. 0363-907X. 45 (5), 6550–6583. doi:10.1002/er.6316.

Yu, E.H., Krewer, U., Scott, K., 2010. Principles and materials aspects of direct alkaline alcohol fuel cells. Energies. 19961073. 3 (8), 1499–1528. Available from http://www.mdpi.com/1996-1073/3/8/1499/pdf. doi:10.3390/en3081499.

Yu, P.S., Liu, C.T., Feng, B., Wan, J.F., Li, L., Du, C.Y., 2015. Highly efficient anode catalyst with a Ni@PdPt core–shell nanostructure for methanol electrooxidation in alkaline media. Int. J. Miner. Metall. Mater. 1869103X. 22 (10), 1101–1107. Available from http://www.springer.com/materials/journal/12613. doi:10.1007/s12613-015-1173-0.

Zenith, F., Krewer, U., 2010. Modelling, dynamics and control of a portable DMFC system. J. Process Control. 09591524. 20 (5), 630–642. doi:10.1016/j.jprocont.2010.02.014.

Zhang, L., Lee, K., Zhang, J., 2007. The effect of heat treatment on nanoparticle size and ORR activity for carbon-supported Pd–Co alloy electrocatalysts. Electrochim. Acta. 00134686. 52 (9), 3088–3094. doi:10.1016/j.electacta.2006.09.051.

Zhang, M., Hao, L., Duan, X., Zou, P., Jeerh, G., Sun, B., Chen, S., Humphreys, J., Walker, M., Xie, K., Tao, S., 2021. An Efficient Symmetric Electrolyzer Based On Bifunctional Perovskite Catalyst for Ammonia Electrolysis. Adv. Sci. 2198-3844. 8 (22), doi:10.1002/advs.202101299.

Zhang, M., Zhang, J., Jeerh, G., Zou, P., Sun, B., Walker, M., Xie, K., Tao, S., 2022b. A symmetric direct ammonia fuel cell using ternary NiCuFe alloy embedded in a carbon network as electrodes. J. Mater. Chem. A. 2050-7488. 10 (36), 18701–18713. doi:10.1039/d2ta04129d.

Zhang, M., Zou, P., Jeerh, G., Sun, B., Walker, M., Tao, S., 2022a. Oxygen Vacancy-Rich La 0.5 Sr 1.5 Ni 0.9 Cu 0.1 O 4–δ as a High-Performance Bifunctional Catalyst for Symmetric Ammonia Electrolyzer. Adv. Funct. Mater. 1616-301X. 32 (38), doi:10.1002/adfm.202204881.

Zhao, T.S., Xu, C., Chen, R., Yang, W.W., 2009. Mass transport phenomena in direct methanol fuel cells. Prog. Energy Combust. Sci. 03601285. 35 (3), 275–292. doi:10.1016/j.pecs.2009.01.001.

Zhao, X., Yin, M., Ma, L., Liang, L., Liu, C., Liao, J., Lu, T., Xing, W., 2011. Recent advances in catalysts for direct methanol fuel cells. Energy Environ. Sci. 17545706. 4 (8), 2736–2753. doi:10.1039/c1ee01307f.

Zhao, Y., Setzler, B.P., Wang, J., Nash, J., Wang, T., Xu, B., Yan, Y., 2019. An Efficient Direct Ammonia Fuel Cell for Affordable Carbon-Neutral Transportation. Joule. 25424351. 3 (10), 2472–2484. Available from: https://www.journals.elsevier.com/joule. doi:10.1016/j.joule.2019.07.005.

Chapter 10

Alkaline fuel cells for transportation applications and beyond

Yifei Wang[a], Xinhai Xu[a], Meng Ni[b], Michael K.H. Leung[c], and Dennis Y.C. Leung[d]

[a]*School of Robotics and Advanced Manufacture, Harbin Institute of Technology, Shenzhen, China*
[b]*Department of Building Environment and Energy Engineering, The Hong Kong Polytechnic University, Hong Kong, China*
[c]*School of Energy and Environment, City University of Hong Kong, Hong Kong, China*
[d]*Department of Mechanical Engineering, The University of Hong Kong, Hong Kong, China*

1 Introduction

Since the British scientist William Grove invented the very first fuel cell in 1839, the real application of this technology did not follow. Nevertheless, the internal combustion engine has taken the lead in power generation for both transportation and stationary applications instead (*https://en.wikipedia.org/wiki/Fuel_cell*, n.d.). This was mainly due to the poor power density of the nascent fuel cell, which used diluted H_2SO_4 as the electrolyte and a Pt rod as the electrode, as shown in Fig. 10.1A. The design was both sluggish in reaction kinetics and limited in reaction surface area. Later, Reid and Noel described a new type of fuel cell with aqueous potassium hydroxide (KOH) electrolyte in the 1900s (Gülzow, 1996). However, it was not until the 1950s that a British engineer, Francis Bacon, successfully developed the first alkaline fuel cell (AFC), which initiated the practical application of fuel cell technology (*https://en.wikipedia.org/wiki/Alkaline_fuel_cell*, n.d.). Compared with Grove's acidic prototype, Bacon's AFC adopted KOH as the electrolyte and double-layer porous Ni coated with Pt as the electrode. It also operated with pressurized gas at temperatures exceeding 200°C. These modifications greatly improved the reaction kinetics and surface area, leading to much higher power density. Moreover, a compact AFC system with multiple thin-film single cells separated by bipolar plates and stacked between two thick end plates was proposed, as shown in Fig. 10.1B. This design further improved the system's power output to 6 kW, enabling it to power forklift trucks, welding machines, and a circular saw (Hamada et al., 2023).

Alkaline Fuel Cells. https://doi.org/10.1016/B978-0-443-36638-3.00014-0

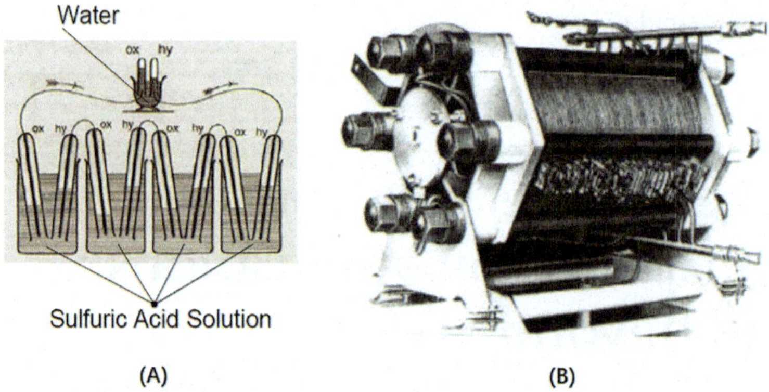

FIG. 10.1 Early fuel cell prototypes. (A) William Grove's acidic fuel cell, invented in 1839; (B) Francis Bacon's alkaline fuel cell, invented in the 1950. *(Panel (A) Data from https://www1.eere.energy.gov/vehiclesandfuels/pdfs/deer_2004/session1/2004_deer_fairbanks.pdf. (n.d.). Panel (B) Data from Varta Fachbuchreihe, Band 6—A Key Historical Reference in Fuel Cell Development.)*

With Bacon's efforts, more and more AFC applications have been implemented in the fields of aerospace, military, and civilian use since the 1960s, particularly in the transportation sector (McLean et al., 2002). For instance, the AFC was very popular in spacecraft because it could provide electricity as well as water from liquid H_2 and O_2, both of which were necessary for astronauts in outer space. As for military missions, the AFC was well known for its distinct advantages of high efficiency, low noise, and zero infrared, and it was successfully applied to air-independent-propulsion (AIP) submarines, mobile radar units, and portable power generators. Furthermore, many pioneering companies have explored the market potential of AFC vehicles in order to reduce carbon emissions, such as AFC tractors for agriculture, AFC cars for road traffic, AFC taxis for city transportation, and AFC ferries for shipping (Gülzow, 2004). However, with the development of the proton exchange membrane fuel cell (PEMFC) later on, an advanced version of Grove's acidic fuel cell, interest in AFC gradually vanished due to several reasons (Wee, 2007). First of all, the alkaline electrolyte is sensitive to CO_2 poisoning from the ambient air, which makes it troublesome for maintenance (Ni et al., 2006). Next, the strong alkaline electrolyte is very corrosive, which brings both durability issues and safety hazards. Last but not least, the hydrogen infrastructure was not mature at that time, making it difficult and expensive to refuel AFC vehicles. Consequently, nowadays only a few AFC companies still exist, focusing instead on stationary power generation, while the transportation sector is largely dominated by the PEMFC (Veeranjaneyulu et al., 2023). Nevertheless, since a noble metal catalyst is inevitable in PEMFCs, researchers have not given up on AFC technology yet, as it can use 100% non-noble catalysts and exhibits significant cost advantages. To resolve the above-mentioned AFC shortcomings, the next-generation AFC based on anion exchange membranes has also been proposed recently, which may bring AFC technology back to the transportation market in the near future (Hren et al., 2021).

This chapter primarily introduces the application history and current status of AFC, particularly in the transportation sector. In addition, the application of AFC for stationary power generation and chemical co-generation will be elaborated upon. Finally, the future perspectives of AEMFC will also be discussed.

2 AFCs for transportation applications

AFCs were initially developed for the transportation sector because of their high energy efficiency and low noise level, which seemed to be a better choice than internal combustion engines. However, they were restricted by the high manufacturing cost as well as poor hydrogen infrastructure at that time, leading to their implementation only in niche applications, particularly in the aerospace and military fields. Nowadays, with the global consensus on the hydrogen economy, interest in fuel cell transportation has been revived, but more in PEMFC technology rather than AFCs.

2.1 Aerospace & military transportation

Considering its high manufacturing cost at the time, AFC was first applied in the fields of aerospace and military, where high performance is of vital importance while manufacturing cost is less of a concern. Fuel cell technology is naturally suitable for aerospace applications because it can obtain fuel and oxidant directly from liquid H_2 and O_2 storage, producing both electricity and water for astronauts in outer space. Meanwhile, since there is no CO_2 in the pure O_2 oxidant, the alkaline electrolyte in AFC is free from CO_2 poisoning, guaranteeing a long operational lifetime. Considering all these benefits, NASA chose the AFC technology developed by Pratt & Whitney and adapted it as the main power supply for their Apollo moon landing spacecraft in the 1960s and 1970s. As shown in Fig. 10.2A, the AFC used in Apollo 11 had a total weight of 181.4 kg and could provide a maximum power of 2.3 kW. Later on, a more advanced version of AFC was adopted in the Space Shuttle Orbiter, with a higher power output of 12 kW. Moreover, the specific weight was also improved from 200 to 20 lb kW^{-1} (Warshay and Prokopius, 1989). This choice of AFC was made because, at the time, PEMFC was not mature enough; its sulfonic acid membrane electrolyte gradually degraded during operation, leading to a short lifetime and contaminated water generation. However, with the invention of more stable proton exchange membrane (PEM), AFC technology was eventually replaced by PEMFC in aerospace missions. In addition to spacecraft, National Aeronautics and Space Administration (NASA) also tested fuel cell technology in their unmanned aerial vehicles (UAVs), such as the Helios prototype with a coupled solar and fuel cell power system, as shown in Fig. 10.2B (*https://en.wikipedia.org/wiki/AeroVironment_Helios_Prototype*, n.d.).

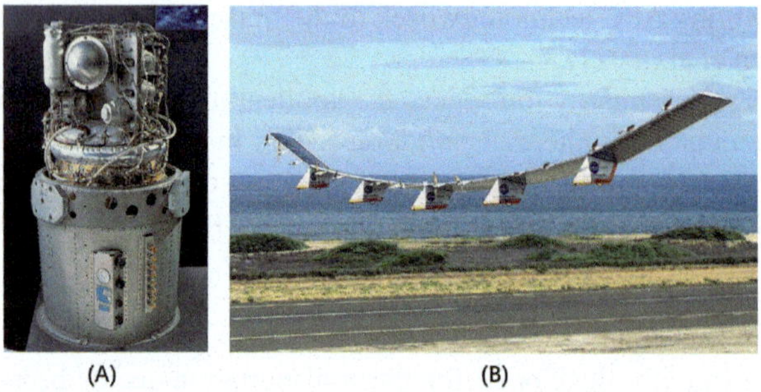

FIG. 10.2 AFC applications in the field of aerospace. (A) The AFC used in the Apollo moon-landing spacecraft; (B) The AFC used in the Helios prototype. *(Panel (A) Data from https://airand-space.si.edu/collection-objects/fuel-cell-apollo/nasm_A19780288000. (n.d.). Panel (B) Data from https://www.nasa.gov/image-article/helios-prototype-flying-wing-2/. (n.d.).)*

Despite the energy efficiency advantage of AFC, PEMFC was eventually selected, considering the requirement for a longer lifetime, while the application potential of AFCs in future UAVs remained much less explored. Furthermore, Rostami et al. (2022) compared the performance of PEMFC and AFC in UAVs through modeling. It was found that for a UAV with a total power of 1.25 kW and a flight duration of 7.4 hours, the required area for PEMFC and AFC was 0.03 m^2 and 0.6 m^2, respectively, indicating a significant advantage of the former choice.

In addition to aerospace, the military is also interested in fuel cell technology, considering its low-noise and zero-infrared operation, which is especially suitable for submarines with AIP. In the 1980s and 1990s, Siemens cooperated with the German Ministry of Defense to develop a 100 kW AFC and tested it on the U-1 type submarine (Strasser, 1992). However, similar to the trend in aerospace, the AFC was later replaced by a 300 kW PEMFC in subsequent 212-type submarines due to its easier electrolyte management and higher tolerance to CO_2. As for unmanned underwater vehicles (UUVs), the United States. Navy once tried AFC in their UUV prototype to test its long-term underwater operation ability in the 2000s, but later switched to PEMFC, such as the Orca prototype developed by Boeing (*https://en.wikipedia.org/wiki/Orca_(AUV)*, n.d.). Similarly, nowadays, PEMFC has become the dominant choice for both military and commercial UUVs.

2.2 Civil transportation

Despite the high cost of AFC at that time, an American agricultural machinery manufacturer, Allis-Chalmers, was very active in developing the world's first AFC-powered farm tractor in 1959, which was even earlier than NASA's Apollo project (*https://americanhistory.si.edu/collections/object/nmah_687671*, n.d.). As shown in Fig. 10.3A, this modified Model-D tractor was powered by an AFC system composed of 1008 single cells arranged in 112 stacks, which could deliver a

FIG. 10.3 AFC applications in the field of civil transportation. (A) Allis-Chalmers' AFC-powered tractor; (B) Karl Kordesch's AFC-powered Austin A40 car; (C) An AFC-powered ferry from the "Hydra" project. *(Panel (A) Data from https://americanhistory.si.edu/collections/object/nmah_687671, n.d. Panel (B) Data from Kordesch, K., Gsellmann, J., Cifrain, M., Voss, S., Hacker, V., Aronson, R. R., Fabjan, C., Hejze, T., Daniel-Ivad, J., 1999. Intermittent use of a low-cost alkaline fuel cell-hybrid system for electric vehicles. J. Power Sources, 80 (1), 190–197. https://doi.org/10.1016/S0378-7753(98)00261-4. Panel (C) Data from https://en.wikipedia.org/wiki/Fuel_cell. n.d.)*

power output of 15 kW and was capable of pulling a weight of about 3000 pounds. With a full tank of liquid H_2, this AFC tractor could work for up to ten hours. Nevertheless, this AFC tractor was not commercialized at that time because it was much more expensive than the diesel engine-powered tractors. By the 1970s, Allis-Chalmers abandoned AFC research and focused on its main products.

In addition to Allis-Chalmers, another company named Union Carbide also experimented with AFC in the 1960s–1970s, adopting high-surface-area Raney nickel as electrode material and choosing lower operating temperatures (60–90°C) to slow down material degradation. At this company, Dr. Karl Kordesch, an Austrian-American chemist and fuel cell pioneer, made groundbreaking advancements to bring AFC into practical consumer applications. Specifically, he built an AFC-NiCd battery hybrid electric motorcycle in 1967. Later, he also adapted his Austin A40 car with an AFC power source, as shown in Fig. 10.3B, which had a driving range of 180 miles (Kordesch et al., 1999). This AFC-powered car was driven on public roads for three years during 1970–1973 (Kordesch et al., 2000), proving the high feasibility and safety level of AFC vehicles. Nevertheless, Union Carbide abandoned AFC research in the 1980s due to technical bottlenecks and economic reasons.

Starting in 1976, ELENCO, a Belgian-Dutch joint venture, began its R&D on AFC technology using air as the oxidant (Van den Broeck et al., 1986). With 30% KOH as the electrolyte, 0.7 mg cm^{-2} of Pt on carbon as the catalyst, and a soda-lime CO_2 scrubber, their 24-cell module could deliver a specific power of 0.1 kW kg^{-1} and operated stably for at least 4800 hours. When applied to a Volkswagen electric van to replace the original lead-acid battery, the van was successfully driven for 100 km in 1983, consuming about 1.8 kg of H2, which was only half of its H_2 storage capacity. This 200 km full driving range was already much higher than that of the lead-acid battery (only 60 km). However, no further progress has been reported since, likely due to the high cost of the Pt catalyst in their AFC design.

When the 2000s arrived, a start-up company called Zero Emission Vehicle Company (ZEVCO) was founded in the UK. The company attempted to apply AFC technology to automotive use, becoming one of the representatives of the early commercial exploration of hydrogen energy transportation (De Geeter et al., 1999). In 2003, ZEVCO launched the first AFC-powered taxi (5 kW) in London. The taxi produced no harmful fumes and made significantly less noise than traditional internal combustion taxis. To avoid the use of pure O_2 tanks while inhibiting CO_2 poisoning, an additional air filtration/scrubbing process was incorporated into the AFC system, thereby prolonging the device's lifetime. Meanwhile, another AFC company called ZeTek Power was also founded in the UK. ZeTek Power developed AFCs for automotive, marine, and distributed power applications (Burchardt et al., 2002). However, as PEMFC technology advanced rapidly during the 2000s, with several PEMFC vehicle prototypes unveiled (e.g., Toyota Mirai, Hyundai Nexo), both companies were eventually shut down due to a lack of funding. Afterward, the focus of AFC applications shifted to stationary power generation, with successor companies such as AFC Energy from the UK and GenCell from Israel continuing the development. In addition, a French company called Eident Energy developed AFCs with circulating aqueous electrolytes, achieving an energy efficiency of 51%, a current density of 137 mA cm^{-2} at 0.67 V, and a lifetime of 2500 hours with a 10% power reduction at that time (Gouérec et al., 2004). Moreover, it was discovered that performance degradation was primarily caused by water flooding rather than irreversible material corrosion, which could be mitigated by applying wet-proof coatings. Nevertheless, no further updates about the company have been found.

In addition to land transportation, AFC technology was also tested in water vehicles. In 2000, the "Hydra" project, led by Germany, successfully built and tested the world's first AFC-powered ferry, as shown in Fig. 10.3C, marking the first commercial attempt at fuel cell technology in the shipbuilding field (*https://en.wikipedia.org/wiki/Fuel_cell*, n.d.). This 22-meter-long passenger ferry was powered by a 5 kW AFC developed by Siemens, with lead-acid batteries as a buffer, and was demonstrated on the Rhineland-Palatinate waterways. However, the project was terminated after 2003 and did not enter the commercialization stage, while subsequent fuel cell ferries and ships generally opted for PEMFC instead.

3 AFCs for other applications

Despite the CO_2 contamination problem, the AFC can still be operated stably with a satisfactory lifetime, provided that air filtration/scrubbing technology is adopted in the system. In this pretreatment process, the 400 ppm CO_2 contaminant in ambient air can be efficiently removed by chemical absorption, physical adsorption, or membrane filtration techniques before entering the AFC stack, ensuring significantly extended KOH electrolyte durability. Moreover, if liquid H_2 is used as fuel, its boil-off process can be utilized to sublime and continuously remove CO_2 from the air (Ahuja and Green, 1996). However, such a pretreatment component also introduces extra system redundancy, making it more suitable for stationary applications rather than transportation.

3.1 Stationary power generation

To date, there are still a few AFC companies active in the global energy market, such as AFC Energy from the UK and GenCell from Israel, both of which focus on stationary power generators for off-grid electricity supply.

AFC Energy provides a series of AFC generator products powered by hydrogen, with a wide range of power outputs up to 200 kW (*https://www.afcenergy.com/*, n.d.). The H-Power Tower is composed of four air-cooled fuel cell modules, which can supply up to 10 kW for powering demonstrations, events, and construction sites. The H-Power Generator can accommodate more modules and further scale up the power output to 30 kW. To achieve even higher power, the air-cooled modules can be replaced with liquid-cooled ones, enabling a maximum power output of 200 kW. Considering the challenges of hydrogen storage and transport, the company is also working on integrating an ammonia cracker or methanol reformer into the AFC system, allowing liquid ammonia or methanol fuel to be used instead of hydrogen. GenCell focuses on AFC generators that use liquid ammonia as fuel (*https://www.gencellenergy.com/*, n.d.). The GenCell REX has a rated power of 5 kW and is designed for utility substation backup power. The GenCell BOX also has a rated power of 5 kW and is intended for long-duration backup power for mission-critical applications, such as telecommunication base stations. To provide even higher power, the GenCell EVOX is a much larger AFC system with a rated power of 382–532 kW, designed for multipurpose, grid-independent, and resilient power applications, including DC fast charging, backup power, and 24/7 emergency operations, such as electric vehicle charging stations and emergency and disaster response. In addition to the two companies mentioned above, a Belgian company called Intensys also worked on AFC generators (Mulder et al., 2008). In collaboration with the Flemish Technological Institute VITO, a 6 kW prototype was developed and demonstrated in a concept building for future technologies in Vilvoorde (Brussels). However, no further updates have been found regarding this company or its AFC products.

The academic community is also interested in AFC technology, particularly in its application to stationary scenarios for combined heat and power (CHP) generation. Staffell and Ingram (2010) conducted a life cycle assessment of a 1 kW AFC-based CHP system powered by natural gas and ambient air. Compared with other fuel cell technologies, it was found that the AFC was more environmentally friendly than the solid oxide fuel cell and phosphoric acid fuel cell, though it was difficult to compare it with the PEMFC. Cox and Treyer (2015) compared the environmental impact and economic value of an ammonia-fed AFC generator and a diesel generator for a telecommunication base transceiver station using life cycle assessment and levelized cost of electricity calculations. They found that power density, energy efficiency, and device lifetime significantly affected the system cost and environmental impact. Specifically, the electricity cost of the AFC generator was 14% higher than that of the diesel generator, but the life cycle CO_2 equivalent emissions were 23% lower. Verhaert et al. (2016) studied the feasibility of small-scale AFC-based CHP systems in building applications, with a particular focus on their control strategy. They confirmed that the inverter and purger are more important than pumps and fans in terms of system performance. Regarding operating conditions, the flow rate of the cooling circuit was critical in controlling water management.

AFC is also highly welcomed in renewable energy systems for converting the generated green hydrogen back to electricity. As shown in Fig. 10.4, Wang et al. (2021) proposed a renewable energy system composed of a wind turbine and solar photovoltaic (PV) for renewable electricity generation, a water electrolyzer for hydrogen production, an AFC for electricity generation from hydrogen, and a Stirling engine for heat recovery from the AFC. By using this system, 10.5 kW of electricity could be provided with an electrical efficiency of 56.9%, of which 9.9 kW was used to produce 221.3 g h^{-1} of hydrogen fuel. By adding the Stirling engine to the AFC, an extra 1.03 kW of power could be recovered. Shen et al. (2021) also developed a renewable energy system based on a wind turbine, AFC, alkaline electrolyzer, and pumped hydro-compressed air electrical energy system. On average, this system could store up to 1.35 kW of power, of which the alkaline electrolyzer consumed 0.63 kW of electricity while the AFC produced 0.07 kW of electricity. Li et al. (2021) also developed a renewable energy system composed of solar PV, AFC, water electrolyzer, Stirling engine, absorption chiller, and pumped hydro-compressed air system. With an AFC power output of 1 kW, the Stirling engine could produce 1.27 kW of electricity, and the absorption chiller could obtain a cooling load of 2.48 kW, resulting in a system energy efficiency of 63.8%. Zeng et al. (2023) compared three solar energy systems in Transient System Simulation Program (TRNSYS), including the solar energy system without electricity storage, the solar system with battery energy storage, and the solar system with hydrogen energy storage with the help

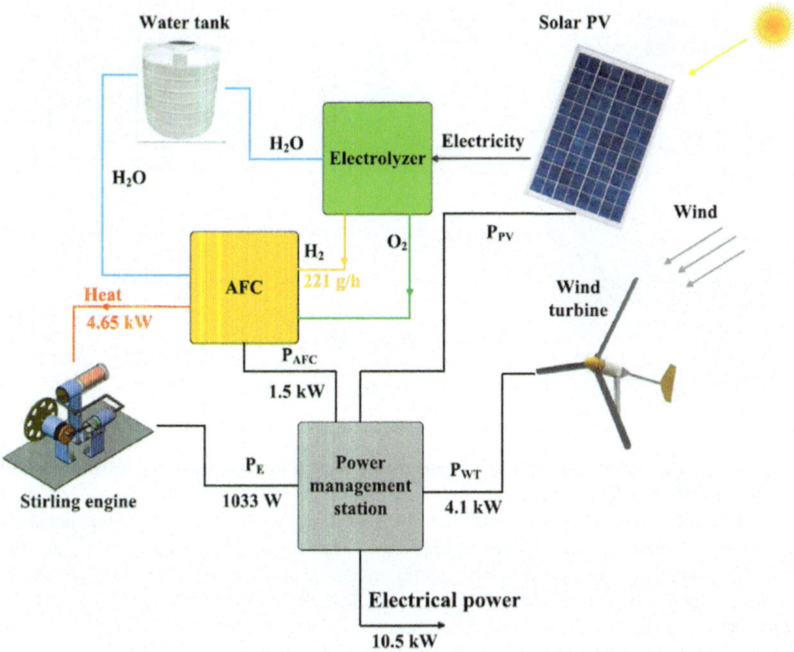

FIG. 10.4 The application of AFC in a renewable energy system. *(Data from Wang, Z., Zhang, X. & Rezazadeh, A., 2021. Hydrogen fuel and electricity generation from a new hybrid energy system based on wind and solar energies and alkaline fuel cell. Energy Rep. 7, 2594–2604.)*

of a water electrolyzer and AFC. The simulation results indicated that the third choice had the lowest life cycle cost, the highest electricity coverage ratio, the lowest annual CO_2 emissions, and the lowest levelized cost of energy, making it more competitive in solar energy storage and application.

To improve the overall energy efficiency of AFC, various techniques have been proposed to better harvest its exhaust heat. As shown in Fig. 10.5A, Yang et al. (2014) studied the performance of an AFC-thermoelectric generator (TEG) hybrid system, in which the waste heat from AFC operation could be harvested by the TEG to further export electricity. By this means, the equivalent maximum power density of the AFC could be improved by 23%, leading to higher energy conversion efficiency. As shown in Fig. 10.5B, Zhu et al. (2022) added a temperature-matching elastocaloric cooler to the AFC in order to harvest its waste heat for cooling purposes. By numerical calculation, it was found that the hybrid system's equivalent peak power density and energy efficiency could be improved by 51.6% and 20.9%, respectively, compared with the single AFC. Zhang and Rahman (2022) employed a thermoradiative device to harvest the waste heat from AFC operation, as shown in Fig. 10.5C, providing extra electricity output. Compared with conventional heat recovery technologies, this hybrid system exhibited better power output performance, with its maximum

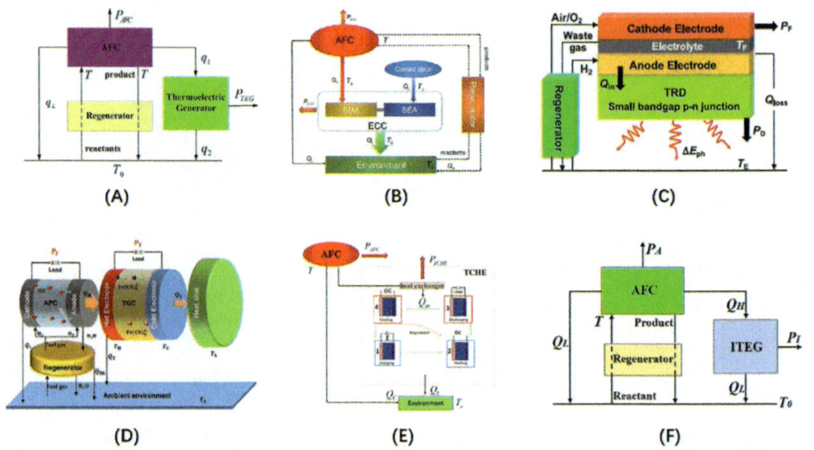

FIG. 10.5 Schematic diagram of different techniques to harvest waste heat from the AFC. (A) an AFC-thermoelectric generator (TEG) hybrid system; (B) an AFC-elastocaloric cooler hybrid system; (C) an AFC-thermoradiative device hybrid system; (d) an AFC-thermogalvanic cell hybrid system; (E) an AFC-thermocapacitive heat engine hybrid system; (F) an AFC-inhomogeneous thermoelectric generator hybrid system. *(Panel (A) data from Yang, P., Zhu, Y., Zhang, P., Zhang, H., Hu, Z., Zhang, J., 2014. Int. J. Hydrogen Energy 39 (22), 11,756–11,762. Panel (B) data from Zhu, H., Li, J., Lai, C., Zhang, H., 2022. International Journal of Hydrogen Energy, 47(63), 27,124–27,138. Panel (C) data from Zhang, X., Rahman, E., 2022. Renew. Energy 200, 1240–1250. Panel (D) data from Wang, M., Ruan, J., Zhang, J., Jiang, Y., Gao, F., Zhang, X., Rahman, E., Guo, J., 2024. Energy, 292, 130,557. Panel (E) data from Zhang, X., Zhao, Q., Wang, J., Zhang, H., 2024. Process Saf. Environ. Protect. 186, 706–714. Panel (F) data from Zhang, C., Li, H., Zhang, X., Shen, M., Jin, X., 2024. Energies, 17 (9), 2066.)*

power density reaching 1.75 times that of the AFC alone. As shown in Fig. 10.5D, Wang et al. (2024) introduced thermogalvanic cells to harvest the waste heat of AFC. Compared with AFC alone, the hybrid system was able to deliver 19.7% more maximum power density, increasing from 247.1 to 295.8 W m^{-2}. Moreover, the energy conversion efficiency was also improved from 10.16% to 10.74%. Zhang et al. (2024b) proposed the addition of a thermocapacitive heat engine to AFC, as shown in Fig. 10.5E, improving both its peak power density and energy/exergy efficiencies. To optimize the hybrid system's performance, it was recommended to sustain a higher working pressure, temperature, electrolyte concentration, and cut-off voltage during operation, while a shorter preheating time and a suitable charge ratio were also beneficial. Furthermore, Zhang et al. (2024a) adopted the inhomogeneous TEG (ITEG) for AFC exhaust heat recovery, as shown in Fig. 10.5F, which could increase the maximum power density by 31.2% and energy efficiency by 54.6%. Specifically, increasing the AFC operation temperature and environmental temperature could further improve the power output. However, the energy efficiency could be improved by increasing the AFC operation temperature or decreasing the environmental temperature and optimizing ITEG geometric characteristics at the same time.

The waste heat from AFC can also be utilized to generate freshwater. As shown in Fig. 10.6A, Zhao et al. (2021) added a direct contact membrane distillation to the AFC in order to utilize its waste heat for water purification. The modeling results revealed that such a hybrid system could improve the equivalent power density and energy efficiency by 44.58% and 44.55%, respectively. To further improve the system's performance, it was recommended to increase the feed water temperature and flow rate, as well as the pore diameter of the hydrodynamic membrane, while the membrane thickness and permeate water temperature should be decreased. In addition, Zhu et al. (2024) employed air gap membrane distillation to obtain freshwater from AFC waste heat, as shown

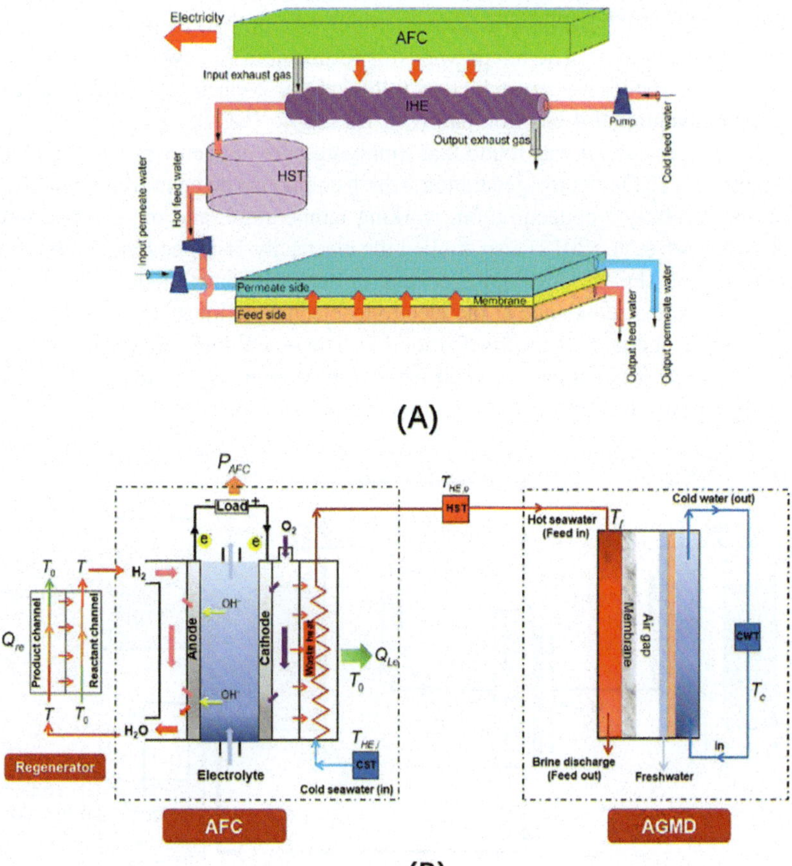

FIG. 10.6 Schematic diagram of different techniques to purify water using the AFC. (A) An AFC-direct contact membrane distillation hybrid system; (B) An AFC-air gap membrane distillation hybrid system. *(Panel (A) Data from Zhao, Q., Zhang, H., Hu, Z., Li, Y., 2021. Energy 225, 120,303. Panel (B) Data from Zhu, H., Xiao, L., Kuang, M., Wang, J., Zhang, H., 2024. Renew. Energy 225, 120,149.)*

in Fig. 10.6B, which was comprehensively studied through energetic, exergetic, economic, and environmental analyses. Compared with AFC alone, the equivalent power density of this hybrid system was 63.3% higher at an operating temperature of 80°C. Consequently, the levelized cost of freshwater was 56.9 USD/m^{-3} and the levelized cost of electricity was about 0.1 USD/kWh.

3.2 Electricity-chemical cogeneration

In addition to electricity and waste heat, the AFC can also output other valuable chemicals such as hydrogen peroxide, leading to an electricity-chemical cogeneration system. To date, only a few research studies have been conducted in this area, and no relevant industrial applications have been identified yet. As shown in Fig. 10.7, Brillas et al. (2002) employed a flow AFC for electricity and H_2O_2 co-generation. The anode was still responsible for hydrogen oxidation, while the cathode oxygen reduction followed the 2-electron pathway instead of the conventional 4-electron pathway, leading to H_2O_2 generation. Through a parametric study, it was found that supplying pure oxygen was beneficial for maximizing H_2O_2 electro-generation, which could also be enhanced by increasing the electrolyte concentration, working temperature, and oxygen pressure. De Schepper et al. (2012) also studied the electricity-H_2O_2 cogeneration capability of the AFC by developing a Tank in Series Reactor model, which considered the charge balance at the electrode–electrolyte interface as well as the species and energy balance inside the AFC. The modeling results indicated that mass transfer intensification could significantly improve reactor performance due to the strong coupling between mass and charge transfer.

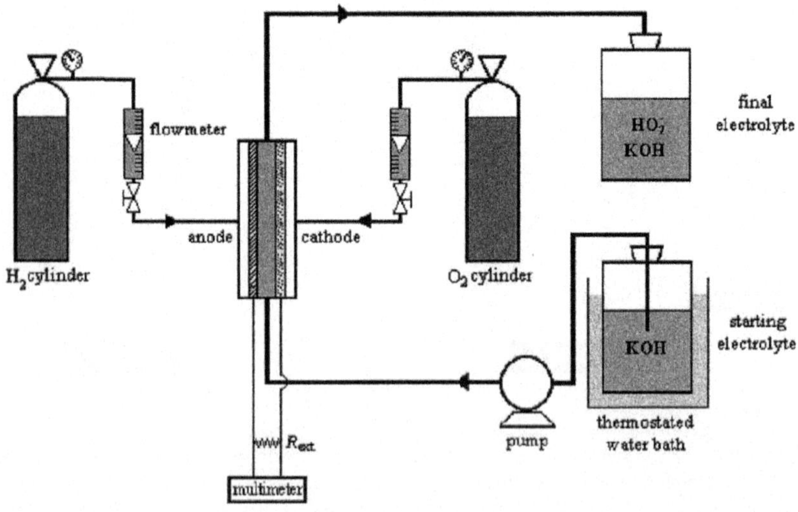

FIG. 10.7 An electricity-H_2O_2 co-generation system based on AFC. *(Data from Brillas, E., Alcaide, F., Cabot, P.-L., 2002. Electrochim. Acta 48 (4), 331–340.)*

4 Application prospect of anion exchange membrane fuel cell

The global interest in AFC is mainly overshadowed by its PEMFC counterpart, which benefits significantly from its solid-state polymer electrolyte membrane. Therefore, inspired by PEMFC, a new generation of AFC called the anion exchange membrane fuel cell (AEMFC) has been proposed recently, which also utilizes a solid polymer electrolyte membrane capable of conducting hydroxide ions (Dekel, 2018). In this way, the CO_2 poisoning issue can be greatly alleviated since there are no cations such as K^+ and Na^+ in the membrane. In addition, the corrosion and leakage hazards are eliminated. More importantly, the alkaline reaction environment is maintained, allowing for 100% nonnoble catalyst utilization, which promises a significant cost advantage over PEMFC. Nevertheless, to date, the AEM conductivity and durability are still not as advanced as the mature PEM product (Das et al., 2022), resulting in only a few pioneering companies focusing on this fuel cell technology.

As one of the major AFC market players, AFC Energy is also deploying research and development (R&D) on next-generation AEMFC technology, such as its high energy-dense fuel cell technology, HydroX-Cell(S)™, which is targeted for heavy motive applications in road, marine, and rail traffic (*https://www.afcenergy.com/news/press-releases/launch-of-world-class-anion-exchange-membrane-hydrogen-fuel-cell-test-facility*, n.d.). The AlkaMem is adopted as the electrolyte, and the modular 10 kW stack is under development at their AEMFC test facility at the company's Surrey headquarters. Another company, called Enapter, is also working on AEM-based technology, but it mainly focuses on AEM water electrolyzers rather than fuel cells (*https://www.enapter.com/*, n.d.). A series of commercial products have been released, ranging from the 2.4 kW AEM Electrolyzer EL 4 to the 1 MW AEM Nexus. Considering the similar materials and structure between AEM water electrolyzers and fuel cells, this company is also highly promising in the future AEMFC market. Furthermore, major AEM manufacturers, such as Fumatech from Germany (*https://www.fumatech.com/en/products/membranes-alkaline-water-electrolysis/*, n.d.) and Tokuyama from Japan (*https://www.tokuyama.co.jp/eng/*, n.d.), are also continuously improving their membrane properties, benefiting the widespread application of AEMFCs in the near future.

5 Conclusion & future perspectives

In this chapter, the application history of AFC since its first appearance in the 1950s is elaborated on in detail, including academic research, practical demonstrations, as well as commercial products. When it was first invented by Francis Bacon, the high power density, low noise, and clean nature of AFC soon attracted wide interest in various fields such as aerospace, military, farm machinery, and road traffic, aiming to replace the conventional internal combustion engine. Specifically, NASA adopted AFC in their Apollo projects during the 1960s–1970s, and Siemens employed AFC in their AIP submarines during the 1980s–1990s. However, due to the high manufacturing cost and immature hydrogen infrastructure at that time, this technology was not widespread for civil transportation in the latter half of the 20th century, with only niche demonstrations implemented by several pioneering AFC companies. When the time came to the 21st century, major AFC companies abandoned the market of transportation and turned to the market of stationary power generation, producing high-power AFC generators for telecommunication base stations, buildings, and emergency power supply. This was indeed a wise choice since PEMFC technology has been developed rapidly in the new century, particularly for electric vehicles, which is more favorable considering its compact structure, freedom from CO_2 poisoning, easy maintenance, and high power density. Nevertheless, when it comes to large-scale stationary power generation, AFC with 100% nonnoble catalysts might be more appealing in terms of cost, which can be widely utilized in various CHP and renewable energy systems. Furthermore, the next-generation AEMFC is also learning quickly from PEMFC by replacing the conventional alkaline solution with solid-state AEM, combining the advantages of both fuel cell technologies.

Currently, large AFC companies such as AFC Energy and GenCell have launched a series of products with power outputs ranging from several kW to several hundred kW, offering excellent energy conversion efficiencies. With the help of efficient CO_2 scrubbers, long-term durability has also been significantly improved compared to before. When facing the global boom in renewable energy system development, the application of AFC generators has very broad prospects, considering their well-balanced combination of device cost, energy efficiency, and operational lifetime among all fuel cell technologies. In the meantime, the next-generation AEMFC is also undergoing rapid development in terms of nonnoble catalysts, AEMs, and system optimization. Once commercial AEMs achieve similar ionic conductivity and long-term durability to PEMs, AEMFCs could quickly penetrate the transportation sector, potentially becoming the dominant force in the future market for fuel cell electric vehicles.

References

Ahuja, V., Green, R.K., 1996. CO_2 removal from air for alkaline fuel cells operating with liquid hydrogen—heat exchanger development. Int. J. Hydrogen Energy. 0360-3199. 21 (5), 415–421.

Brillas, E., Alcaide, F., Cabot, P.-L., 2002. A small-scale flow alkaline fuel cell for on-site production of hydrogen peroxide. Electrochim. Acta. 0013-4686. 48 (4), 331–340.

Burchardt, T., Gouérec, P., Sanchez-Cortezon, E., Karichev, Z., Miners, J.H., 2002. Alkaline fuel cells: contemporary advancement and limitations. Fuel. 0016-2361. 81 (17), 2151–2155.

Cox, B., Treyer, K., 2015. Environmental and economic assessment of a cracked ammonia fuelled alkaline fuel cell for off-grid power applications. J. Power Sources. 0378-7753. 275, 322–335.

Das, G., Choi, J.-H., Nguyen, P.K.T., Kim, D.-J., Yoon, Y.S., 2022. Anion exchange membranes for fuel cell application: a review. Polymers. 2073-4360. 14 (6), 1197.

De Geeter, E., Mangan, M., Spaepen, S., Stinissen, W., Vennekens, G., 1999. Alkaline fuel cells for road traction. J. Power Sources. 0378-7753. 80 (1-2), 207–212.

De Schepper, P., Danilov, V., Denayer, J.F.M., 2012. A tank in series model for alkaline fuel cell in cogeneration of hydrogen peroxide and electricity. Int. J. Hydrogen Energy. 0360-3199. 37 (22), 17258–17267.

Dekel, D.R., 2018. Review of cell performance in anion exchange membrane fuel cells. J. Power Sources. 0378-7753. 375, 158–169.

Gouérec, P., Poletto, L., Denizot, J., Sanchez-Cortezon, E., Miners, J.H., 2004. The evolution of the performance of alkaline fuel cells with circulating electrolyte. J. Power Sources. 0378-7753. 129 (2), 193–204.

Gülzow, E., 1996. Alkaline fuel cells: a critical view. J. Power Sources. 0378-7753. 61 (1), 99–104. Available from: https://www.sciencedirect.com/science/article/pii/S0378775396023440. doi:10.1016/S0378-7753(96)02344-0.

Gülzow, E., 2004. Alkaline Fuel Cells. Fuel Cells. 1615-6846. 4 (4), 251–255. doi:10.1002/fuce.200400042.

Hamada, A.T., Orhan, M.F., Kannan, A.M., 2023. Alkaline fuel cells: status and prospects. Energy Rep. 2352-4847. 9, 6396–6418. Available from: https://www.sciencedirect.com/science/article/pii/S235248472301020X. doi:10.1016/j.egyr.2023.05.276.

Hren, M., Božič, M., Fakin, D., Kleinschek, K.S., Gorgieva, S., 2021. Alkaline membrane fuel cells: anion exchange membranes and fuels. Sustain. Energy Fuels 5 (3), 604–637. https://american-history.si.edu/collections/object/nmah_687671.

https://en.wikipedia.org/wiki/AeroVironment_Helios_Prototype.

https://en.wikipedia.org/wiki/Alkaline_fuel_cell.

https://en.wikipedia.org/wiki/Fuel_cell.

https://en.wikipedia.org/wiki/Orca_(AUV).

https://www.afcenergy.com/.

https://www.afcenergy.com/news/press-releases/launch-of-world-class-anion-exchange-membrane-hydrogen-fuel-cell-test-facility.

https://www.enapter.com/.

https://www.fumatech.com/en/products/membranes-alkaline-water-electrolysis/.

https://www.gencellenergy.com/.

https://www.tokuyama.co.jp/eng/.

Kordesch, K., Gsellmann, J., Cifrain, M., Voss, S., Hacker, V., Aronson, R.R., Fabjan, C., Hejze, T., Daniel-Ivad, J., 1999. Intermittent use of a low-cost alkaline fuel cell-hybrid system for electric vehicles. J. Power Sources. 0378-7753. 80 (1), 190–197. Available from: https://www.sciencedirect.com/science/article/pii/S0378775398002614. doi:10.1016/S0378-7753(98)00261-4.

Kordesch, K., Hacker, V., Gsellmann, J., Cifrain, M., Faleschini, G., Enzinger, P., Fankhauser, R., Ortner, M., Muhr, M., Aronson, R.R., 2000. Alkaline fuel cells applications. J. Power Sources. 0378-7753. 86 (1–2), 162–165.

Li, D., Guo, J., Zhang, J., Zhan, L., Alizadeh, M., 2021. Numerical assessment of a hybrid energy generation process and energy storage system based on alkaline fuel cell, solar energy and Stirling engine. J. Energy Storage. 2352-152X. 39 102631.

McLean, G.F., Niet, T., Prince-Richard, S., Djilali, N., 2002. An assessment of alkaline fuel cell technology. Int. J. Hydrogen Energy. 0360-3199. 27 (5), 507–526. Available from: https://www.sciencedirect.com/science/article/pii/S0360319901001811. doi:10.1016/S0360-3199(01)00181-1.

Mulder, G., Coenen, P., Martens, A., Spaepen, J., 2008. The development of a 6 kW fuel cell generator based on alkaline fuel cell technology. Int. J. Hydrogen Energy. 0360-3199. 33 (12), 3220–3224.

Ni, M., Leung, M., Leung, K.H., Dennis, Y.C., 2006. Technological development and prospect of alkaline fuel cells. In: Proceedings of 16th World Hydrogen Energy Conference. 33–39.

Rostami, M., Manshadi, M.D., Afshari, E., 2022. Performance evaluation of two proton exchange membrane and alkaline fuel cells for use in UAVs by investigating the effect of operating altitude. Int. J. Energy Res. 0363-907X. 46 (2), 1481–1496.

Shen, Y., XueQiang, L., Wang, N., Li, L., Hoseyni, A., 2021. Introducing and investigation of a pumped hydro-compressed air storage based on wind turbine and alkaline fuel cell and electrolyzer. Sustain. Energy Technol. Assess. 2213-1388. 47 101378.

Staffell, I., Ingram, A., 2010. Life cycle assessment of an alkaline fuel cell CHP system. Int. J. Hydrogen Energy. 0360-3199. 35 (6), 2491–2505.

Strasser, K., 1992. Mobile fuel cell development at Siemens. J. Power Sources. 0378-7753. 37 (1-2), 209–219.

Van den Broeck, H., Adriaensen, L., Alfenaar, M., Beekman, A., Blanchart, A., Van Bogaert, G., Vanneste, G., 1986. Status of ELENCO's alkaline fuel cell development. Int. J. Hydrogen Energy. 0360-3199. 11 (7), 471–474.

Veeranjaneyulu, K., Joshi, S., Devalla, V., Kiran, K.S., Khushal, 2023. Recent advancements of PEMFC in transport applications. Adv. Aeromechan. Mater. Manuf. 0094-243X. 2492 (1), 050019.

Verhaert, I., Mulder, G., De Paepe, M., 2016. Evaluation of an alkaline fuel cell system as a micro-CHP. Energy Convers. Manag. 0196-8904. 126, 434–445.

Wang, M., Ruan, J., Zhang, J., Jiang, Y., Gao, F., Zhang, X., Rahman, E., Guo, J., 2024. Modeling, thermodynamic performance analysis, and parameter optimization of a hybrid power generation system coupling thermogalvanic cells with alkaline fuel cells. Energy. 0360-5442. 292 130557.

Wang, Z., Zhang, X., Rezazadeh, A., 2021. Hydrogen fuel and electricity generation from a new hybrid energy system based on wind and solar energies and alkaline fuel cell. Energy Rep. 2352-4847. 7, 2594–2604.

Warshay, M., Prokopius, P.R., 1989. Grove Anniversary (1839–1989) Fuel Cell Symposium E-5084 the Fuel Cell in Space: Yesterday, Today and Tomorrow.

Wee, J.-H., 2007. Applications of proton exchange membrane fuel cell systems. Renew. Sustain. Energy Rev. 1364-0321. 11 (8), 1720–1738. Available from: https://www.sciencedirect.com/science/article/pii/S1364032106000281. doi:10.1016/j.rser.2006.01.005.

Yang, P., Zhu, Y., Zhang, P., Zhang, H., Ziyang, H., Zhang, J., 2014. Performance evaluation of an alkaline fuel cell/thermoelectric generator hybrid system. Int. J. Hydrogen Energy. 0360-3199. 39 (22), 11756–11762.

Zeng, R., Tang, X., Deng, Y., Zhang, X., Hongqiang, L., Yin, W., Zhang, G., 2023. Design and optimization of solar energy system with hydrogen energy storage and alkaline fuel cell. Energy Convers. Manag. 0196-8904. 295 117628.

Zhang, C., Hanqi, L., Zhang, X., Shen, M., Jin, X., 2024a. The Performance Evaluation of a Hybrid System Combining an Alkaline Fuel Cell with an Inhomogeneous Thermoelectric Generator. Energies. 1996-1073. 17 (9), 2066.

Zhang, X., Rahman, E., 2022. Thermodynamic analysis and optimization of a hybrid power system using thermoradiative device to efficiently recover waste heat from alkaline fuel cell. Renew. Energy. 0960-1481. 200, 1240–1250.

Zhang, X., Zhao, Q., Wang, J., Zhang, H., 2024b. Synergizing alkaline fuel cell and thermocapacitive heat engine for hydrogen energy efficient utilization. Process Saf. Environ. Prot. 0957-5820. 186, 706–714.

Zhao, Q., Zhang, H., Ziyang, H., Yangyang, L., 2021. An alkaline fuel cell/direct contact membrane distillation hybrid system for cogenerating electricity and freshwater. Energy. 0360-5442. 225 120303.

Zhu, H., Jiarui, L., Lai, C., Zhang, H., 2022. Recycling alkaline fuel cell waste heat for cooling production via temperature-matching elastocaloric cooler. Int. J. Hydrogen Energy. 0360-3199. 47 (63), 27124–27138.

Zhu, H., Xiao, L., Kuang, M., Wang, J., Zhang, H., 2024. Innovative use of air gap membrane distillation to harvest waste heat from alkaline fuel cell for efficient freshwater production: a comprehensive 4E study. Renew. Energy. 0960-1481. 225 120149.

Chapter 11

Market dynamics of alkaline fuel cells: A global overview

Muhammad Mubeen
The Islamia University of Bahawalpur, Bahawalpur, Pakistan

1 Introduction

Fuel cells are becoming a central technology in the global push toward clean and reliable energy. They offer efficient, flexible power generation with water as the only by product. Among the different types of fuel cells, such as proton exchange membrane, solid oxide, molten carbonate, phosphoric acid, direct methanol, and alkaline, alkaline fuel cells (AFCs) have a special role. They operate at moderate temperatures between 60 and 90 degrees Celsius, provide reliable performance, and can handle some fuel impurities. These characteristics made AFCs suitable for use in the National Aeronautics and Space Administration (NASA)'s Gemini and Apollo missions during the 1960s. Today, the same features are driving renewed interest in AFCs for both gridsupport and offgrid power uses. In 2023, the global fuel cell market was valued at 3.3 billion United States dollars. It is expected to grow to 8.7 billion United States dollars by 2028, with an average annual growth rate of 21.7%. This rise is supported by strong policy backing, lower technology costs, and growing demand for zero-emission vehicles and decentralized power systems. Within this growing market, AFCs are once again gaining attention. They are especially useful in situations that need fast startup, high fuel efficiency, and simpler system design. Countries like Japan and South Korea are actively supporting fuel cell development through national hydrogen strategies, making them important regions for AFC growth (Marketsandmarkets, 2023). The higher projected compound annual growth rate (CAGR) for AFCs, estimated at 18 to 20% over the period from 2025 to 2033, is attributed to anticipated cost reductions through the scale up of alkaline membrane production, lower catalyst loadings, and government incentives that support next-generation fuel cell technologies (Orion Market Research Pvt. Ltd, 2026).

Alkaline Fuel Cells. https://doi.org/10.1016/B978-0-443-36638-3.00013-9

1.1 Market drivers

Several interrelated factors propel AFC market growth:

1.1.1 Stringent emissions regulations:

As transportation and stationary power sectors face increasingly ambitious CO_2 targets, policymakers are mandating or incentivizing low and zero emission alternatives. Fuel cell electric vehicles, or FCEVs, which emit only water during operation, directly support these goals. Global initiatives such as EV30@30, which aims to have 30% of new vehicle sales be zero-emission by 2030, are driving the deployment of FCEVs across Europe, North America, and the Asia Pacific region.

1.1.2 Government subsidies and R&D incentives:

To de risk early stage technologies, many governments offer subsidies for fuel cell manufacturing, hydrogen refueling infrastructure, and demonstration projects. In the United States, the Department of Energy's H2@Scale initiative and Solid State Energy Conversion Alliance have collectively invested over USD 2 billion in fuel cell and hydrogen research since 2005; similar programs in Japan and South Korea underpin large fleets of AFC powered buses and forklifts.

1.1.3 Hydrogen infrastructure expansion

AFCs' reliance on hydrogen rather than reformate gases means that investments in high pressure storage, distribution pipelines, and on site electrolyzers directly enhance the technology's viability. Governments and private consortia are now deploying hydrogen corridors alongside major highways and urban centers, lowering refueling costs and improving system uptime.

1.1.4 System cost reductions

Advances in bipolar plate coatings, gasket materials, and potassium carbonate management have driven down both capital expenditures and maintenance intervals. Modern AFC stacks exhibit higher power densities at lower precious-metal loadings than earlier generations, narrowing the cost gap with proton exchange membrane systems.

1.2 Market restraints

Despite favorable tailwinds, several challenges temper near term AFC adoption:

1.2.1 Hydrogen storage & logistics:

Compressed hydrogen systems add weight and volume to vehicles, constraining range relative to liquid fuel alternatives. Safety standards require rigorous component qualification and up to 1,500 high-cycle durability tests for onboard tanks, increasing upfront costs.

1.2.2 Electrolyte carbonation

Exposure to atmospheric CO_2 gradually converts potassium hydroxide (KOH) electrolyte into potassium carbonate, reducing ionic conductivity. While recirculation and membrane separators can mitigate carbonate buildup, they introduce additional system complexity and parasitic power losses.

1.2.3 Liquid-electrolyte handling

Fuel-cell packages must incorporate robust seals and corrosion resistant materials to contain the caustic electrolyte, complicating stack assembly and maintenance compared to solid electrolyte designs.

1.3 Opportunities

1.3.1 Backup and uninterruptible power

Data centers, hospitals, and critical infrastructure demand rapid-start, high reliability power systems. AFC based generators offer minimal noise, low vibration, and sustained output under variable loads, making them ideal for mission critical applications.

1.3.2 Microgrids & remote sites

In offgrid communities and islanded microgrids, AFCs paired with renewables (solar, wind) and storage can deliver dispatchable, low emission power with a minimal environmental footprint. Hybrid configurations, where excess renewables produce hydrogen via electrolyzers—enhance energy security and resilience.

1.3.3 Marine & heavy-duty transport

AFC's tolerance for reformate impurities and rapid dynamic response position them as ideal for auxiliary power units in ships, rail locomotives, and long haul trucks, where onboard space and refueling flexibility are critical.

1.4 Competitive landscape

AFC technology is manufactured by both well established industrial companies and innovative startups. Prominent firms such as Doosan Fuel Cell in South Korea, Aisin Seiki in Japan, and Ballard Power Systems in Canada provide a wide range of products across stationary, transportation, and portable applications. Collaborations between major energy companies like Shell and Eneco and fuel-cell manufacturers are enabling large-scale hydrogen projects, including offshore wind farms integrated with hydrogen-to-power systems.

2 Global hydrogen market momentum and implications for alkaline fuel cells

The global energy transition has propelled hydrogen to the forefront as a key vector for decarbonization across multiple sectors. This momentum is driven by extensive policy commitments, technological advancements, and substantial capital flows to scale hydrogen production, storage, and utilization infrastructure. AFCs, as a mature hydrogen conversion technology, stand to benefit significantly from this expanding hydrogen ecosystem. Understanding the dynamics of the global hydrogen market is, therefore, essential to contextualize current and future growth trajectories for AFC technologies.

2.1 Expanding hydrogen project pipeline and investment landscape

As of October 2023, 1418 hydrogen projects have been announced globally, sppanning a wide range of applications, including production, transport, storage, and end-use conversion technologies. Notably, over 1000 of these projects are targeted for full or partial deployment by 2030, signaling a rapid scale-up phase aligned with international net-zero targets and hydrogen strategies. The total announced capital investment across these projects reaches approximately USD 570 billion, reflecting a remarkable 30% increase within nine months, underscoring the heightened confidence and strategic prioritization that governments and the private sector place in hydrogen development. However, it is critical to note that while announced investments signal strong future activity, only USD 39 billion, or about 26% of the total capital, has advanced to the final investment decision (FID) stage. This discrepancy illustrates a common challenge in the hydrogen sector: the substantial gap between project announcements and actual financial closure or construction commencement. This gap directly impacts AFC manufacturers and developers, as the timely availability of green hydrogen feedstock hinges on successful project maturation and commissioning. Within the announced projects, clean hydrogen supply capacity is expected to reach 45 million tonnes per annum (Mt p.a.) by 2030. The composition of this supply is anticipated to be predominantly renewable hydrogen (approximately 70%) produced via water electrolysis powered by renewable energy sources, complemented by low-carbon hydrogen (30%) derived from fossil fuels with carbon capture and storage (CCS). This robust pipeline of clean hydrogen capacity is integral for AFCs, which require high-purity hydrogen fuel to maximize efficiency and operational durability.

2.2 Regional trends driving hydrogen and AFC market growth

The geographical distribution of hydrogen projects and investments reveals distinct regional leadership and strategic priorities. Europe currently leads with

540 projects and 193 billion United States dollars invested, benefiting from comprehensive policy frameworks such as the European Union's Hydrogen Strategy and the European Green Deal. These policies incentivize the development of hydrogen infrastructure and technologies, fostering favorable market conditions for AFC adoption. The European hydrogen ecosystem also emphasizes renewable hydrogen production, which aligns well with the zero-emission potential of AFCs. North America follows as a dynamic market, with 248 projects and significant investment growth catalyzed by policy instruments such as the United States Inflation Reduction Act (IRA). The IRA's tax credits and subsidies for clean energy technologies have stimulated private-sector investment, reinforcing the commercial viability of hydrogen production and fuel-cell deployment. This policy-driven surge enhances AFC prospects in power generation, transportation, and industrial sectors. Latin America's hydrogen market is characterized by fewer projects but comparatively large investment volumes, primarily due to gig-scale renewable hydrogen projects leveraging abundant solar and wind resources. While direct AFC market activity in this region remains nascent, the development of large-scale hydrogen supply infrastructure creates potential entry points for AFC systems in the medium term. In Asia, China and India are rapidly expanding their hydrogen initiatives. China alone has committed USD 12.5 billion to hydrogen projects, focusing on the northern and central regions where industrial hydrogen demand and renewable energy potential converge. These developments complement AFC market expansion, especially given China's emphasis on clean energy technologies and the deployment of hydrogen fuel cell. India's growing hydrogen strategy, though still in the early stages, adds further momentum to the regional hydrogen economy.

2.3 Project maturity, deployment challenges, and market implications

Despite the promising pipeline, a critical bottleneck persists in converting announced projects into operational assets. Globally, only 7% of hydrogen investments have reached FID, with Europe at 4%, indicating slower project maturity. China leads with 35% of investments achieving FID, reflecting a more advanced development cycle. This divergence in project maturity highlights the complexities of financing, permitting, technology validation, and supply chain constraints that continue to challenge the hydrogen sector. For AFC manufacturers, this delayed project maturation presents both challenges and strategic opportunities. The slow conversion of hydrogen projects to operational status necessitates AFC companies to engage proactively with project developers, aligning product development and commercialization timelines with evolving hydrogen supply availability. Early partnerships can mitigate risks and facilitate the integration of AFC systems into hydrogen infrastructure from inception. Moreover, scaling electrolyzer capacity is fundamental to achieving the anticipated hydrogen production volumes required by AFC markets. Current installed electrolysis capacity stands at approximately 1.1 GW globally, with projections to reach 305 GW by 2030—representing a 250-fold increase. This dramatic scale-up requires accelrating manufacturing capabilities, strengthening supply chain resilience, and deploying capital. As electrolysis capacity expands, the resultant increase in clean hydrogen availability will underpin AFC market growth by ensuring reliable, low-carbon fuel supplies necessary for large-scale adoption.

2.4 Strategic outlook for AFC market dynamics

The rapid growth in hydrogen production capacity, driven by renewable energy integration and substantial investment flows, creates an enabling environment for the deployment of AFC technology. AFCs, known for their high efficiency, low operating temperature, and cost-effectiveness compared to other fuel cell types, are well-positioned to capitalize on increasing low-cost hydrogen availability. Renewable hydrogen, which constitutes the majority of the projected supply, further enhances the sustainability profile of AFC applications, reinforcing their appeal in power generation, industrial processes, and the transport sector. Nonetheless, the hydrogen market's current developmental stage underscores the importance of policy support, technological innovation, and coordinated industry efforts to bridge the gap between announced projects and operational reality. AFC market stakeholders must navigate regulatory landscapes, incentivize investment, and advance technological improvements to optimize performance and cost competitiveness. In conclusion, the global hydrogen market's rapid scale-up and evolving investment landscape offer critical context for understanding the market dynamics shaping AFC development. Regions with mature hydrogen project pipelines, particularly Europe, China, and North America, are poised to lead the adoption of AFCs. Concurrently, the ambitious expansion of electrolyzer capacity will be a pivotal enabler of AFC growth, securing the hydrogen supply chain needed for widespread commercialization and advancing broader decarbonization goals.

3 Market dynamics of AFCs in the evolving hydrogen economy

As the global energy system undergoes a major transformation toward sustainability, hydrogen technologies have emerged as essential components of decarbonization strategies across various sectors. Among the diverse hydrogen energy technologies, AFCs have attracted attention for their high efficiency, cost-effective materials, and suitability for both stationary and transportation applications. This chapter explores the evolving market dynamics of AFCs by situating them within the broader hydrogen economy, with a particular focus on pricing strategies, collaboration models, and market penetration influences that shape their commercial adoption and growth trajectory.

4 AFCs and their role in clean hydrogen utilization

AFCs operate using an alkaline electrolyte, typically potassium hydroxide, to convert hydrogen and oxygen into electricity, with water and heat as the only byproducts. Their high energy conversion efficiency and the use of non-noble metals in electrode construction make them attractive for early commercialization in niche markets, such as military power systems, space applications, and backup power supplies. However, for AFCs to capture a larger market share, the availability and affordability of clean hydrogen are critical. This interdependence necessitates evaluating broader hydrogen market dynamics to understand the strategic opportunities and challenges for AFC deployment.

4.1 Hydrogen pricing and demand implications for AFC markets

The cost of hydrogen remains a central determinant in the competitiveness of fuel cell technologies. AFCs, like other fuel cell systems, are highly sensitive to hydrogen prices, which fluctuate based on production methods, policy subsidies, and delivery logistics. Currently, gray hydrogen produced from fossil fuels without carbon capture is the cheapest option; however, its environmental footprint contradicts the sustainability goals that AFCs aim to support. The shift toward green hydrogen, produced via electrolysis using renewable energy sources, is imperative for AFCs to realize their full environmental benefits. However, the high cost of green hydrogen production remains a barrier to its widespread use (Gulli et al., 2024). Under Canada's Transformative Hydrogen Scenario, the hydrogen and fuel cell industry is expected to earn up to CAD 50 billion each year by 2050. This growth is driven by rising demand for hydrogen and by support from local manufacturing, services, and infrastructure. The estimate is based on an average hydrogen price of 2 Canadian dollars (CAD) per kilogram. It includes direct income from hydrogen production, as well as additional benefits to related industries such as steam-methane-reforming (SMR), carbon capture, utilization, and storage (CCUS), and hydrogen pipeline development (Natural Resources Canada, 2020). These cost reductions are expected to enhance the commercial viability of AFCs, particularly in applications where sustainability and operational reliability are prioritized. In addition, the impact of alternative fuel prices and the substitution effect are crucial. If hydrogen remains priced higher than conventional fuels, the adoption of AFCs may be delayed despite their superior efficiency (Harichandan and Kar, 2023).

4.2 Market penetration and consumer adoption of AFCs

Market penetration refers to the extent to which a new product gains acceptance relative to its total potential market. For AFCs, this involves not only technological readiness but also consumer awareness, cost parity with incumbent systems, and availability of supporting infrastructure. As the article under review highlights, demand for hydrogen is price-sensitive and affected by factors such as delivery lead time and the environmental credentials of the hydrogen source

(Gulli et al., 2024). These factors apply equally to AFCs, whose success depends on timely hydrogen supply and the assurance of low-carbon credentials. The study identifies penetration effects and consumer demand elasticity as central variables in shaping hydrogen adoption. AFC producers and system integrators must consider these when entering the market. Higher lead times for hydrogen delivery could diminish consumer trust in AFC reliability, especially in time-sensitive applications such as logistics and critical infrastructure (Harichandan and Kar, 2023). Therefore, market strategies should include improved delivery mechanisms, possibly through localized hydrogen hubs, to enhance AFC uptake.

4.3 Collaboration models and strategic alliances

One of the study's core contributions is the evaluation of three collaboration scenarios: (1) market-based pricing, (2) cooperative integration, and (3) coordinated decision-making. Each model has distinct implications for AFC market expansion:

4.3.1 Market-based pricing (competition)

In this decentralized model, hydrogen producers act independently, leading to competitive pricing but often insufficient alignment with sustainability goals. In AFC markets, this could lead to price volatility and inconsistent hydrogen quality, thereby hindering consumer confidence and long-term planning (Ding et al., 2021).

4.3.2 Cooperative integration (collusion model)

Here, producers form strategic alliances to optimize shared outcomes, such as cost reduction and sustainable production. This scenario benefits AFC deployment by stabilizing supply chains and potentially lowering hydrogen prices, enabling more affordable AFC systems (Madio and Pignataro, 2024).

4.3.3 Coordinated decision-making (tariff contract model)

This model features centralized collaboration facilitated by hydrogen hubs, where pricing and delivery logistics are managed collectively. The study found that this structure not only improves profitability but also aligns with environmental goals by enabling broader adoption of green hydrogen. For AFCs, this model presents the most promising path to scale, especially when supported by policy incentives and infrastructure investment.

4.4 Policy, incentives, and hydrogen hubs: Opportunities for AFC growth

Policy alignment plays a pivotal role in the diffusion of AFCs. Government interventions, such as carbon taxes, green hydrogen subsidies, and fuel cell deployment mandates, can significantly influence AFC market dynamics. The concept of hydrogen hubs, highlighted in the study, offers a coordinated infrastructure platform for integrating green hydrogen production, storage, and distribution. These hubs can be tailored to meet the needs of AFCs in specific sectors, such as transportation corridors or industrial clusters, thereby creating localized markets that reduce delivery lead times and improve reliability. Moreover, the study underscores the importance of balancing commission fee structures within hubs to avoid eroding profitability. For AFC manufacturers and service providers, active participation in hub governance could create favorable conditions for the adoption of their technology (Gulli et al., 2024).

4.5 Challenges and future outlook

Despite the positive outlook, the market for AFCs faces several challenges. These include:

- Technological limitations in handling CO_2 and impurities in non-pure hydrogen streams
- Limited deployment of hydrogen infrastructure
- High upfront capital costs compared to battery-based systems

However, as clean hydrogen becomes more accessible and delivery mechanisms more efficient, AFCs are expected to play a vital role in distributed and offgrid energy systems. The integration of AFCs with renewable energy sources and microgrid technologies could further enhance their value proposition in a decarbonized energy landscape.

5 Electrolytic hydrogen cost dynamics and implications for AFC markets

The commercial viability and market penetration of AFCs hinge not only on advances in cell design and stack durability but also on the availability of low-cost, low-carbon hydrogen feedstock. Although AFC technology traditionally relies on alkaline electrolysis for on-site hydrogen generation, the broader hydrogen economy encompasses multiple production pathways, most notably polymer-electrolyte membrane (PEM) electrolysis. The Department of Energy's Hydrogen Program Record No. 24,005 (May 20, 2024) presents a comprehensive, peer-reviewed techno-economic assessment of PEM-based clean hydrogen production, offering critical benchmarks for the levelized cost of hydrogen (LCOH) under varying renewable electricity scenarios. While PEM and alkaline electrolysis differ in membrane chemistry and operating conditions, the resulting hydrogen commodity flows into common infrastructure and markets; thus, the PEM cost projections inform the upstream supply economics that AFC systems ultimately depend upon.

5.1 Levelized cost of hydrogen from renewable PEM electrolysis

Applying the H2A-Lite discounted cash flow model developed by the National Renewable Energy Laboratory and vetted by industry, the U.S. Department of Energy (DOE) analysis evaluates hydrogen production across four illustrative electricity supply cases: a United States grid average case, a midwestern hydropower case, a best-in-class land based wind case, and a hybrid wind plus solar photovoltaic (PV) case optimized for capacity factor (National Laboratory of the Rockies, U.S. Department of Energy, 2026). Under current capital cost assumptions, with installed PEM electrolyzers priced at \$2000 per kilowatt and excluding subsidies or tax credits, the grid-average scenario results in an LCOH of approximately \$7.50 per kilogram of hydrogen. This reflects high electricity prices of 8.3 cents per kilowatt-hour and near constant utilization at a 97% capacity factor. In contrast, electrolyzers coupled with a hybrid wind and solar photovoltaic installation in Texas, leveraging complementary daily and seasonal resource profiles, achieve an LCOH as low as \$5.20 per kilogram of hydrogen, with electricity costs of 3.3 cents per kilowatt-hour and a 74% capacity factor. Intermediate scenarios, such as hydropower with 50% capacity and an electricity cost of 3.4 cents per kilowatt-hour, and class-1 wind with 51% capacity and and an electricity cost of 2.9 cents per kilowatt-hour, produce LCOHs of \$6.70 and \$6.40 per kilogram, respectively (Bloomberg New Energy Finance, 2022). These cost differentials underscore the pivotal roles of electricity prices and capacity factors in determining hydrogen production economics. For AFC market development, regions where renewable resources can support capacity factors above 60% at low wholesale rates (below 3 ¢/kWh) stand to benefit most, yielding hydrogen costs approaching the \$5/kg threshold that significantly enhances fuel cell competitiveness.

5.2 Technical and financial parameters influencing LCOH

The robustness of these LCOH estimates stems from carefully selected system parameters and financial assumptions designed to reflect early market PEM installations. Key electrolyzer specifications include an average system electricity consumption of roughly 57 kWh kg^{-1} H_2, based on a stack operating point of 2 A cm^{-2} at 1.9 V per cell, plus balance-of-plant losses (Pivovar and Ahluwalia, n.d.), (Wang et al., 2023). A centralized plant capacity of 50,000 kg of hydrogen per day is assumed. Electrolyzer replacement costs are estimated at 11% of the installed capital every 40,000 operating hours, while fixed annual operations and maintenance costs are set at 5% of the total installed capital. Financial parameters include a real return on equity of 10.89%, a debt-to-equity ratio of 0.62, an interest rate of 5%, and a combined tax rate of 25.74%. These inputs yield a nominal weighted-average cost of capital of approximately 10%. Sensitivity analyses reveal that a 10% reduction in electrolyzer capital cost (\$1 800/kW vs. \$2000/kW) or a 5% increase in capacity factor can each lower LCOH by roughly \$0.20–\$0.30 per kilogram (National Laboratory of the Rockies, U.S. Department of Energy, 2026).

5.3 Trends in electrolyzer capital costs

Published cost surveys indicate that total installed costs for PEM electrolyzer systems currently range between $1400 and $2500 per kilowatt, inclusive of module, balance-of-plant, and installation expenditures (Badgett et al., 2024; Tengler, 2024). While earlier estimates placed these costs closer to $1000–$1800/kW, inflation and supply-chain pressures have driven mid-range values upward (Lazard, 2021; Bhashyam, 2023). Ongoing R&D under the Hydrogen from Next-generation Electrolyzers of Water (H2NEW) Consortium and the DOE's Clean Hydrogen Electrolysis Program targets a reduction to $1000/kW by 2030, supported by manufacturing scale-up, standardized module designs, and streamlined installation practices (Satyapal et al., Clean Hydrogen Electrolysis Program, n.d.; Randolph et al., 2022). Each $100/kW decrease in installed capital cost corresponds to an LCOH decline of approximately $0.15–$0.20/kg at baseline capacity factors, emphasizing the critical importance of cost-down innovations for market expansion.

5.4 Implications for AFC market dynamics

AFCs, while electrochemically distinct, consume the same commodity hydrogen as PEM-based systems. Therefore, the LCOH benchmarks established for PEM electrolysis directly influence the wholesale price that AFC operators can expect to pay. In markets where renewable supply chains and grid interconnections permit LCOH near or below $6/kg, AFC systems in stationary power, backup, and mobility applications can achieve competitive levelized cost of energy (LCOE) metrics relative to diesel generators and batteries, particularly when factoring in the high round-trip efficiency potential of AFCs (50–60%) and low operating temperatures. Conversely, in regions where the LCOH remains above $7 per kilogram due to high electricity tariffs, limited renewable energy integration, or restricted grid access, the adoption of AFCs is likely to be limited to specialized applications. These may include offgrid installations or high-value industrial sites, where AFCs offer unique advantages despite higher hydrogen costs.

5.5 Pathway to sub-$2/kg hydrogen and AFC scale-up

The DOE Hydrogen Shot initiative sets an ambitious target of reducing electrolytic hydrogen costs to $1/kg by 2031, with an intermediate goal of $2/kg by 2026. Achieving these milestones will require synchronized progress across multiple fronts: improvements in electrolyzer stack durability and performances, enhanced balance-of-plant integration efficiencies, rapid scale-up of manufacturing to leverage economies of scale, and continued deployment of high-capacity-factor renewable projects (including hybrid wind + solar and emerging marine or geothermal resources). For AFC markets, securing reliable hydrogen at or below $2–$3/kg will be transformative, enabling widespread adoption in decentralized energy systems and transportation applications, where fuel cell capital and operating costs remain a barrier today. In sum, the DOE's PEM-electrolysis cost projections furnish a vital reference point for stakeholders planning AFC deployments worldwide. By mapping the anticipated evolution of hydrogen production economics under varied renewable scenarios, this analysis informs strategic decisions on plant siting, offtake contracting, and technology choice, ultimately shaping the global market dynamics of AFC systems. As electrolyzer capital costs decline and renewable electricity becomes ever cheaper, AFCs stand poised to capitalize on abundant, affordable hydrogen, unlocking a new era in clean, efficient power generation.

6 AFCs in transport decarbonization

The urgent need to curb global warming to 1.5°C has placed road transport squarely in the crosshairs of decarbonization efforts. Today, this sector consumes over 90% of global liquid fuels and accounts for nearly 7.7 Gt CO_2 in 2021 alone (International Energy Agency (IEA), 2024; International Energy Agency (IEA), 2021). While battery electric vehicles (BEVs) and polymer electrolyte membrane fuel cell electric vehicles (PEMFC-FCEVs) lead the development landscape, AFCs provide a combination of rapid refueling, high electrical efficiency, and lower catalyst costs that is suited to certain targeted mobility sectors.

6.1 Environmental and policy drivers

International climate agreements, most notably the Paris Accord and the United Nations' Sustainable Development Goals (SDGs 7 and 13), have set a clear trajectory: deep greenhouse gas (GHG) cuts across all sectors by mid-century (United Nations, 2016), Intergovernmental Panel on Climate Change (IPCC), 2022). The International Energy Agency's "Net Zero by 2050" roadmap specifies over 400 critical milestones, including a 20% reduction in transport emissions by 2030 (International Energy Agency (IEA), 2021). In Europe, the Fit for 55 package codifies this direction, banning the sale of new petrol and diesel vehicles by 2035 (Erbach and Jensen, 2024; European Parliament, 2022). At the same time, national and regional hydrogen mobility initiatives, such as California's Fuel Cell Partnership, are coordinating public and private incentives to develop the refueling infrastructure required for FCEVs (Hydrogen Fuel Cell Partnership, 2025). AFCs, which operate at moderate temperatures (50–90°C) using aqueous potassium hydroxide electrolytes, are well suited to these policy aims. Their tolerance for nonprecious metal catalysts (e.g., nickel) minimizes system expenditures, while their high reaction kinetics contribute to superior electrical efficiency. As governments tighten emissions standards, AFCs' combination of cost-effective materials and robust performance positions them as a compelling complement to proton exchange membrane fuel cells (PEMFCs) in both light duty fleets and stationary backup power systems.

6.2 Well-to-wheel emissions and carbon intensity

Well-to-wheel (WTW) studies consistently demonstrate that hydrogen FCEVs powered by SMR hydrogen emit 15–45% less GHG than conventional ICE vehicles—even before accounting for CCUS. When CCUS is applied, lifecycle emissions can decline significantly from 247 g CO_2e/mi to 105 g CO_2e/mi (Liu et al., 2020). AFCs demonstrate a comparative advantage in this context: their alkaline electrolyte system is compatible with low-level CO_2 impurities, allowing direct integration of hydrogen from methanol reforming or biomass-gasification processes. Such integration with emerging carbon-capture-use hubs can further reduce overall emissions (Quarton and Samsatli, 2020; Al-Qahtani et al., 2020a). However, the carbon intensity of the electricity used for hydrogen electrolysis remains a pivotal factor: FCEVs refueled via grid-powered electrolyzers in regions with fossil-dominated power mixes can produce higher WTW emissions than ICEs or BEVs charged from the same grid (Wang, 2020; Yoo et al., 2018). Thus, strategic AFC deployment should target areas with available renewable energy or dedicated green hydrogen corridors.

6.3 Cost and infrastructure dynamics

Currently, green hydrogen is more expensive than hydrogen produced by steam methane reforming, but forecasts indicate a significant reduction in electrolyzer and renewable energy costs, especially from solar and wind, by 2030 (International Energy Agency (IEA), 2022). AFC stacks use affordable nickel catalysts and robust alkaline membranes, making it feasible to achieve system costs under £800 per kilowatt by the end of the decade, comparable to the goals set for advanced PEMFCs. Their flexible design enables easy scaling, with smaller units suited for remote sites like telecom towers or data centers, and larger systems ideal for supporting public transport hubs such as bus depots. Yet, widespread market adoption hinges on a parallel expansion of hydrogen refueling stations (HRS). Today, just over 1020 HRS exist globally (International Energy Agency (IEA), 2023), a fraction of the network needed to sustain both light-duty and heavy-duty FCEV fleets. Coordinated investment in "hydrogen highways" and interoperable station designs will be essential to unlocking the potential of the AFC market.

6.4 Heavy-duty and fleet applications

AFC technology demonstrates strong performance in heavy duty and fleet applications, where rapid refueling and predictable routing align with the inherent advantages of fuel cell systems. In contrast to BEVs—which require substantial battery mass and extended charging durations—AFC-powered trucks and buses can refuel within 3–5 minutes and achieve driving ranges exceeding 600 km, closely paralleling diesel performance benchmarks (International Energy Agency (IEA), 2023). In the UK, pilot demonstrations involving AFC-based shuttle buses and airport ground-support vehicles have demonstrated robust operational reliability and considerable efficiency under real-world conditions. Fleet operators use centralized hydrogen production and storage systems alongside AFC units designed for durability, capable of withstanding thousands of start-stop cycles without degrading non-platinum catalysts. As the shift to electric urban logistics and public transport accelerates, AFC systems offer a practical way to connect zero-emission goals with real-world feasibility.

7 Regional market analysis of AFCs

The global advancement of hydrogen energy is significantly influencing the development of clean energy technologies, with AFCs emerging as a critical component in this transition. Owing to their high efficiency, relatively low cost, and compatibility with high purity green hydrogen, AFCs are increasingly being integrated into national energy strategies. As countries aim to reduce carbon emissions and achieve long term climate targets, AFCs are gaining attention for their potential in stationary power, industrial use, and transport applications. This section provides an in depth analysis of key regional markets, including Europe, China, South Korea, Japan, and Australia, highlighting how each region's hydrogen policies, infrastructure development, and technological innovation are shaping the demand and deployment of AFCs. Through this comparative analysis, the chapter explores both the enabling conditions and the barriers influencing the global expansion of AFC technologies.

7.1 Europe: hydrogen infrastructure development and its implications for the AFC market

The rapid evolution of Europe's hydrogen strategy has catalyzed substantial shifts in the continent's energy landscape, with far reaching implications for downstream technologies, including the AFC market. Amid an urgent push for decarbonization and energy independence, hydrogen is increasingly recognized as a cornerstone of the European Union's (EU) net-zero transition, especially following geopolitical disruptions such as the energy crisis triggered by the Russia-Ukraine conflict (European Commission Directorate-General for Energy, 2019). These macro-level developments are accelerating hydrogen infrastructure planning and, in turn, reshaping market dynamics highly relevant to the AFC sector. The European Commission (EC) laid the groundwork for renewable hydrogen deployment through the "Clean Energy for All Europeans" initiative in 2016 (European Commission Directorate-General for Energy, 2019), which has since evolved with enhanced ambition. Under the "Fit-for-55" legislative package, the EU committed to a 55% reduction in GHG emissions by 2030, while simultaneously setting a goal of producing 6.7 million tons of renewable hydrogen by that year (Council of the European Union, 2025). This target was revised upward in the "REPowerEU" communication in

2022, increasing the domestic production goal to 10 million tonnes and adding another 10 million tonnes of imported hydrogen, collectively equivalent to 666 TWh (*REPowerEU: A Plan to Rapidly Reduce Dependence on Russian Fossil Fuels and Fast Forward the Green Transition, 2022*). These figures imply a future market in which large volumes of hydrogen are produced and distributed across the continent, an essential precursor to expanding applications such as AFCs. Despite strong policy signals and national hydrogen strategies, Europe currently lacks a cohesive infrastructure to suppor the envisaged hydrogen economy. The European Hydrogen Backbone initiative has proposed a pan-European hydrogen transmission network that utilizes both new and repurposed gas pipelines (Wang et al., 2020; European Hydrogen and Backbone, 2022). This network seeks to link production hubs, primarily located on the edges of Europe, with demand centers in Central and Western Europe. These demand centers include industrial clusters and urban transportation systems. Such connectivity is particularly important for fuel cell technologies, which depend on the reliable and timely supply of highpurity hydrogen across different locations. To identify optimal transition pathways, (Kountouris et al., 2024) employ a techno-economic energy system model (Balmorel) to simulate multiple hydrogen infrastructure scenarios across the EU27, the UK, Norway, Switzerland, and the Balkans. Three scenarios are analyzed: Hydrogen Europe (H2E), which allows imports and all production technologies (electrolysis, SMR, and SMR with CCS); Green Hydrogen Europe (GH2E), which restricts production to electrolysis but permits imports; and Self-Sufficient Green Hydrogen Europe (SSGH2E), which bans imports and relies solely on European electrolysis capacity. The latter two scenarios hold particular significance for the AFC market, as they prioritize green hydrogen, the primary fuel type compatible with AFCs, which require high-purity hydrogen and benefit from cost-effective production. A critical insight from this modeling is that green hydrogen, derived from electrolysis using renewable electricity, is central to a resilient European hydrogen economy. By 2050, a self-sufficient Europe would require approximately 518 GW of electrolyzer capacity, a staggering increase from the 4 GW installed in 2022, representing a 127-fold expansion. Given that alkaline electrolyzers are among the most mature and cost-effective technologies available today (Nami et al., 2022), their widespread deployment would inherently favor the parallel expansion of the AFC market. AFCs, known for their efficiency and suitability for stationary and mobility applications, align naturally with such green hydrogen pathways (Kountouris et al., 2024). However, the path forward is complex. The study identifies a potential "lock-in" effect of blue hydrogen—produced from natural gas via SMR with CCS—which may delay or even undermine green hydrogen scalability if deployed too extensively in the medium term. The associated risk is not only technological inertia but also a continued dependency on methane, which could have adverse implications for the purity and carbon intensity of hydrogen used in AFC systems. This is particularly relevant as AFCs are sensitive to CO_2 contamination in their electrolyte, thus underscoring the preference for green hydrogen. Infrastructure expansion is another decisive factor. The Balmorel model demonstrates that cost-optimal development favors peripheral hydrogen production centers, with strategic pipeline build-outs to Western and Central European demand hubs. These findings are supported by complementary modeling studies that emphasize the importance of sector coupling and multi-vector energy integration (Pickering et al., 2022; Wetzel et al., 2023; Frischmuth and Härtel, 2022; Seck et al., 2022). For the AFC sector, this geographical distribution suggests that large-scale deployment may be concentrated in industrial and urban clusters with strong pipeline or storage connectivity, rather than in isolated or rural settings. Moreover, the development of hydrogen storage facilities, a core recommendation of the study, would directly support AFC applications that require buffering and load balancing to cope with intermittent renewable generation. From a cost perspective, the study finds that a self-sufficient hydrogen economy, while eliminating imports, would raise total system expenses by only about 3%, further reinforcing the economic feasibility of green hydrogen scenarios. This marginal increase is justified, especially given the long-term benefits of avoiding fossil-fuel lock-in and promoting energy sovereignty. For the AFC market, this is a critical signal: if green hydrogen becomes available at scale in Europe at a relatively low incremental cost, it creates a stable and sustainable foundation for fuel cell deployment across sectors, including power backup, offgrid systems, and even electric mobility. Technological uncertainty plays a dual role. On one hand, future increases in CO_2 transportation and storage costs may render blue hydrogen less competitive, thereby accelerating green hydrogen adoption and, by extension, the AFC market (Smith et al., 2021). On the other hand, the actual ramp-up of electrolysis capacity hinges on advances in manufacturing scale and cost reduction. Here again, alkaline electrolyzers are particularly well-positioned due to their lower capital costs and simpler design relative to other technologies such as PEM or SOEC (International Renewable Energy Agency (IRENA), 2020). In summary, the study provides a robust analytical foundation for understanding the spatial, temporal, and technological dimensions of Europe's hydrogen transition. It shows that a coherent, sector-coupled hydrogen infrastructure, with a strong emphasis on electrolysis and renewable power integration, is not only achievable but also economically prudent. This scenario bodes well for the AFC market, as it enhances the availability of high-purity green hydrogen, reduces systemic reliance on fossil-based hydrogen, and ensures the logistical infrastructure needed for widespread AFC deployment. The alignment of policy, infrastructure, and technology thus positions Europe as a potential leader in both green hydrogen and its applications, including the advanced fuel cell sectors.

7.1.1 Europe's renewable energy landscape

Europe has witnessed a remarkable transformation in its energy sector, particularly through the integration of renewable energy sources such as solar, wind, hydroelectric, and green hydrogen. The EU and the UK have made substantial progress, significantly increasing the share of renewables in their energy mix. From 2015 to 2021, the contribution of renewables rose from 23.6% to 34.1%, reflecting strategic investments, technological advancements, and favorable policies. Among the EU-27 and the UK, Germany has emerged as a leader in renewable energy production, followed closely by Austria and Belgium. Germany not only leads in solar and wind energy capacity but also ranks first in total renewable energy output. Regionally, Northern Sweden, Germany, and France are identified as renewable energy hubs due to their favorable geographical and climatic conditions (Eurostat, 2025). A major trend identified is the correlation between high electricity consumption and renewable energy adoption. Countries with higher energy demands are also the most aggressive in transitioning to clean energy, suggesting that energy-intensive economies are driving the green shift. One of the most critical developments in the European energy landscape is the increasing role of green hydrogen, a zero-carbon fuel produced via electrolysis powered by renewable sources. Germany and France have taken the lead in this area, showcasing substantial production capacities that position them at the forefront of this emerging sector (Eurostat, 2025). The EC's Hydrogen Strategy, launched in July 2020, sets ambitious goals: installing 6 GW of electrolyzer capacity for renewable hydrogen and producing up to 1 million tons of renewable hydrogen by 2024. Green hydrogen is pivotal in decarbonizing hard-to-abate sectors such as steel manufacturing, chemicals, aviation, and maritime transport (Commission, E., 2020). Furthermore, as renewable energy sources like wind and solar are intermittent, green hydrogen serves as a key solution for energy storage and grid balancing. Surplus electricity can be stored as hydrogen and converted back to electricity during periods of low renewable generation, thereby enhancing grid stability (Kakoulaki et al., 2021). In the broader context of energy security, Europe currently imports over half (about 55%) of its energy. The development of domestic green hydrogen infrastructure can significantly reduce this dependency. The EU aims to produce up to 10 million tons of renewable hydrogen by 2030, aligning energy independence with environmental goals (Rodríguez-Fernández et al., 2022). Moreover, the green hydrogen economy is anticipated to drive economic growth and job creation. According to estimates, clean hydrogen could meet 13–14% of the EU's energy demand by 2050, potentially developing into a €140 billion annual industry and creating numerous employment opportunities (H2cluster Finland, 2023). Significant financial commitments have been made to support this transition. Under the EU's Next Generation Recovery Plan, €150 billion is allocated to the "Power Up" initiative, which includes hydrogen technologies (European Commission, 2023). In addition, the Clean Hydrogen Partnership, a public-private alliance with a €1 billion budget, is spearheading research and development in the field (Clean Hydrogen Partnership, 2021). Europe's renewable energy transition is multi-faceted and robust, with clear policy direction, substantial investment, and increasing deployment of key technologies such as green hydrogen. These developments position the EU and UK as global leaders in the pursuit of a sustainable, secure, and decarbonized energy future.

7.1.2 UK market dynamics for AFCs

The UK's Net Zero Strategy initiates a substantial change in its transport and power sectors, creating a strong foundation for AFC market adoption. With a mandatory goal to achieve net zero GHG emissions by 2050, the UK has recognized road transport as a major source of emissions and a key priority for decarbonization (Dixon et al., 2022; Chaudry et al., 2022). The planned ban on new petrol and diesel vehicles from 2030 (and the requirement that all new cars must be zero-emission by 2035) highlights the need for multiple low-carbon pathways, including FCEVs powered by AFCs, which offer rapid refueling and long range compared to BEVs (Penn et al., 2022; Haugen et al., 2021). AFCs are especially well-suited to the UK's energy landscape because of their mature technology, high energy efficiency, and flexibility to run on pure hydrogen or hydrogen mixed with reformate from domestic bio-sources. Unlike proton exchange membrane fuel cells, AFCs can tolerate low levels of carbon monoxide. This tolerance allows AFCs to work with small-scale methanol reforming or bio-methane streams, which is a significant advantage given the UK's increasing investments in carbon capture and renewable methanol produced from captured CO_2 (Quarton and Samsatli, 2020; Al-Qahtani et al., 2020b). Recent techno-economic studies demonstrate that coupling alkaline electrolyzers with AFC stacks can achieve power-to-liquid efficiencies up to 0.71, making them competitive for both stationary backup power in remote regions and fleet applications in logistics hubs (Lonis et al., 2021). Several UK initiatives underscore AFC market momentum. The UK Hydrogen Strategy allocated £500 million to promote the deployment of electrolyzers and fuel cells, with preliminary projects in Scotland and North East England leveraging locally produced green hydrogen to supply AFC-powered backup generators at telecom and data center sites (Quarton and Samsatli, 2020). Moreover, cooperative research and development programs between universities (e.g., Imperial College London) and industry consortia are improving low-cost, durable alkaline membranes to reduce system costs below £800/kW by 2030—in line with advanced PEMFC objectives (Haugen et al., 2022; Burrin et al., 2021). As the UK scales up its hydrogen refueling infrastructure, AFCs are poised to capture a significant share of both the on-road FCEV market and the offgrid power segment, driving the country's transition to a resilient, zero-emission energy landscape.

7.2 North American market dynamics for AFCs

The North American fuel cell market is undergoing rapid transformation, driven by increasing investments in hydrogen infrastructure and the urgent need for low-emission power solutions. Within this evolving landscape, AFCs are emerging as a viable and strategic technology, particularly in stationary and backup power applications. Characterized by their mature operating principles, low-cost catalysts, and compatibility with high-purity hydrogen, AFCs are increasingly deployed in critical infrastructure such as telecommunications, data centers, and microgrids. Market forecasts indicate strong growth across various fuel cell segments in the region, with AFCs playing a significant role in meeting the growing demand for decentralized, efficient, and clean energy systems. As the sector diversifies into portable and offgrid applications, AFCs are expected to benefit from both public and private-sector efforts to advance hydrogen-based energy solutions.

7.2.1 Market growth and segmentationmarket growth and segmentation

The North American fuel-cell market is poised for robust expansion, with AFCs playing a key role in stationary and backup power applications. According to Mordor Intelligence, the North America Fuel Cell Technology Market is forecast to grow at a CAGR of 14.97% through 2029, driven by post-pandemic recovery and renewed investment in hydrogen infrastructure. AFCs, valued for their mature technology, low catalyst cost, and efficient hydrogen utilization, are increasingly deployed in telecom backup systems, data centers, and microgrids. In the portable and direct-methanol segments, high energy density and the convenience of liquid fuels underpin a projected spike: the direct methanol fuel cell market alone is expected to exceed USD 900 million by 2034, while the broader portable fuel-cell segment may grow at over 27% CAGR through 2034, supporting offgrid and distributed power needs (Future Market Insights, Inc, 2024).

7.2.2 Key policy drivers

United states

The United States Department of Energy's Hydrogen and Fuel Cell Technologies Office (HFTO) has catalyzed AFC advancement through programs such as H2@ Scale. A flagship demonstration, in collaboration with Caterpillar, Microsoft, and Ballard Power Systems, showcased a megawatt-scale AFC backup system at Microsoft's Wyoming data center—delivering 99.999% uptime over a simulated 48-hour outage (U.S. Department of Energy, 2024). Federal support under the Infrastructure Investment and Jobs Act commits several billion dollars to hydrogen production, storage, and distribution infrastructure (International Energy Agency (IEA), 2026). Furthermore, HFTO's Technical Targets for Liquid Alkaline Electrolysis aim to push electrolyzer performance to 1 A/cm^2 at 1.8 V and 48 kWh/kg H_2 by 2026, with capital costs down to USD 100/kW—pivotal enablers for cost-competitive AFC deployments (U.S. Department of Energy, 2022).

Canada

Canada's 2020 Hydrogen Strategy Progress Report targets hydrogen to supply up to 18% of the national energy demand by 2050, focusing especially on heavy-duty transport, steelmaking, and commercial buildings, which are sectors well-suited for AFCs (Canada Energy Regulator, 2025). With clean hydrogen production expected to reach 6–20 million tonnes annually by mid-century, AFCs are poised to support stationary power, industrial processing, and heavy-duty vehicle applications. The sector generated CAD 527 million in revenue in 2021, with CAD 125 million invested in R&D, underscoring the country's focus on innovation (Canadian Hydrogen and Fuel Cell Association (CHFCA), 2022).

7.2.3 Investment landscape

Venture capital, corporate, and government funding are accelerating the commercialization of AFCs. According to Global Market Insights, North America accounts for nearly half of the global fuel-cell market, and the regional market is projected to exceed USD 9.21 billion by 2034, growing at a CAGR of 20.47%. This creates an environment in which AFCs are becoming increasingly competitive with combustion turbines in peaking applications (Huang and Li, 2022). Major players such as Plug Power, Ballard Power Systems, and Loop Energy are scaling up AFC systems—for example, Loop's 120 kW AFC modules—while driving down levelized costs to approximately USD 0.06 to 0.09 per kWh for grid-connected backup applications (Global Market Insights Inc, 2024).

7.2.4 Outlook and challenges

Despite momentum, challenges remain:

- **Infrastructure gaps:** Hydrogen refueling and high-pressure storage networks are nascent, limiting long-range AFC mobility use-cases.
- **Supply chain:** Electrolyzer costs and catalyst availability must decrease further to reach DOE's sub-$2/kg hydrogen targets.
- **Regulatory alignment:** Streamlined codes for the handling of liquid and gaseous handling will be critical to accelerating AFC installations.

Continued public–private partnerships, coupled with aggressive electrolyzer performance targets (U.S. Department of Energy, 2022) and cross-border hydrogen hubs under the Clean Hydrogen Partnership, will be essential to sustaining AFC market growth. As North America advances its decarbonization goals, AFCs, supported by their mature technology and cost benefits, are well-placed to provide reliable, low-carbon power for stationary, portable, and transportation applications.

7.3 Australia: strategic positioning in the emerging hydrogen and AFC economy

Australia is actively establishing itself as a strategic player in the global hydrogen economy, driven by robust government support, industrial interest, and international partnerships. These developments create a strong foundation for the commercialization of AFCs, especially in niche areas such as offgrid power generation, mining operations, and future export partnerships.

7.3.1 National strategy and infrastructure development

The Australian government's National Hydrogen Strategy, first launched in 2019 and updated in 2024, sets a clear roadmap for achieving a globally competitive hydrogen industry by 2030. The strategy envisions large-scale investments in hydrogen infrastructure, with targeted support for emerging technologies, including fuel cells for stationary, transport, and industrial applications (*Department of Climate Change, Energy, the Environment and Water [DCCEEW], 2019*; DCCEEW, 2024a). These policy actions are creating a favorable environment for AFC deployment by ensuring a reliable supply of high-purity hydrogen and supporting infrastructure such as hydrogen hubs, pipelines, and storage networks (DCCEEW, 2024b). According to the National Hydrogen Infrastructure Assessment (2024), over 100 billion Australian dollars (AUD) in potential investments have been identified across infrastructure sectors, including electrolyzers, refueling stations, and hydrogen storage. This is significant for AFCs, which require clean hydrogen and stable supply channels for optimal performance (DCCEEW, 2024c).

7.3.2 Research capacity and innovation pathways

Australia's national science agency, the Commonwealth Scientific and Industrial Research Organisation (CSIRO), is central to driving hydrogen innovation. Its Hydrogen Roadmap outlines research priorities for hydrogen generation and fuel cell integration, aiming to enhance technological readiness across applications. CSIRO's 2024 electrolyzer manufacturing report highlights Australia's growing manufacturing capabilities, which could reduce hydrogen production costs and make AFC systems more viable in distributed energy markets (Csiro, 2024). Scientific research also underscores the country's increasing contribution to AFC technologies. A recent peer-reviewed study in Materials Today Sustainability discusses solid alkaline membrane development as a pathway to improving AFC efficiency and life cycle sustainability, with Australian research institutions among the contributors (Hamada et al., 2023). A study by Queensland University of Technology highlights AFCs as a scalable solution for powering Australia's remote communities and mining operations, which have high energy demands and limited access to the grid (Smith, 2024).

7.3.3 Industrial uptake and export outlook

Australia's energy-intensive industries, particularly mining and mineral processing, offer fertile ground for the deployment of AFCs. AFCs, with their high energy efficiency and relatively simple operational requirements, are well-suited to remote and rugged environments. Industry stakeholders are already exploring hybrid systems combining solar energy, hydrogen storage, and AFCs to provide uninterrupted offgrid power (Smith, 2024). Internationally, Australia's hydrogen export strategy focuses on key AFC markets like Japan and South Korea, both of which are making significant investments in AFC infrastructure. Strategic alignment with these countries increases the likelihood of bilateral projects that combine Australian hydrogen exports with AFC deployment initiatives in East Asia (*Department of Climate Change, Energy, the Environment and Water (DCCEEW), 2019*; DCCEEW, 2024b). Australia's hydrogen economy is rapidly evolving through coordinated government policies, targeted R&D, and infrastructure expansion. These developments directly benefit the AFC sector by reducing hydrogen costs, expanding use cases, and encouraging innovation in fuel cell materials and systems. With sustained investment, Australia is well-positioned to integrate AFCs into its clean energy landscape.

7.4 Asia-pacific markets

7.4.1 South Korea: Market dynamics of AFCs

The South Korean AFC market remains at an emergent stage compared to the well-established PEMFC and phosphoric acid fuel cell (PAFC) segments. However, focused development in the hydrogen vehicle sector has driven notable forecasted growth: the AFC segment in the South Korean hydrogen fuel-cell vehicle market is projected to expand from a modest base in 2025 to approximately USD 600 million by 2035, at a CAGR of 23.3% over 2025–2035 (Bhoyar and Akre, 2026). In the broader context, Asia-Pacific is expected to register the highest regional CAGR in the global AFC market, reflecting strong policy support and industrialization, with South Korea among its fastest-growing national markets (Reed Intelligence Research Team, 2026).

Key sectors driving demand for AFCs

AFCs are increasingly in demand across key sectors such as transportation, stationary power, and specialized industries due to their high efficiency, low catalyst costs, and strong performance in varied operating conditions.

Transportation applications

South Korea's passenger hydrogen fuel-cell vehicle fleet is primarily dominated by PEMFC technologies. However, several specialized vehicle programs, especially in the heavy-duty and micro-mobility sectors, have started pilot trials using AFCs due to their robustness and easier electrolyte management in subzero conditions (Wikipedia, 2025). Government incentives under the Ministry of Trade, Industry and Energy hydrogen roadmap prioritize demonstrations of next-generation fuel cells, including AFCs, for public transit and commercial fleets (UK's Department for International Trade, 2021).

Stationary and backup power: Telecom base stations and remote backup power installations have shown early adoption of small–scale AFC units, leveraging their high tolerance to fuel impurities and lower platinum loading than PEMFCs. Recent trials in Jeju Island's offgrid telecom network achieved over 5000 hours of continuous operation with ruggedized AFC stacks, validating their suitability for demanding, unmanned applications (HTF Market Intelligence, 2025).

Specialized industrial uses: In petrochemical and semiconductor fabrication facilities, where reactor purity and continuous power are essential, AFCs have been tested as uninterruptible power supply (UPS) systems. Their alkaline electrolyte allows easier management of by-product CO_2 than acid-based systems, reducing auxiliary purification costs.

Policy and regulatory framework

South Korea's Hydrogen Economy Roadmap 2040 (announced in 2021) earmarks KRW 1.5 trillion (≈USD 1.1 billion) over 2021–2025 for diversifying fuel-cell R&D beyond PEMFCs, explicitly mentioning alkaline membrane and liquid-alkaline systems. Alongside direct R&D subsidies (covering up to 70% of project costs), the government provides up to 80% capital-expenditure subsidies for demonstration installations of non-PEMFC technologies in public facilities (UK's Department for International Trade, 2021). New safety and quality infrastructure standards under the Korean Agency for Technology and Standards (KATS) are being updated to include AFC-specific operating conditions, such as pressure, temperature, and electrolyte management, to accelerate certification and market entry (Government of the Republic of Korea, Ministry of Trade, Industry and Energy, 2019).

Competitive landscape and industry players

The competitive landscape for AFCs in South Korea is shaped by a dynamic mix of research institutions, innovative startups, major industrial conglomerates, and strategic international partnerships, all working to advance AFC technology and expand its commercial deployment across diverse applications.

Research Institutes & Startups: Incheon National University's Polymer Energy Materials Center has pioneered cross-linked anion-exchange membranes that deliver >1 A cm^{-2} at 0.7 V in single-cell tests with alkaline liquid electrolyte circulation, highlighting the technological potential of AEMFC hybrids (Eurekalert, 2023). Startups such as H2AlkaTech (Seoul) are scaling lab-proven AEMFC stacks (5 kW modules) toward pilot deployment in telecom stations by 2026.

Large conglomerates: While Doosan Fuel Cell and Hyundai Motor Group continue to focus primarily on PEMFCs for stationary and automotive uses, Hyundai is also expanding its expertise in core hydrogen technologies. In January 2024, Hyundai and Kia signed an agreement with W. L. Gore & Associates to co-develop advanced PEMs for next-generation hydrogen fuel cell systems, aiming to improve performance and durability for commercial vehicles (Corporation, H. M. C. K., 2024).

International partnerships

South Korean companies like Korea Hydro & Nuclear Power (KHNP) have formed strategic partnerships with international AFC and hydrogen technology leaders such as Norway's Nel and the United States' FuelCell Energy. These collaborations combine South Korea's strong manufacturing and nuclear power expertise with advanced alkaline electrolysis and fuel cell technologies to develop clean hydrogen solutions for domestic use and for export to global markets (Čučuk, 2025).

Technology performance, variation, and innovation

AFC technology has advanced from static, batch refilled liquid-electrolyte stacks to circulating-electrolyte systems with real-time pH control and impurity removal. The use of flowing or mobile electrolytes has been shown to mitigate issues such as carbonate formation and thermal management challenges, thereby enhancing system stability and durability (Hamada et al., 2023). Standardization efforts, led by KATS in collaboration with the German–Korea Hydrogen Standards Working Group, aim to reduce cell-to-cell variation by defining uniform protocols for alkaline electrolyte composition, membrane-electrode assembly fabrication, and stack assembly tolerances.

Market barriers and future outlook

Key barriers include the lack of established supply chains for large-volume alkaline electrolytes and the need to develop durable, low-cost anion-exchange membranes for long-term operation. The government's upcoming Hydrogen Innovation Cluster in Ulsan (2025) will host a dedicated AFC pilot line to overcome scale-up challenges and validate integrated production, aiming to cut module costs by 40% by 2030. If achieved, South Korea is poised to become a leading exporter of AFC technology for niche applications in cold climates, remote power, and specialty vehicles—complementing its global leadership in PEMFC and fuel-cell electric vehicles.

7.4.2 China: market dynamics of AFCs

AFCs occupy a distinctive niche within China's hydrogen energy roadmap, leveraging mature chemical processing capabilities and favorable cost structures to support both stationary power and distributed energy applications.

Unlike PEMFCs, which rely on costly platinum-group catalysts, AFCs use an aqueous KOH electrolyte and non-precious-metal electrodes, typically nickel or silver. This leads to substantially lower stack costs and simpler manufacturing processes. China's mature nickel refining and KOH production industries offer a well-integrated supply chain, allowing AFC system costs to decrease below 1000 USD per kilowatt at scale, roughly 20 to 30% lower than similar PEMFC systems (International Energy Agency (IEA), 2024).

Application segments of AFCs in China's energy and industrial landscape

In China, AFCs are being used across many sectors as the country works to reduce carbon emissions and increase the use of clean energy. These fuel cells are now found in places like data centers, factories, and remote areas where reliable electricity is needed. They also help store and reuse energy from wind and solar power, especially in regions with an excess of renewable energy that often goes unused. By using AFCs, China is discovering new ways to support clean energy, improve power reliability, and reduce pollution in both cities and industrial zones.

Market Segmentation and Application Domains To date, AFCs represent roughly 15% of China's cumulative fuel cell capacity (estimated at 2 GW total installed by the end of 2024), with the remaining share dominated by PEMFCs and emerging solid oxide fuel cells (SOFCs).

Stationary and backup power: Over 200 MW of AFC systems have been deployed across telecom towers, data centers, and government facilities in Jiangsu, Zhejiang, and Guangdong provinces, where reliable offgrid power is critical during extreme weather events and grid instability (Han et al., 2022).

Renewable energy integration: In regions such as Inner Mongolia, Hebei, and Xinjiang, where over 230 billion kWh of wind and solar power was curtailed in 2022, AFCs function as the generation element in power-to-gas-to-power (P2G2P) microgrid projects. Through alkaline electrolysis, excess renewable electricity is converted to hydrogen at efficiencies approaching 75%, stored, and later reconverted to power via AFC stacks at round-trip efficiencies of 45%–50%, effectively mitigating curtailment and providing peaking capacity (Liao et al., 2024).

Industrial backup in heavy manufacturing: In the steel and petrochemical zones of Shandong and Shanxi, AFC backup units (5–100 kW each) replace diesel generators to maintain process continuity during outages, reducing CO_2 emissions by over 60% on a life-cycle basis when paired with byproduct hydrogen streams from chlor-alkali plants.

Policy and incentive framework

China's 14th Five-Year Plan and the recently released "Hydrogen Energy Industry Development Plan (2021–2035)" explicitly identify alkaline electrolysis and AFC technology as strategic priorities. Key policy measures include:

R&D subsidies: Up to 50% cost-sharing grants for pilot plants with alkaline electrolysis capacity exceeding 1 MW, aimed at driving hydrogen production costs below 3 USD/kg by 2025.

Demonstration funding: Preferential loans and tax incentives for AFC installations in grid-stabilization projects, covering up to 30% of capital expenditure for systems above 500 kW.

Standards and certification: National standards for KOH purity, electrode manufacturing, and stack durability, which require at least 40,000 hours of operational lifetime, help ensure quality and reliability throughout the supply chain. These measures have catalyzed a rapid expansion of AFC pilot projects, with annual installation rates growing at a CAGR of 32% from 2021 to 2024.

Key industry players and supply chain dynamics

China's AFC ecosystem comprises over 50 specialized firms spanning catalyst suppliers, membrane producers, stack integrators, and system integrators. Prominent enterprises include:

- Tenglong Corporation: Leading producer of alkaline membranes and electrode coatings, boasting an annual capacity of 500 MW of electrolyzer components
- Snowman Group: Specialists in modular AFC stack assembly, with over 150 MW of stacks delivered for stationary and microgrid projects
- Guangdong Guohong Corporation: Integrated system provider offering turnkey AFC plants (100 kW–5 MW) with proprietary KOH circulation and thermal management technologies

The domestic supply chain achieves a "make in-China" ratio of over 85% for essential AFC components, which helps lower costs and shorten lead times. In addition, strategic collaborations between chemical conglomerates such as Sinopec and Binhua Group, and fuel cell companies promote the use of industrial hydrogen byproduct streams, including coke oven gas and chlor-alkali offgases, as affordable feedstocks for AFC systems (Gulli et al., 2024).

Economic and environmental impact

Cost modeling indicates that grid-connected AFC systems can deliver electricity at 0.45–0.60 CNY/kWh (0.06–0.09 USD/kWh) when operated with electrolytic hydrogen sourced at 3 USD/kg, making them competitive with peaking gas turbines (0.08–0.10 USD/kWh) under current tariff structures. Lifecycle analysis also demonstrates that AFCs powered by green or byproduct hydrogen can achieve CO_2 emission reductions of up to 90% relative to diesel generators and 70% relative to natural-gas-fired turbines.

Future outlook

Looking toward 2030, AFC technology is poised to scale further as China targets 5 GW of cumulative fuel cell capacity and 2 million tons per year of electrolytic hydrogen production under the "1 + N" policy framework. Continued improvements in alkaline electrolysis efficiency (targeting > 80%) and stack durability (>60,000 h) will lower the LCOE for AFC systems to below 0.05 USD/kWh, unlocking new markets in rural electrification, data centers, and heavy-duty transportation. Through sustained policy support, industry collaboration, and integration with renewable energy assets, AFCs will play a vital role in China's transition to a low-carbon, hydrogen-based energy economy by 2060.

7.4.3 Japan: market dynamics of AFCs

Japan has long been at the forefront of fuel-cell development, propelled by energy security concerns, an aging population seeking reliable distributed power, and a strong industrial base capable of high-precision manufacturing. While PEMFCs and SOFCs have dominated both R&D and commercial deployments, AFCs are now emerging as a compelling alternative for specific niche applications. This section presents a comprehensive analysis of Japan's AFC market, covering size and growth trends, key demand sectors, policy frameworks, competitive landscape, technological innovations, and the barriers that must be overcome for AFCs to achieve widespread adoption.

Market size, growth trends, and forecasts

According to IMARC Group, Japan's total fuel-cell market was valued at USD 451.5 million in 2024 and is expected to reach USD 1,592.3 million by 2033 (CAGR of 15.0% between 2025 and 2033; IMARC Group, 2026). Although PEMFCs, especially in passenger vehicles and residential micro-CHP systems (known as Ene-Farm), make up the majority of installations, AFCs are steadily gaining a larger share in Japan's distributed power and specialty vehicle markets. Cognitive Market Research projects that, within the Asia-Pacific AFC market, Japan will represent nearly 10% of shipments by 2030, driven by three primary factors:

- Back up and offgrid power for critical infrastructure
- Specialized industrial UPS, where grid reliability and power quality are paramount
- Pilot transport fleets targeting low-speed, high-duty-cycle applications

The higher projected CAGR for AFCs, estimated to be between 18% and 20% from 2025 to 2033, is attributed to expected cost reductions driven by increased production of alkaline membranes, reduced catalyst loadings, and government incentives supporting next-generation fuel cells (Orion Market

Research Pvt. Ltd, 2026). Notably, Japan's early AFC market base remains small, with approximately 5 megawatts of cumulative capacity as of 2024; however, it is projected to surpass 150 megawatts by 2030 if current deployment plans and subsidy programs are successfully implemented.

Key sectors driving demand for AFCs

Stationary and backup power (Micro-CHP & telecom): Japan's dense urban fabric and high reliance on telecommunications underpin strong demand for resilient, emissions-free backup power. AFC configurations (5–20 kW) offer two compelling advantages over PEMFCs:

Fuel-impurity tolerance: Liquid-alkaline electrolytes can absorb trace CO_2 without rapid performance degradation, reducing the need for upstream fuel conditioning in field deployments.

Lower precious-metal loading: AFC cathode catalysts typically use silver or lower platinum-group-metal (PGM) loadings, reducing stack costs by 20%–30% at current commodity prices.

Industrial UPS: Semiconductor fabs, pharmaceutical clean rooms, and data-center colocation facilities demand sub-second power-quality continuity. AFCs have been explored as rack-mounted UPS units capable of rapid transition from grid power to fuel-cell operation. Preliminary trials have demonstrated that these systems can endure thousands of switching cycles with minimal degradation in output voltage, meeting the reliability specifications of high-value manufacturing operations.

Transport and specialty vehicles: While hydrogen buses and passenger cars in Tokyo and Osaka use PEMFC stacks for high power density and rapid transient response, AFC technology is being evaluated for slower, duty-cycle-intensive applications where water management simplicity and cost take precedence. Demonstration fleets of port logistics vehicles and campus shuttles in cities like Kobe and Yokohama are operating with alkaline-electrolyte stacks. These vehicles offer modest ranges (~80 km) at low speeds, leveraging AFCs' tolerance to sub-ambient temperatures and simpler humidification requirements to reduce system complexity.

Several Japanese mobile operator base stations in rural areas have piloted AFC modules, demonstrating continuous operation exceeding 8000 hours. These trials have reported significant reductions in diesel generator runtimes and lower total cost of ownership compared to conventional backup systems. In addition, micro-CHP pilots at small commercial facilities, such as hotels and office annexes, are utilizing 10 kW circulating-electrolyte AFC assemblies that achieve electrical efficiencies of 25%–30% and thermal utilization above 60%.

Policy and regulatory framework

National Energy Strategy & Funding: Japan's sixth Strategic Energy Plan (2021) codifies a target of 300,000 t-H_2/year of domestic production and imports by 2030, supported by a hydrogen R&D and deployment budget exceeding JPY 300 billion (Re-*examining* Renewable Energy Institute, 2022). Although proton exchange membrane fuel cells (PEMFCs) dominate existing demonstration and subsidy programs, the plan explicitly mandates support for next-generation fuel-cell technologies. These include AFCs and hydroxide-exchange membrane systems, aimed at reducing PGM loadings and lowering overall system costs.

Japan's sixth Strategic Energy Plan (2021) codifies a target of 300,000 t-H_2/year of domestic production and imports by 2030, supported by a hydrogen R&D and deployment budget exceeding JPY 300 billion (Re-*examining* Renewable Energy Institute, 2022). Although PEMFCs dominate current demonstration and subsidy programs, the plan explicitly mandates support for next-generation fuel-cell technologies such as AFCs and hydroxide-exchange membrane systems to reduce platinum-group metal loadings and lower system costs.

NEDO roadmap and subsidies: Under the Strategic Roadmap for Hydrogen and Fuel Cells, the New Energy and Industrial Technology Development Organization (NEDO) administers R&D grants (TRL 3–5) and demonstration subsidies (TRL 6–8), covering up to 70% of capital costs for AFC pilot production lines, grid-independent AFC power parks, and specialty-vehicle fleet trials (Nedo, 2021). Since 2022, NEDO has allocated 18 billion Japanese yen (JPY) specifically to AFC membrane development and pilot stack fabrication, fostering collaboration between universities and startups.

Regulatory standards and quality infrastructure: Japan has undertaken efforts to update safety guidelines and technical standards for alkaline fuel-cell systems, focusing on leak containment, electrolyte handling, and waste neutralization. AFC systems are often exempt from certain high-pressure hydrogen regulations because of their inherently lower operating pressures. These regulatory refinements are intended to streamline certification procedures and reduce approval timelines, forming part of broader quality-infrastructure improvements to support wider fuel-cell adoption.

Japan has undertaken efforts to update safety guidelines and technical standards for alkaline fuel-cell systems, focusing on leak containment, electrolyte handling, and waste neutralization. AFC systems are often exempted from certain high-pressure hydrogen regulations due to their inherently lower operating pressures. These regulatory refinements are intended to streamline certification procedures and reduce approval timelines, forming part of broader quality-infrastructure improvements to support wider fuel-cell adoption.

Technology performance, variation, and innovation: Japan's AFC R&D ecosystem targets two critical performance levers:

Membrane durability: Cross-linked AEMs with sterically hindered quaternary sites now withstand 80°C alkaline operation for >10,000 h without significant loss in ion-exchange capacity. This advance addresses the historical challenge of membrane degradation in concentrated KOH environments.

Stack modularity and standardization Industry consortia are converging on a 5 kW module form factor with interchangeable graphite bipolar plates, standardized gasket materials, and unified electrolyte-circulation manifolds. This modularity reduces manufacturing variation, simplifies maintenance, and shortens lead times—critical for scaling beyond pilot volumes.

Roadmap targets: By 2028, the Japanese AFC roadmap aims to achieve:

- Cell cost < JPY 10,000 per kW
- System capital cost < JPY 300,000 per kW

Achieving these targets would place AFC systems on roughly cost-parity with diesel gensets on a levelized-cost-of-electricity basis in remote applications, unlocking new markets in rural electrification and telecom backup.

AI disclosure

During the preparation of this work, the author(s) used ChatGPT (GPT-4-turbo, OpenAI, May 2024 version) in order to improve sentence structure, enhance readability, and refine academic language expression. After using this tool, the author(s) reviewed and edited the content as needed and take(s) full responsibility for the content of the publication.

References

Al-Qahtani, A., González-Garay, A., Bernardi, A., Galán-Martín, Á., Pozo, C., Dowell, N.M., Chachuat, B., Guillén-Gosálbez, G., 2020b. Electricity grid decarbonisation or green methanol fuel? A life-cycle modelling and analysis of today's transportation-power nexus. Appl. Energy. 03062619. 265, 114718. doi:10.1016/j.apenergy.2020.114718.

Al-Qahtani, A., González-Garay, A., Bernardi, A., Galán-Martín, A., Pozo, C., Mac Dowell, N., Chachuat, B., Guillén-Gosálbez, G., 2020a. Electricity grid decarbonisation or green methanol fuel? A life-cycle modelling and analysis of today's transportation-power nexus. Appl. Energy 265, 114718.

Badgett, A., Brauch, J., Thatte, A., Rubin, R., Skangos, C., Wang, X., Ahluwalia, R., Pivovar, B., Ruth, M., 2024. Updated manufactured cost analysis for proton exchange membrane water electrolyzers. No. NREL/TP-6A20-87625. National Renewable Energy Laboratory (NREL), Golden, CO (United States).

Bhashyam, A., 2023. 2023 Hydrogen Levelized Cost Update: Green Beats Gray. Bloomberg New Energy Finance.

Bhoyar, T., Akre, S., 2026. South Korea Hydrogen Fuel Cell Vehicle Market Research Report by Type and by End User: Growth & Industry Forecast 2025 to 2035. M. R. Future.

Bloomberg New Energy Finance, 2022. Wind and solar corporate PPA prices rise up to 16.7% across Europe.

Burrin, D., Roy, S., Roskilly, A.P., Smallbone, A., 2021. A combined heat and green hydrogen (CHH) generator integrated with a heat network. Energy Convers. Manag. 01968904. 246, 114686. doi:10.1016/j.enconman.2021.114686.

Canada Energy Regulator, 2025. Market snapshot: hydrogen update—a small molecule's role in Canada's energy system. https://www.cer-rec.gc.ca/en/data-analysis/energy-markets/

market-snapshots/2025/market-snapshot-hydrogen-update-a-small-molecules-role-in-canadas-energy-system.html.

Canadian Hydrogen and Fuel Cell Association (CHFCA), 2022. Canadian Hydrogen and Fuel Cell Sector Profile: Final Report. MNP LLP.

Chaudry, M., Jayasuriya, L., Blainey, S., Lovric, M., Hall, J.W., Russell, T., Jenkins, N., Jianzhong, W., 2022. The implications of ambitious decarbonisation of heat and road transport for Britain's net zero carbon energy systems. Appl. Energy. 03062619. 305, 117905. doi:10.1016/j.apenergy.2021.117905.

Clean Hydrogen Partnership, 2021. Clean Hydrogen Joint Undertaking. European Commission.

Commission, E., 2020. A Hydrogen Strategy for a Climate-Neutral Europe. E. Commission.

Corporation, H. M. C. K., 2024. Hyundai Motor and Kia to Develop Polymer Electrolyte Membrane with Gore for Hydrogen Fuel Cell Systems. Hyundai Global Newsroom.

Council of the European Union, 2025. Fit for 55: How the EU will turn climate goals into law.

Csiro, 2024. Hydrogen Electrolyser Manufacturing Report.

Čučuk, A., 2025. Nel and KHNP looking into development of clean hydrogen production technology Offshore Energy.

Dcceew, 2024a. Australia's National Hydrogen Strategy - 2024 Update.

Dcceew, 2024b. Hydrogen Policy Portal.

Dcceew, 2024c. National Hydrogen Infrastructure Assessment – Final Report.

Department of Climate Change, Energy, the Environment and Water (DCCEEW), 2019. DCCEEW.

Ding, S., Huang, J., Zhang, X., Zhong, M., 2021. Game equilibrium based control analysis on the sustainable market structure of rare metal mineral resources – evidence from China. J. Environ. Eng. Landsc. Manag. 18224199. 29 (2), 73–84. Available from http://journals.vgtu.lt/index.php/JEELM/about. doi:10.3846/jeelm.2021.14186.

Dixon, J., Bell, K., Brush, S., 2022. Which way to net zero? a comparative analysis of seven UK 2050 decarbonisation pathways. Renew. Sustain. Energy Transit. 2667095X. 2, 100016. doi:10.1016/j.rset.2021.100016.

Erbach, G., Jensen, L., 2024. Towards Climate Neutrality: Fit for 55 Package (EPRS | European Parliamentary Research Service, PE 733.513). European Parliament.

Eurekalert, 2023. New elastic and durable AEM for AEMFC.

European Commission Directorate-General for Energy, 2019. Clean Energy for all Europeans. Publications Office of the European Union.

European Commission, 2023. Next Generation EU: Recovery plan for Europe. European Commission.

European Hydrogen and Backbone, 2022. A European Hydrogen Infrastructure Vision Covering 28 Countries.

European Parliament, 2022. EU Ban on the Sale of New Petrol and Diesel Cars from 2035 Explained. European Parliament.

Eurostat, 2025. Energy statistics - an overview.

Frischmuth, F., Härtel, P., 2022. Hydrogen sourcing strategies and cross-sectoral flexibility trade-offs in net-neutral energy scenarios for Europe. Energy. 03605442. 238, 121598. doi:10.1016/j.energy.2021.121598.

Future Market Insights, Inc, 2024. Portable Fuel Cells Market Analysis by AFC and Methanol: Historical Data 2014–2022, Base Year 2023, Forecast 2025–2034. Future Market Insights, Inc.

Global Market Insights Inc, 2024. North America proton exchange membrane (PEM) fuel cell market size & share 2024 to 2032: Market size by application, country analysis & forecast (Report ID: GMI8699).

Government of the Republic of Korea, Ministry of Trade, Industry and Energy, 2019. Hydrogen Economy Roadmap of Korea: National Strategy for Hydrogen Economy Development. Government of the Republic of Korea.

Gulli, C., Heid, B., Noffsinger, J., Waardenburg, M., Wilthaner, M., 2024. Global Energy Perspective 2023: Hydrogen Outlook. McKinsey & Co. https://www.mckinsey.com/industries/oil-and-gas/our-insights/global-energy-perspective-2023-hydrogen-outlook.

H2cluster Finland, 2023. A Hydrogen Economy Strategy for Finland.

Hamada, A.T., Orhan, M.F., Kannan, A.M., 2023. Alkaline fuel cells: Status and prospects. Energy Rep. 9, 6396–6418.

Han, Y., Luoyi, L., Chen, W., Hou, Y., Zhang, J., 2022. A bi-cyclic co-optimization method for sizing of electricity-hydrogen hybrid energy storage microgrid. Sustain. Energy Fuels. 2398-4902. 6 (17), 4048–4061. doi:10.1039/d2se00556e.

Harichandan, S., Kar, S.K., 2023. Financing the hydrogen industry: exploring demand and supply chain dynamics. Environ. Sci. Pollut. Res. 16147499. Available from https://www.springer.com/journal/11356. doi:10.1007/s11356-023-30262-9.

Haugen, M.J., Flynn, D., Greening, P., Tichler, J., Blythe, P., Boies, A.M., 2022. Electrification versus hydrogen for UK road freight: conclusions from a systems analysis of transport energy transitions. Energy Sustain. Dev. 23524669. 68, 203–210. Available from http://www.elsevier.com. doi:10.1016/j.esd.2022.03.011.

Haugen, M.J., Paoli, L., Cullen, J., Cebon, D., Boies, A.M., 2021. A fork in the road: which energy pathway offers the greatest energy efficiency and CO2 reduction potential for low-carbon vehicles? Appl. Energy. 03062619. 283, 116295. doi:10.1016/j.apenergy.2020.116295.

HTF Market Intelligence, 2025. Alkaline Fuel Cells Market: Global Share, Size & Changing Dynamics 2024–2030 (Report ID: HTF1767800). HTF Market Intelligence.

Huang, W., Li, H., 2022. Game theory applications in the electricity market and renewable energy trading: a critical survey. Front. Energy Res. 2296598X. 10. Available from http://www.frontiersin.org/Energy_Research/. doi:10.3389/fenrg.2022.1009217.

Hydrogen Fuel Cell Partnership, 2025. California Hydrogen Mobility Vision & Roadmap. Hydrogen Fuel Cell Partnership.

IMARC Group, 2026. Japan fuel cell market size, share, trends and forecast by type, application, and region, 2026–2034 (Report ID: SR112026A17484).

Intergovernmental Panel on Climate Change (IPCC), 2022. Climate change 2022: mitigation of climate change. Contribution of Working Group III to the Sixth Assessment Report of the Intergovernmental Panel on Climate Change.

International Energy Agency (IEA), 2021. Net Zero by 2050: A Roadmap for the Global Energy Sector. International Energy Agency.

International Energy Agency (IEA), 2021. Net Zero by 2050. International Energy Agency (IEA).

International Energy Agency (IEA), 2022. Global Hydrogen Review.

International Energy Agency (IEA), 2023. Global EV Outlook 2023: Catching up With Climate Ambitions. International Energy Agency.

International Energy Agency (IEA), 2024. CO_2 emissions in 2023: a new record high, but is there light at the end of the tunnel.

International Energy Agency (IEA), 2026. Infrastructure and Jobs Act: Clean Hydrogen Initiatives – United States. International Energy Agency.

International Renewable Energy Agency (IRENA), 2020. Green Hydrogen Cost Reduction: Scaling up Electrolysers to Meet the 1.5°C Climate Goal. International Renewable Energy Agency (IRENA).

Kakoulaki, G., Kougias, I., Taylor, N., Dolci, F., Moya, J., Jäger-Waldau, A., 2021. Green hydrogen in Europe – a regional assessment: substituting existing production with electrolysis powered by renewables. Energy Convers. Manag. 01968904. 228, 113649. doi:10.1016/j.enconman.2020.113649.

Kountouris, I., Bramstoft, R., Madsen, T., Gea-Bermúdez, J., Münster, M., Keles, D., 2024. A unified European hydrogen infrastructure planning to support the rapid scale-up of hydrogen production. Nat. Commun. 20411723. 15 (1). Available from https://www.nature.com/ncomms/. doi:10.1038/s41467-024-49867-w.

Lazard, 2021. Levelized Cost of Hydrogen Analysis - Version.

Liao, M., Hu, W., Wang, C., Yu, Y., Zhou, H., Song, L., 2024. Research on synergistic development of green hydrogen energy and different industries in new power system—Taking Xinjiang as an example. In: 6th International Conference on Power and Energy Technology (ICPET 2024), July 2024. Institute of Electrical and Electronics Engineers (IEEE), China, pp. 1009–1014.

Liu, X., Reddi, K., Elgowainy, A., Lohse-Busch, H., Wang, M., Rustagi, N., 2020. Comparison of well-to-wheels energy use and emissions of a hydrogen fuel cell electric vehicle relative to a conventional gasoline-powered internal combustion engine vehicle. Int. J. Hydrogen Energy 45 (1), 972–983.

Lonis, F., Tola, V., Cau, G., 2021. Assessment of integrated energy systems for the production and use of renewable methanol by water electrolysis and CO2 hydrogenation. Fuel. 00162361. 285, 119160. doi:10.1016/j.fuel.2020.119160.

Madio, L., Pignataro, A., 2024. Collusion sustainability with a capacity-constrained firm. Oxf. Econ. Pap. 14643812. 1, 227–249. http://oep.oxfordjournals.org/. doi:10.1093/oep/gpac054.

Marketsandmarkets, 2023. Global Fuel Cell Market Size, Share & Growth Analysis.

Nami, H., Rizvandi, O.B., Chatzichristodoulou, C., Hendriksen, P.V., Frandsen, H.L., 2022. Techno-economic analysis of current and emerging electrolysis technologies for green hydrogen production. Energy Convers. Manag. 01968904. 269, 116162. doi:10.1016/j.enconman.2022.116162.

National Laboratory of the Rockies, U.S. Department of Energy, 2026. H_2A: Hydrogen Analysis Production Models. U.S. Department of Energy, National Laboratory of the Rockies.

Natural Resources Canada, 2020. Hydrogen Strategy for Canada: Seizing the Opportunities for Hydrogen – A Call to Action. Government of Canada.

Nedo, 2021. Japan's Action to Promote Hydrogen.

Orion Market Research Pvt. Ltd, 2026. Alkaline Fuel Cell Market Size, Share & Trends Analysis Report by Electrolyte (Immobilized or Static and Flowing) and End-User (Spacecraft and Space Shuttles, Defense and Military, Communication Towers, Automotive, and Others), Forecast Period 2025–2035 (Report Code: OMR2027450).

Penn, A.S., Bartington, S.E., Moller, S.J., Hamilton, I., Levine, J.G., Hatcher, K., Gilbert, N., 2022. Adopting a Whole Systems Approach to Transport Decarbonisation, Air Quality and Health: an Online Participatory Systems Mapping Case Study in the UK. Atmosphere. 20734433. 13 (3). Available from https://www.mdpi.com/2073-4433/13/3/492/pdf. doi:10.3390/atmos13030492.

Pickering, B., Lombardi, F., Pfenninger, S., 2022. Diversity of options to eliminate fossil fuels and reach carbon neutrality across the entire European energy system. Joule. 25424351. 6 (6), 1253–1276. Available from https://www.journals.elsevier.com/joule. doi:10.1016/j.joule.2022.05.009.

Pivovar, B., M., R.A., n.d. H2NEW: hydrogen from Next-Generation Electrolyzers—LTE Task 3c: system and Techno-economic Analysis.

Quarton, C.J., Samsatli, S., 2020. The value of hydrogen and carbon capture, storage and utilisation in decarbonising energy: insights from integrated value chain optimisation. Appl. Energy. 03062619. 257, 113936. Available from https://www.journals.elsevier.com/applied-energy. doi:10.1016/j.apenergy.2019.113936.

Randolph, K., Vickers, J., Peterson, D., Hubert, M., Miller, E., 2022. Program Record 22002: Historical Cost Reduction of PEM Electrolyzers. DOE Hydrogen Program Record. U.S. Department of Energy, Hydrogen and Fuel Cell Technologies Office.

Reed Intelligence Research Team, 2026. Asia Pacific Alkaline Fuel Cell Market Size & Outlook, 2026–2034 (Base Year 2025; Forecast Period 2026–2034). Reed Intelligence.

Renewable Energy Institute, 2022. Re-examining Japan's hydrogen strategy: moving beyond the "hydrogen society" fantasy.

REPowerEU: A Plan to Rapidly Reduce Dependence on Russian Fossil Fuels and Fast Forward the Green Transition, 2022. European Commission, European Commission.

Rodríguez-Fernández, L., Carvajal, A.B.F., de Tejada, V.F., 2022. Improving the concept of energy security in an energy transition environment: application to the gas sector in the European Union. Extr. Ind. Soc. 2214790X. 9. Available from http://www.journals.elsevier.com/the-extractive-industries-and-society/. doi:10.1016/j.exis.2022.101045.

Satyapal, S., Thomas, S., Marcinkoski, J., Schrecengost, R., Miller, E., Stetson, N., Papageorgopoulos, D., Adams, J., Rustagi, N., Melaina, M., Harting, K., Parker, K., Osvatics, C., Wieliczko, M., Taie, Z., Arjona, V., 2021. Department of Energy Update on Hydrogen Shot, RFI Results, and Summary of Hydrogen Provisions in the Bipartisan Infrastructure Law. United States Department of Energy, Hydrogen and Fuel Cell.

Seck, G.S., Hache, E., Sabathier, J., Guedes, F., Reigstad, G.A., Straus, J., Wolfgang, O., Ouassou, J.A., Askeland, M., Hjorth, I., Skjelbred, H.I., Andersson, L.E., Douguet, S., Villavicencio, M., Trüby, J., Brauer, J., Cabot, C., 2022. Hydrogen and the decarbonization of the energy system in Europe in 2050: a detailed model-based analysis. Renew. Sustain. Energy Rev. 13640321. 167, 112779. doi:10.1016/j.rser.2022.112779.

Smith, E., Morris, J., Kheshgi, H., Teletzke, G., Herzog, H., Paltsev, S., 2021. The cost of CO_2 transport and storage in global integrated assessment modeling. Int. J. Greenh. Gas Control 109, 103367.

Smith, J., 2024. Deploying Alkaline Fuel Cells in Off-Grid Australia: A Feasibility Assessment.

Tengler, M., 2024. Hydrogen: 10 Things to Watch for 2024. Bloomberg New Energy Finance.

U.S. Department of Energy, 2022. Technical targets for liquid alkaline electrolysis. Clean Energy Manufacturing Initiative, U.S. Department of Energy.

U.S. Department of Energy, 2024. H_2@Scale. Office of Energy Efficiency & Renewable Energy.

UK's Department for International Trade, 2021. The hydrogen economy South Korea: Market intelligence report. UK's Department for International Trade.

United Nations, 2016. The Sustainable Development Goals Report 2016. United Nations.

Wang, A., 2020. European hydrogen backbone – how a dedicated hydrogen infrastructure can be created.

Wang, Q., Xue, M., Lin, B.L., Lei, Z., Zhang, Z., 2020. Well-to-wheel analysis of energy consumption, greenhouse gas and air pollutants emissions of hydrogen fuel cell vehicle in China. J. Clean. Prod. 275, 123061.

Wang, X., Star, A.G., Ahluwalia, R.K., 2023. Performance of polymer electrolyte membrane water electrolysis systems: configuration, stack materials, turndown and efficiency. Energies. 19961073. 16 (13). Available from http://www.mdpi.com/journal/energies/. doi:10.3390/en16134964.

Wetzel, M., Gils, H.C., Bertsch, V., 2023. Green energy carriers and energy sovereignty in a climate neutral European energy system. Renew. Energy. 18790682. 210, 591–603. Available from http://www.journals.elsevier.com/renewable-and-sustainable-energy-reviews/. doi:10.1016/j.renene.2023.04.015.

Wikipedia, 2025. Ulsan Green Hydrogen Town. Wikimedia Foundation.

Yoo, E., Kim, M., Song, H.H., 2018. Well-to-wheel analysis of hydrogen fuel-cell electric vehicle in Korea. Int. J. Hydrogen Energy. 03603199. 43 (41), 19267–19278. Available from http://www.journals.elsevier.com/international-journal-of-hydrogen-energy/. doi:10.1016/j.ijhydene.2018.08.088.

Chapter 12

Government policies: Impacts and future opportunities

Liuke Duan and Zhefei Pan
Key Laboratory of Low-grade Energy Utilization Technologies and Systems, Ministry of Education, Chongqing University, Chongqing, China
School of Energy and Power Engineering, Institute of Engineering Thermophysics, Chongqing University, Chongqing, China

1 Introduction

In the current context of increasingly severe global climate change, achieving carbon neutrality has become a common goal for countries around the world. To reach this objective, nations have been vigorously promoting hydrogen energy as a crucial area for energy transformation, engaging in intense strategic competition. Alkaline fuel cells (AFCs), with their high efficiency and clean energy characteristics, hold a pivotal position in the hydrogen energy utilization system and are regarded as one of the important technological paths toward sustainable energy development. Fig. 12.1 shows the reaction schematic diagram of a hydrogen-oxygen AFC.

Government policies, as a significant external driving force for industrial development, have a profound impact on the technological research and development (R&D) direction, industrial development scale, and market application scope of AFCs. Policies can not only provide direct incentives, such as financial support and tax preferences for technological R&D, but also guide the healthy development of the industry and shape the technological roadmap by formulating industry standards and regulating market access. For example, certain policies may favor specific technological R&D paths, influencing the resource investment directions of enterprises and research institutions, and thus determining the development trajectory of AFC technology.

This chapter analyzes how national policies drive AFC development through diverse policy tools and how future policy trends may reshape the global hydrogen energy landscape, supporting the interpretation of current drivers and future industrial trends.

Alkaline Fuel Cells. https://doi.org/10.1016/B978-0-443-36638-3.00003-6

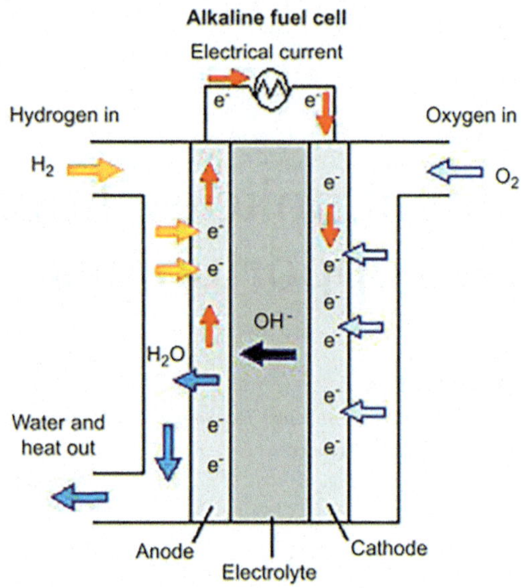

FIG. 12.1 The alkaline fuel cell (Breeze, 2017). *(Modified from Breeze, P., 2017. Chapter 3—The alkaline fuel cell. In P. Breeze (Ed.), Fuel Cells, pp. 23–32. Academic Press. https://doi.org/10.1016/B978-0-08-101039-6.00003-0.)*

2 Global policy framework and strategic competition

2.1 Comparison of policy tools among major economies

Table 12.1 compares policy tools for AFCs across major economies. In terms of core policy tools, the United States provides tax credits through the Inflation Reduction Act (IRA); the European Union (EU) combines the Carbon Border Adjustment Mechanism (CBAM) with green hydrogen auctions; Japan has introduced a 15-year operational subsidy scheme; China has issued the 14th Five-Year Plan for the Modern Energy System; South Korea has released the Hydrogen Economy Activation Roadmap; and the Middle East, represented by Saudi Arabia and the United Arab Emirates (UAE), has promoted Saudi Arabia's NEOM (New Future) Green Hydrogen Project and the UAE National Hydrogen Strategy 2050. For ammonia-related support, the EU has established certification requirements for Renewable Fuels of Non-Biological Origin (RFNBOs), Japan provides ammonia-fired power-generation support, and China has issued the Guiding Opinions on Vigorously Implementing the Renewable Energy Substitution Action. Overall, these economies use different combinations of tax credits, carbon pricing, direct subsidies, public R&D funding, industrial planning, and direct investment to support hydrogen, ammonia, and AFC development.

Since the oil crisis in the 1970s, the United States has recognized the importance of energy diversification. The subsequent Energy Policy Act laid the

TABLE 12.1 Comparison of policy tools for alkaline fuel cells in major economies.

Dimension	United States	European Union	Japan	China	South Korea	Middle East (Saudi Arabia and the UAE)
Core Policy Tools	Tax credits under the Inflation Reduction Act (IRA) (United States Congress, 2022)	Carbon Border Adjustment Mechanism (CBAM) + green hydrogen auctions (European Parliament and Council, 2023)	15-year operational subsidy scheme (Japan Organization for Metals and Energy Security (JOGMEC), 2025)	14th Five-Year Plan for the Modern Energy System (National Development and Reform Commission, National Energy Administration, 2022)	"Hydrogen Economy Activation Roadmap" (Ministry of Trade, Industry and Energy, 2019)	Saudi Arabia's NEOM (New Future) Green Hydrogen Project (ACWA Power, n.d.) and the UAE National Hydrogen Strategy 2050 (UAE Ministry of Energy and Infrastructure, 2023)
Ammonia Energy Support	None	Renewable Fuels of Non-Biological Origin (RFNBO) certification (European Commission, 2023)	Ammonia-fired power-generation support (3 JPY/kWh) (Ichiryu and Takano, 2024)	Guiding opinions on vigorously implementing the renewable energy substitution action (National Development and Reform Commission et al., 2024)	None	None
Types of Policy Tools	R&D tax credits	Carbon pricing linkage and direct subsidies (green hydrogen auctions)	Direct subsidies	Direct subsidies	Direct subsidies	Direct investment and industrial planning guidance

foundation for clean energy policies. With changes in the global energy landscape and the intensification of climate change, the United States has continuously adjusted its energy policies. The clean energy plan proposed by the Biden administration aims to achieve a 100% clean energy economy and net-zero emissions by 2050, creating a favorable policy environment for the development of AFCs. Special R&D funding programs, such as the Energy Frontier Research Centers (EFRCs) and the Advanced Research Projects Agency-Energy (ARPA-E), established by the US Department of Energy (DOE), provide financial support for research on key AFC technologies (U.S. Department of Energy, n.d.a, n.d.b). For example, supported by these programs, the Center for Alkaline-Based Energy Solutions (CABES) at Cornell University has conducted extensive research on catalyst materials and developed novel nonprecious metal catalysts, reducing reliance on precious metals while enhancing catalytic activity (Zhang et al., 2025). As shown in Fig. 12.2, Cornell University studied a cobalt-manganese oxide catalyst using advanced X-ray techniques. The tax credit policy under the IRA allows enterprises to claim tax credits for qualified R&D investment in AFCs, reducing corporate R&D costs and increasing funds available for innovation (United States Congress, 2022). During the production stage, enterprises can also obtain tax credits for purchasing special equipment for AFC production, which stimulates capacity expansion and production efficiency. In addition, the DOE supports R&D activities of small enterprises in the hydrogen energy and fuel cell sectors through the Small Business Innovation Research and Small Business Technology Transfer programs, thereby promoting technological innovation, facilitating commercialization, and strengthening the industrial ecosystem (U.S. Department of Energy, 2024). The three-level funding system comprising Thematic, Open, and SCALEUP plans, established by ARPA-E, has played an important role in technology transfer. In 2023, the SCALEUP Plan invested 47 million US dollars (USD) to support seven

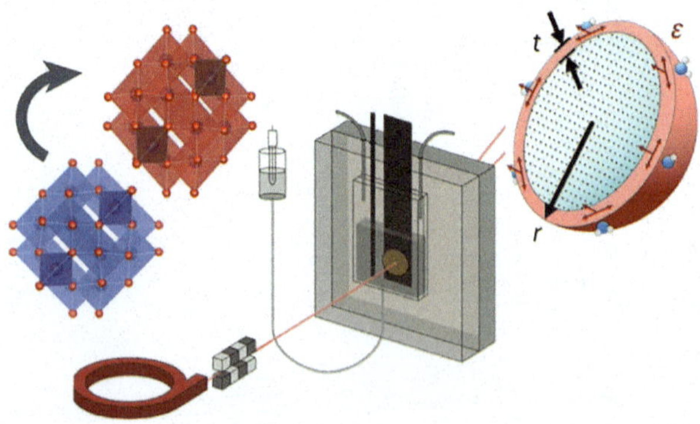

FIG. 12.2 Schematic diagram of Cornell University's achievements. *(Modified from Zhang, J., et al., 2025. Multimodal in situ X-ray mechanistic studies of a bimetallic oxide electrocatalyst in alkaline media. Nat. Catal. https://doi.org/10.1038/s41929-025-01289-7.)*

transformative projects. The project manager system implemented by ARPA-E maintains team innovation through a 3–5 year tenure system. In 2023, its budget increased to USD 700 million, and funded projects generated an average of 1.2 patents per USD 1 million of support.

The EU has systematically promoted breakthroughs and industrialization in AFC technology through a four-dimensional policy framework of "strategic planning, funding support, market incentives, and industrial synergy." Between 2019 and 2024, the European Green Deal established the 2050 carbon neutrality target and guided the EU Hydrogen Strategy to set a 40 gigawatt (GW) electrolyzer installation goal by 2030, while the Fit for 55 legislative package raised the renewable energy share in transportation to 29%, creating rigid market demand for AFCs. The 2022 REPowerEU strategy further doubled the green hydrogen production target to 10 million tons per year, directly stimulating demand for alkaline electrolyzers. Calculations by the EU Joint Research Centre show that this policy mix will drive the core materials market for AFCs to reach 4.8 billion euros (EUR) by 2030 (Vezzoni, 2023). Fig. 12.3 shows the structure of the REPowerEU plan. In R&D investment, a tiered funding system was constructed: basic research relies on the Horizon Europe program (EUR 15 billion allocated to hydrogen out of a total EUR 95.5 billion budget for 2021–2027), pilot-scale conversion is supported by the Innovation Fund (AFC funding increased to 22% in 2024), and commercialization is incentivized through the European Hydrogen Bank with contracts-for-difference subsidies of up to EUR 4/kg (Vezzoni, 2023). This "basic research, applied development, and market promotion" full-chain mechanism has yielded significant results. For example, the EU consortium's AEM-4E project, which focuses on anion-exchange membrane (AEM) technology, modified membrane molecular structures and increased conductivity from 50 mS/cm in 2020 to 80 mS/cm in 2024, surpassing the US DOE's 2025 target ahead of schedule. In market regulation, the CBAM and green certification systems form a coordinated approach: CBAM requires imported steel and chemical products to account for full lifecycle carbon emissions, forcing companies such as ThyssenKrupp's Duisburg steel plant to adopt green hydrogen processes, where AFC electrolysis achieves 78% efficiency on a lower heating value basis and reduces emissions by 64% compared to traditional methods (European Parliament and Council, 2023). Meanwhile, the Renewable Energy Directive (RED III) accelerated the "additionality" requirement for green hydrogen from 2028 to 2027, compelling companies to rapidly deploy alkaline electrolyzers to meet the 3000-hour-per-year renewable power supply standard (European Hydrogen Observatory, 2024). Industrial cluster policies, such as Hydrogen Valley North Rhine-Westphalia in Germany, gathered 30 upstream and downstream firms, reducing membrane electrode production costs by 38%. The European Commission further approved the Hy2Tech project under the Important Projects of Common European Interest mechanism, injecting EUR 5.4 billion in public funds to drive companies such as Bosch and Siemens Energy to build GW-scale alkaline electrolyzer production lines,

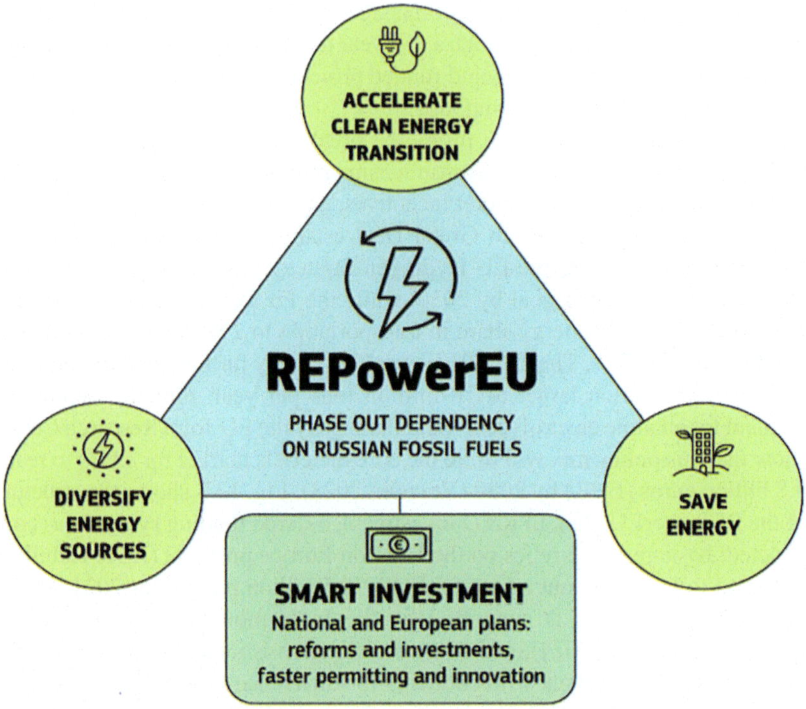

FIG. 12.3 Structure of the REPowerEU plan. *(Modified from Vezzoni, R., 2023. Green growth for whom, how and why? The REPowerEU Plan and the inconsistencies of European Union energy policy. Energy Res. Soc. Sci. 101, 103134. https://doi.org/10.1016/j.erss.2023.103134.)*

forming a "policy guidance, corporate leadership, and cluster synergy" development model (European Hydrogen Observatory, n.d.a). By 2023, EU AFC patents surged to 1572, a 217% increase from 2019, establishing a full-industry-chain technological barrier spanning material R&D (e.g., BASF's novel cobalt-based catalysts), equipment manufacturing (e.g., NEL's proton exchange membrane (PEM) electrolyzer retrofitting), and system integration (e.g., McKinsey's hydrogen island design) (European Hydrogen Observatory, n.d.b).

The Japanese government has systematically shaped the development of AFCs through comprehensive policy design and financial support, establishing a strategic framework centered on national priorities, supported by industrial chain coordination, and extended through international cooperation. Fig. 12.4 shows a schematic diagram of existing energy sources in Japan and the proposed hydrogen economy scenario (Wen and Aziz, 2024). The 2017 Basic Hydrogen Strategy first positioned hydrogen as a national strategic industry, with the 2023 revision setting specific targets to reduce alkaline electrolysis system costs to 52,000 Japanese yen (JPY)/kW by 2030 while planning JPY 15 trillion in hydrogen investments over 15 years (White & Case LLP, 2023).

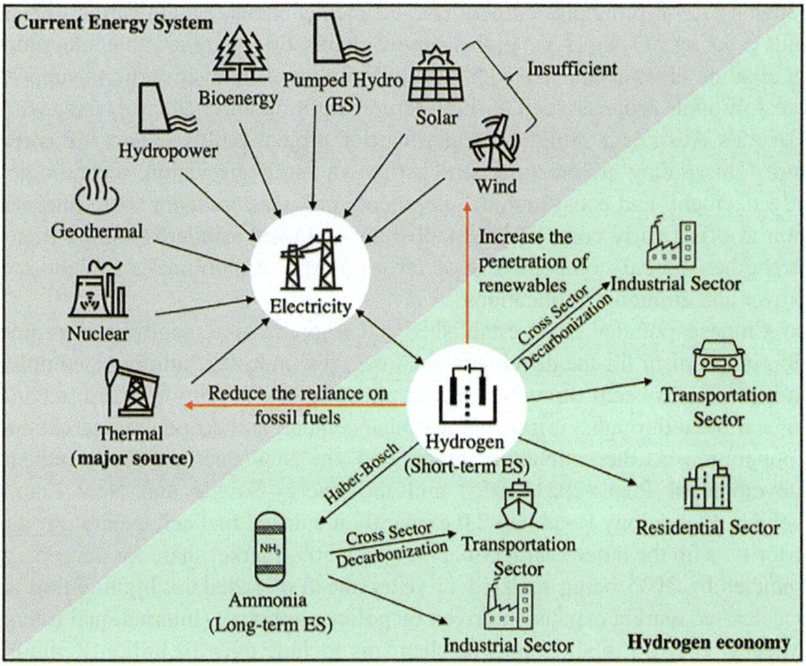

FIG. 12.4 Schematic of existing energy sources in Japan and the proposed hydrogen economy scenario. *(Modified from Wen, D., Aziz, M., 2024. Perspective of staged hydrogen economy in Japan: a case study based on the data-driven method. Renew. Sustain. Energy Rev. 189, 113907. https://doi.org/10.1016/j.rser.2023.113907.)*

These measures directly stimulated core material R&D, as seen in Asahi Kasei's JPY 35 billion Kawasaki plant for alkaline electrolysis membranes, a technology choice closely tied to JPY 11.4 billion in government subsidies (Hydrogen Council, 2024). Policy guidance manifests through multidimensional planning, with the Hydrogen Industry Strategy of Japan's Ministry of Economy, Trade and Industry prioritizing sectors demonstrating "rapid deployment, large markets, and technological advantages." While PEM fuel cells dominate passenger vehicles, the cost benefits of AFCs in industrial applications have made AFCs preferred solutions for stationary power and commercial vehicle systems. Honda's JPY 14.7 billion subsidy supports AFC deployment in data center backup power, targeting 2000 annual units by the mid-2020s, while Toyota's JPY 11.2 billion grant funds 75,000 commercial vehicle fuel cell systems by 2030, leveraging the durability and power advantages of AFCs (Piper, 2024). The Green Transformation (GX) Supply Chain Program integrates the industrial chain, complemented by Hydrogen Safety Strategy standards enabling large-scale AFC adoption. Japan's inclusion of ammonia as a hydrogen carrier (targeting JPY 10/Nm^3 by 2030) creates unique AFC applications in ammonia plant energy systems through ammonia cracking compatibility. Internationally,

Japan's 3E + S principles embed AFC efficiency standards and carbon thresholds (<3.4 kg CO_2/kg H_2) in global supply chains, driving renewable integration in alkaline electrolysis. The JPY 800 billion Green Innovation Fund supports breakthrough projects such as high-temperature alkaline electrolyzers, while Toyota's Australian liquid hydrogen carrier project builds export infrastructure. This policy architecture demonstrates a "strategic vision, technological advancement, and ecosystem development" progression, using state intervention to offset early costs while establishing technical standards and patents to strengthen global competitiveness, creating AFC opportunities in industrial power and ammonia applications.

Chinese policies have established a comprehensive, multitiered promotion mechanism for the development of the AFC industry, building a complete industrial ecosystem through strategic planning, fiscal reinforcement, technological breakthroughs, infrastructure enhancement, and deepened international cooperation. At the national strategic level, the New Energy Vehicle Industry Development Plan (2021–2035) and the Energy-Saving and New Energy Vehicle Technology Roadmap 2.0 explicitly identified fuel cell technology as a priority, with the latter's target of achieving a 50% market share for new energy vehicles by 2035 being realized 11 years ahead of schedule, highlighting the accelerated market expansion driven by policy alignment (International Energy Agency, 2020). Fiscal support mechanisms include over 10 billion Renminbi (RMB) allocated by the central government from 2018 to 2020 for new energy vehicle adoption, complemented by local initiatives such as Shanghai's 5 billion RMB fund for fuel cell R&D and Shenzhen's subsidies of up to 100,000 RMB for fuel cell vehicle purchases, collectively reducing corporate R&D costs by 20%–30% through coordinated central-local financial frameworks (Ministry of Finance, Ministry of Industry and Information Technology, Ministry of Science and Technology, National Development and Reform Commission, 2020). Technological advancements, supported by national programs such as the 863 Plan, have driven breakthroughs in core materials such as catalysts and PEMs: Beijing Hydrogenium Energy's high-power-density systems are now deployed in logistics vehicles, Xinergy Power's metal bipolar plate stacks achieve 3.0 kW/L power density, and State Power Investment Corporation's "Hydrogen Rise" system lifespan surpassing 20,000 hours reflects the efficacy of performance-based subsidy mechanisms. Infrastructure development has seen rapid growth, with hydrogen refueling stations expanding from 128 in 2020 to support Beijing's 2000 fuel cell buses, forming a closed-loop "technology-product-application" ecosystem. Regional policy innovations include Guangdong's industrial chain integration initiatives, Suzhou's specialized incentives for domestic membrane electrode production, and Xiong'an's "Hydrogen Highway" pilot project exploring heavy-duty transport commercialization, creating regionally differentiated development models. Internationally, China's participation in global standardization through partnerships such as the China-Europe New Energy Vehicle Alliance with Daimler and Siemens, and Great Wall Motor's collaboration with

Germany's H2 MOBILITY on hydrogen refueling networks, has facilitated technical standardization and market integration. Sustained policy efforts have elevated the industry chain's completeness from 40% in 2015 to 85% by 2022, fostering globally competitive firms such as SinoHytec (whose fuel cell systems achieve a driving range of 500 km) and Vision Group, while Contemporary Amperex Technology Co., Limited has reduced stack costs by 40% compared with imported alternatives (China.org.cn, 2024). From a market perspective, policy-driven growth has expanded the AFC market from 10 billion RMB in 2019 to a projected 200 billion RMB by 2025, corresponding to a compound annual growth rate of 40%. This growth has broadened applications from public transportation to distributed power generation and energy storage, as exemplified by Beijing's 2000 fuel cell buses and Suzhou's membrane electrode localization projects. To address technical challenges, government-funded R&D has advanced nonprecious metal catalysts, with nickel-based alternatives nearing platinum-level performance, while the Hydrogen Industry Standardization Framework established 53 industry standards to enhance safety and market access. Future policies prioritize global competitiveness, as seen in Great Wall Motor's 2023 European hydrogen network partnerships and 2024 fuel cell vehicle exports surpassing 5000 units, signaling China's shift from technological follower to rule-maker. This systemic integration of policy tools positions AFCs as a cornerstone of China's dual-carbon goals and cements its innovative leadership in the global hydrogen economy.

South Korea's "Hydrogen Economy Activation Roadmap" sets an ambitious goal of reaching 6.2 million fuel cell vehicles by 2040 and provides a 60% subsidy based on the vehicle selling price for hydrogen refueling stations (Ministry of Trade, Industry and Energy, 2019). This measure has significantly promoted the popularization of fuel cell vehicles and the construction of hydrogen refueling station infrastructure. Hyundai Motor's cooperation with Saudi Arabia to develop a green hydrogen-ammonia supply chain takes full advantage of the resources and technical strengths of both sides and promotes the international development of the industry. The South Korean government has also increased R&D subsidies for direct ammonia fuel cells (DAFCs), planning to invest 20 billion won by 2025, which promotes the development of DAFC technology and injects new vitality into the AFC industry.

In the Middle East, Saudi Arabia's NEOM Green Hydrogen Project is a major initiative in the global green hydrogen industry. As one of the world's largest green hydrogen plants, it is planned to export up to 1.2 million tonnes of green ammonia annually (ACWA Power, n.d.). This project not only generates substantial demand for AFC electrolyzers and related equipment manufacturing but also promotes the development of the local hydrogen energy industry chain. The UAE National Hydrogen Strategy 2050 leverages the country's solar resources and aims to strengthen its position in low-carbon hydrogen production and export markets (UAE Ministry of Energy and Infrastructure, 2023). Through large-scale investment in photovoltaic hydrogen production projects, the UAE is promoting the development of related green ammonia industries and contributing to the global green ammonia market.

2.2 Geopolitical and technological standard competition

The "IRA" of the United States sets requirements for the local supply chain, such as a minimum 50% domestic content rate for electrolyzers (United States Congress, 2022). This policy, to a certain extent, squeezes the market space of Chinese and South Korean enterprises. On the one hand, it raises the threshold for Chinese and South Korean enterprises to enter the US market, restricting the sales of their products in the United States. On the other hand, it prompts US local enterprises to increase investment in the R&D and production of electrolyzers, enhancing the competitiveness of local enterprises and intensifying global market competition.

The EU's CBAM requires imported products to meet certain carbon emission standards. This forces export-oriented enterprises to use green hydrogen, indirectly promoting the application of AFCs in industries such as steel and chemical engineering (European Parliament and Council, 2023). To meet the EU's carbon emission standards, relevant enterprises need to seek cleaner energy sources. As an important application technology for green hydrogen, AFCs have seen an increase in application demand in these industries, which promotes the popularization of AFC technology and the development of the industry.

There is a competition for the right to speak in electrolyzer efficiency certification between Europe and the United States. The EU requires electrolyzer efficiency to be at least 60%, while the US DOE aims for 65%. Higher efficiency standards are conducive to promoting enterprises to increase R&D investment and improve electrolyzer technology. However, they may also create technical barriers, restricting the development of some enterprises with lower technical levels. At the same time, different standards may lead to market segmentation and increase the market entry costs for enterprises.

Patent layouts for AEMs differ among China, Japan, and South Korea. Japan holds 43% of global AEM patents and, therefore, has a leading edge in this field. China and South Korea are also continuously increasing R&D investment and actively laying out AEM patents. These differences reflect the competitive situation of technological R&D among countries. Countries with more patents have greater advantages in technological innovation and industrial development and can better promote the progress of AFC technology and industrial development.

3 Policy-driven industrial transformation

3.1 Technological innovation: From laboratory to commercialization

Special R&D funding programs, such as the EFRCs and the ARPA-E, established by the US DOE, provide long-term and stable financial support for research institutions and enterprises (U.S. Department of Energy, n.d.a, n.d.b). With the help of these funds, researchers have studied catalysts and membrane materials for AFCs. Taking the CABES at Cornell University as an example, Fig. 12.5 shows the synthesis of high-surface-area carbon-supported metal nitrides (MxN) by ammonia nitridation (Zeng et al., 2022). Supported by these programs and equipped with advanced research facilities, researchers explore ways to enhance catalytic activity at the atomic and molecular levels. They have investigated new materials such as transition-metal compounds and carbon-based composites and developed nonprecious metal catalysts with improved catalytic activity, reducing dependence on precious metals and offering possibilities for cost reduction. Substantial investment has also promoted research on polymer membranes and nanocomposites. Through molecular structure design and nanotechnology modification, researchers have developed membrane materials with higher ion conductivity and stronger chemical stability, improving overall cell performance. These advances have increased the number of related publications and patents, promoting the transfer of AFC technology from the laboratory to commercialization.

The EU invests substantially through the Horizon Europe R&D funding framework and the Clean Hydrogen Joint Undertaking to support technological R&D covering the hydrogen energy value chain. In the catalyst R&D field, funded projects are dedicated to developing more efficient and stable catalysts to improve the energy conversion efficiency of AFCs. Research institutions and enterprises collaborate to explore new catalytic mechanisms and materials and adopt advanced preparation processes to continuously optimize catalyst performance. In membrane material research, the EU has promoted the development of new AEMs. By optimizing the chemical structure, functional group

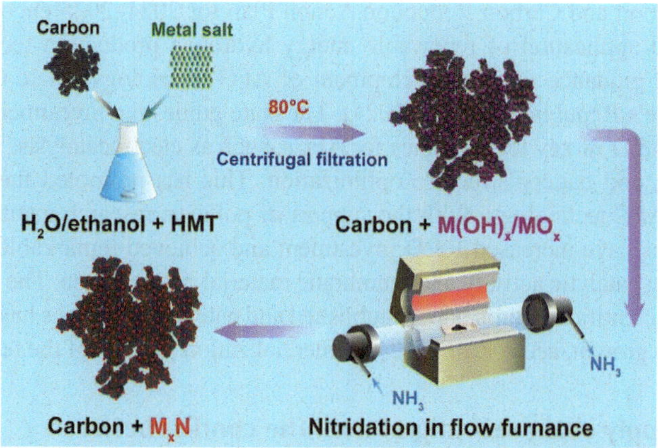

FIG. 12.5 Schematic synthesis procedure of metal nitrides (MxN) supported on high-surface-area carbon via nitridation with ammonia. *(Modified from Zeng, R., et al., 2022. Nonprecious transition metal nitrides as efficient oxygen reduction electrocatalysts for alkaline fuel cells. Sci. Adv. 8, eabj1584. https://doi.org/10.1126/sciadv.abj1584.)*

design, and preparation process of membrane materials, the ion-conduction performance and stability of the membranes have been improved. The EU has continuously achieved breakthroughs in AFC technology R&D, with a large number of scientific research achievements and an increasing number of relevant papers and patents. These results provide a solid technical foundation for the commercial application of AFCs and accelerate the transformation of technology from the laboratory to the market. Some enterprises have developed high-performance AFC systems using the EU's scientific research achievements and have conducted demonstration applications in transportation, distributed power generation, and other fields.

The Japanese government guides the investment of funds in R&D through policies such as the Hydrogen Energy Basic Strategy. In the field of hydrogen production by electrolysis, setting price targets for mainstream alkaline water electrolysis equipment encourages enterprises and research institutions to optimize technologies and reduce costs. Enterprises improve electrolysis efficiency and reduce energy consumption by improving the structural design of electrolyzers, optimizing electrode materials, and refining electrolyte formulations. Research institutions, on the other hand, start with basic theoretical research to explore new electrolysis reaction mechanisms and provide theoretical support for technological innovation. Japan actively explores emerging hydrogen production technologies such as AEM electrolysis, increases R&D investment, and strives to achieve technological breakthroughs in this field to provide more efficient and low-cost hydrogen production methods for AFCs. Japan has also developed large-scale and high-efficiency hydrogen storage and transportation technologies, ensuring a stable hydrogen supply for AFCs and promoting the commercial application of the technology.

China encourages innovation through a series of policies. The Energy Conservation and Carbon Reduction Action Plan for 2024–2025 encourages the R&D and application of renewable energy hydrogen production technologies, providing guidance for the development of AFC technology (State Council of the People's Republic of China, 2024). The state guides the investment of funds into the R&D of key technologies for AFCs, such as electrocatalysts, membrane materials, and battery structure optimization. This has promoted the improvement of AFC technology. With the support of policies, research institutions and enterprises have increased R&D investment and achieved remarkable results in improving catalytic activity and membrane material performance. The number of relevant scientific research papers published and patent applications has witnessed explosive growth, accelerating the commercialization process of the technology.

3.2 Supply chain reconfiguration: The conflict between regionalization and globalization

The tax credit policy in the United States has had a profound impact on the AFC industry. It has acted as a powerful incentive for upstream and downstream enterprises in the industry chain to develop and expand. As a result, a comprehensive and complete industry chain has emerged, spanning from the supply of raw materials, such as metals and polymers, to the manufacturing of key components like electrodes and membranes, and finally to the system integration of AFCs.

The industry cluster effect around Cornell University is a prime example of the positive influence of this policy. A large number of related enterprises, including small-scale startups and large-scale manufacturing companies, have gathered in this area. These enterprises benefit from their proximity to each other in several ways. Firstly, they can share resources more efficiently. For example, they can jointly invest in expensive research equipment, such as advanced material testing facilities, which would be difficult for individual companies to afford. Secondly, they can engage in extensive technical exchanges. Regular seminars, workshops, and industry conferences are held in the area, where researchers and engineers from different companies can share their latest findings and ideas. This cross-fertilization of knowledge often leads to new technological breakthroughs and product improvements.

However, the "IRA" in the United States has introduced a new set of challenges. The act's requirements for the local supply chain have pushed the supply chain toward regionalization. This shift has the potential to disrupt the well-established global supply chain. Many international enterprises that have been supplying components and materials to the US market are now facing difficult decisions. Some may find it unprofitable to meet the local content requirements of the act, especially if they have to relocate their production facilities or source materials from more expensive local suppliers. As a result, they may reduce their supply to the US market and instead focus on other markets where the regulatory environment is more favorable. This could lead to shortages of certain components in the US market and increase the cost of production for US-based enterprises. Moreover, the weakening of the global supply chain synergy may slow down the overall pace of technological innovation and product development in the AFC industry.

In the EU, the government's public funds for hydrogen-related projects and the subsidy policies of the European Hydrogen Bank have been instrumental in promoting the development of AFCs. These financial incentives have attracted a large number of enterprises, ranging from established energy companies to innovative startups, to enter the AFC market.

The EU is also actively promoting the construction of AFC industry clusters across European countries. These clusters are designed to foster collaborative cooperation among upstream and downstream enterprises in the industry chain. For example, in some clusters, research institutions, component manufacturers, and system integrators are located in close proximity to one another. This allows for seamless communication and cooperation throughout the product development process. Component manufacturers can work closely with research institutions to develop new and improved materials and components, while system integrators can provide valuable feedback on the performance requirements of the final products.

However, the EU's CBAM poses a significant challenge to the global supply chain. The CBAM is intended to prevent carbon leakage by imposing a carbon tax on imported goods based on their carbon footprint. Although the policy aims to promote environmental sustainability, many export-oriented enterprises, especially those from developing countries, may find it difficult to meet the strict carbon emission standards. These enterprises may face higher costs if they have to invest in new technologies and processes to reduce carbon emissions. As a result, some may reduce exports to the EU market, which could disrupt the global supply chain and lead to higher prices for AFC products in the EU. Moreover, the CBAM may trigger trade frictions between the EU and other countries, further complicating the global trade environment. Fig. 12.6 shows the representation of various countries in global hydrogen governance (Lentschig et al., 2025).

In Japan, the government's support for the AFC industry through financial subsidies and policy guidance has been crucial for its growth. In 2024, the government's decision to allocate funds for the construction of gigawatt-scale electrolyzers and fuel cell production capacity was a significant step forward.

Companies like Asahi Kasei Corporation have benefited greatly from this government support. With the received funds, Asahi Kasei has been able to upgrade its production facilities and increase the production capacity of its alkaline electrolyzers. This not only helps to meet the growing domestic demand for hydrogen production but also positions Japan as a major player in the global alkaline electrolyzer market.

The policies have also led to the rapid growth of upstream and downstream enterprises in Japan's AFC industry chain. A complete industrial ecosystem has

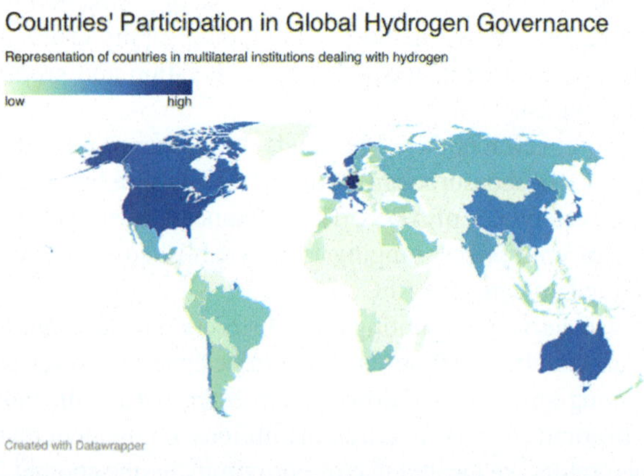

FIG. 12.6 Countries' representation in global hydrogen governance. *(Modified from Lentschig, H., Patonia, A., Quitzow, R., 2025. Multilateral governance in a global hydrogen economy: An overview of main actors and institutions, key challenges and future pathways. Int. J. Hydrogen Energy 97, 76–87. https://doi.org/10.1016/j.ijhydene.2024.11.393.)*

been formed, with strong linkages between raw material suppliers, component manufacturers, and system integrators. The industry clusters in Japan are constantly expanding and strengthening. These clusters have several advantages, such as reduced logistics costs, improved communication and cooperation, and the ability to attract and retain talent.

Japan's active promotion of green ammonia imports and related technology R&D is also having a significant impact on the global hydrogen energy supply chain. Green ammonia is considered a promising hydrogen carrier, as it can be easily transported and stored. Japan's large-scale imports of green ammonia are changing the global trade flow of green ammonia. It is also strengthening regional cooperation between Japan and green ammonia-producing countries. For example, Japan is collaborating with countries in the Middle East and Australia to develop green ammonia production projects. These projects involve joint investment, technology transfer, and long-term supply contracts, which are promoting the development of a more stable and sustainable global hydrogen energy supply chain.

In China, local governments are actively responding to national policies by introducing a series of industrial support policies for the AFC industry. For example, in 2023, Anyang City provided 1:1 matching funds with central funds for fuel cell vehicle-related projects. This financial support has encouraged local enterprises to invest in the research, development, and production of AFCs for vehicles.

In 2024, the accelerated introduction of green hydrogen subsidies in many places has further promoted the development of upstream and downstream enterprises in the AFC industry chain. As a result, a complete and collaborative industrial ecosystem has been formed. The industry cluster effect is becoming increasingly prominent, with enterprises in different regions specializing in different aspects of the industry chain. For example, some regions are focusing on the production of raw materials, while others are specializing in the manufacturing of key components or the system integration of AFCs. As shown in Fig. 12.7, this illustrates the distribution of the hydrogen fuel cell technology industrial chain in the five urban agglomerations (Wang et al., 2025).

China is actively participating in global industry chain cooperation. Chinese enterprises are exporting AFC products and technologies to many countries around the world. At the same time, they are also importing advanced technologies and components from other countries to improve their own product quality and performance. However, China also faces supply chain risks brought about by geopolitics. Tensions in international relations, trade disputes, and sanctions can disrupt the supply of key materials and components, which can have a negative impact on the production and development of the AFC industry in China.

To address these challenges, China is taking a proactive approach. While continuing to expand overseas markets, China is also strengthening regional cooperation with neighboring countries. Through initiatives such as the "Belt and Road Initiative," China is promoting the interconnection of the AFC

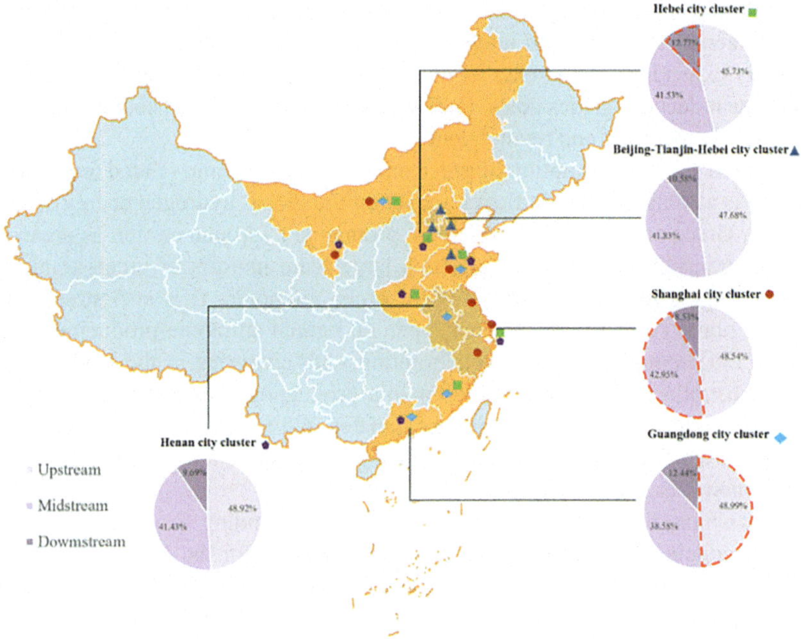

FIG. 12.7 Hydrogen fuel cell technology industrial chain distribution in five city clusters. *(Modified from Wang, X., Fan, L.W., Zhang, H., Zhou, P., 2025. Hydrogen fuel cell technology development in China: technology evolution, city-cluster network and industry chain distribution. Energy 322, 135606. https://doi.org/10.1016/j.energy.2025.135606.)*

industry chain. This includes building transportation infrastructure, establishing trade agreements, and promoting technology transfer and cooperation. By doing so, China aims to ensure the stability of the supply chain and promote the coordinated development of regionalization and globalization of the AFC industry.

4 Future opportunities: Hydrogen-ammonia synergy and emerging markets

4.1 Policy synergy in the hydrogen-ammonia industry chain

Japan has incorporated ammonia energy into its Hydrogen Energy Basic Strategy and issued the Green Ammonia Roadmap. By importing green ammonia, Japan has reduced the cost of hydrogen energy, providing a more economical hydrogen source for AFCs. Japan's subsidy of 3 JPY/kWh for ammonia-based power generation has promoted the application of ammonia energy in the energy sector and driven the coordinated development of the hydrogen-ammonia industry chain. At the same time, Japan has increased investment in related R&D, promoting the development of technologies such as DAFCs and creating new opportunities for the AFC industry. With the advancement of DAFC technology,

AFCs may use ammonia as a fuel in future application scenarios, further expanding their market space.

The EU's "North Sea Green Hydrogen Corridor" plan aims to convert green hydrogen into liquid ammonia for export. This plan has driven the growth in demand for alkaline electrolyzers. The EU's green ammonia certification mechanism (RFNBO standard) has standardized the green ammonia market and promoted the production and application of green ammonia. Through policy guidance, the EU has promoted the coordinated development of the hydrogen-ammonia industry chain, creating a broader market space for the AFC industry. For example, under the impetus of this plan, relevant enterprises have increased investment in alkaline electrolyzers, driving the development of the upstream equipment manufacturing industry in the AFC industry chain. The ammonia fuel demands for the shipping routes are shown in Fig. 12.8 (Wang et al., 2023).

The South Korean government has increased R&D subsidies for DAFCs, planning to invest 20 billion won in 2025, which has promoted the development of DAFC technology. The goals for fuel cell vehicles and the subsidy policy for hydrogen refueling stations set in South Korea's Hydrogen Economy Activation Roadmap have promoted the development of the fuel cell industry and provided support for the coordinated development of the hydrogen-ammonia industry chain. With the popularization of fuel cell vehicles and the construction of hydrogen refueling stations, demand for hydrogen and ammonia will increase, promoting further coordination across the hydrogen-ammonia industry chain.

China has launched hydrogen-ammonia coupling demonstration projects, such as the Ningdong Renewable Hydrogen Coupling Coal-to-Ammonia Project, to explore the collaborative development model of hydrogen and ammonia. Through policy guidance, China has promoted the coordinated development of the hydrogen-ammonia industry chain, providing practical experience and

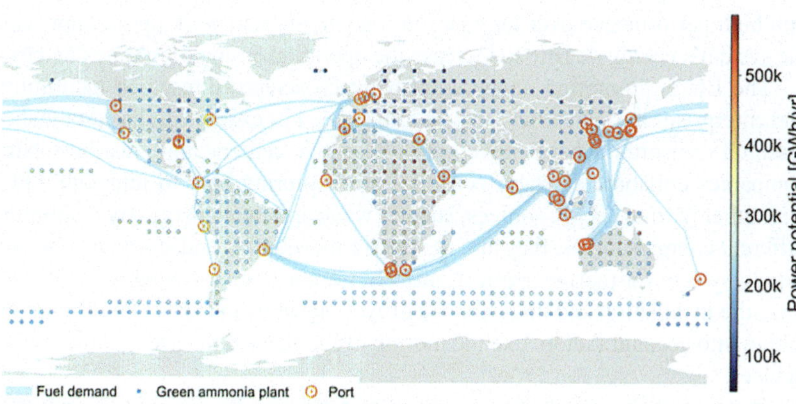

FIG. 12.8 Candidate ammonia production sites for global shipping routes. *(Modified from Wang, H., Daoutidis, P., Zhang, Q., 2023. Ammonia-based green corridors for sustainable maritime transportation. Digit. Chem. Eng., 6, 100082. https://doi.org/10.1016/j.dche.2022.100082.)*

technical accumulation for the application of the AFC industry in the hydrogen-ammonia field. For example, during the operation of this project, technical data and management experience regarding the mixed use of hydrogen and ammonia have been accumulated, providing valuable references for the application of AFCs in hydrogen-ammonia collaborative scenarios.

4.2 Expansion of applications in marine energy and aerospace

In the marine energy field, with the policy support of countries and regions such as the United States, the EU, and Japan, AFCs are on the cusp of wide-scale application in marine devices. Marine buoys, which are crucial for ocean monitoring, require a reliable and long-lasting power source. AFCs can provide stable and efficient electricity, enabling these buoys to continuously collect data on ocean temperature, salinity, and water quality. This data is essential for climate research, fisheries management, and early-warning systems for natural disasters.

Underwater vehicles, including remotely operated vehicles and autonomous underwater vehicles, also stand to benefit greatly from AFCs. These vehicles are used for a variety of tasks, such as deep-sea exploration, pipeline inspection, and marine biodiversity research. Traditional power sources, like batteries, often have limited endurance, but AFCs can offer longer operation times, allowing underwater vehicles to travel greater distances and perform more complex tasks.

In the United States, companies like Boeing are leveraging government policies to conduct in-depth research on the application of AFCs in the marine energy field. Boeing is using its expertise in aerospace and materials science to develop fuel cell systems that can withstand the harsh marine environment. The company is focusing on issues such as corrosion resistance, as the high salt content in seawater can damage the components of the fuel cell. They are testing new materials for the electrodes and membranes that can resist corrosion and maintain high performance over long periods. Boeing is also working on improving the system's energy density to increase the operational time of marine devices.

The EU is promoting cooperation between scientific research institutions and enterprises through policy-driven initiatives. For example, it has established research consortia where universities, research centers, and marine-related companies collaborate. These consortia are exploring how to integrate AFCs with other marine energy sources, such as wave and tidal energy. By combining different energy sources, they aim to create a more stable and sustainable power supply system for marine applications. They are also researching ways to optimize the installation and maintenance of AFC systems in marine environments, taking into account factors like the motion of the sea and the accessibility of the devices.

Japan, with its strong foundation in materials and technology, is actively exploring the use of AFCs in various marine applications. Japanese researchers are developing lightweight and compact AFC systems for small-scale marine vessels, such as fishing boats and coastal patrol boats. These systems can not only reduce reliance on fossil fuels but also lower the emissions of harmful pollutants. In addition, Japan is studying the use of AFCs in ocean-based energy storage systems. These systems can store excess energy generated from renewable sources during peak production times and release it when needed, providing a more stable power supply for offshore facilities.

China is also making significant progress in the application of AFCs in the marine energy field. Chinese researchers are working on developing AFC systems specifically designed for China's unique marine environment. For example, they are considering the impact of the high-sediment content in some coastal waters on the performance of the fuel cell. They are developing filtration and pretreatment systems to ensure the smooth operation of the fuel cell. China is also exploring the use of AFCs in large-scale marine energy farms, where multiple fuel cell systems can be integrated to provide power for desalination plants, underwater communication systems, and other marine-related facilities.

In the aerospace field, Japan's advantages in materials and technology, combined with government support, are driving the exploration of AFCs in aerospace applications. Japanese research teams are focusing on developing high-performance and lightweight fuel cell systems for unmanned aerial vehicles (UAVs) and satellites. For UAVs, these fuel cell systems can significantly extend the flight time compared to traditional battery-powered UAVs. This is particularly important for applications such as long-range surveillance, agricultural monitoring, and environmental research.

For satellites, AFCs can provide a more reliable and long-lasting power source. In space, where the environment is harsh and the power supply is critical, the stability and high energy density of AFCs make them an attractive option. Japanese researchers are working on improving the radiation resistance of the fuel cell components, as satellites are constantly exposed to high-energy radiation in space. They are also developing advanced control systems to optimize the power output of the fuel cells in different space conditions.

China is also actively involved in the R&D of AFCs for aerospace applications. Chinese scientists are conducting in-depth studies on the design and integration of fuel cell systems for high-altitude, long-endurance UAVs. These UAVs can be used for a variety of purposes, such as high-altitude atmospheric research, remote sensing, and communication relay. The development of AFC systems for UAVs in China focuses on improving energy efficiency and reducing the weight of the system. Scientists are exploring new materials and manufacturing techniques to achieve these goals.

In addition, China is looking into the application of AFCs in satellite power systems. By using AFCs, satellites can have a more stable and long-lasting power supply, which is crucial for the operation of complex satellite payloads. China is also collaborating with international partners in aerospace research, sharing knowledge and resources to accelerate the development of AFC technology for space applications.

The United States is also actively exploring the use of AFCs in the aerospace field. American research institutions and companies are conducting research on using AFCs in next-generation satellite power systems. They aim to improve the energy density and lifespan of the fuel cells to meet the increasing power demands of modern satellites. These satellites are equipped with more advanced communication, imaging, and scientific instruments, which require a reliable and high-power energy source. American researchers are also studying the integration of AFCs with other power sources, such as solar panels, to create a hybrid power system that can provide continuous power under different space conditions. This hybrid system can take advantage of the high-power generation of solar panels during sunlight periods and the stable power output of AFCs during eclipse or low-light conditions.

5 Conclusion: The tripartite interaction among policy, technology, and market

The policies of major global economies have played a crucial role in the development of the AFC industry, reshaping the industrial ecosystem through institutional incentives and constraints. Policies guide the flow of funds, encouraging scientific research institutions and enterprises to focus on key technologies, such as the R&D of new catalyst materials and the improvement of membrane material performance. By leveraging economies of scale and technology spillover, marginal cost reduction is achieved, improving battery efficiency and reducing costs. Significant achievements have been made in technological innovation, and there is great potential for future innovation (Fig. 12.9).

In terms of industrial cooperation, international cooperation and industry-university-research cooperation have become increasingly close under the encouragement of policies. Each country, leveraging its own advantages, cooperates in all links of the industry chain, achieving resource sharing and technological complementarity. This has accelerated the transformation of scientific research achievements into productive forces and promoted the gradual

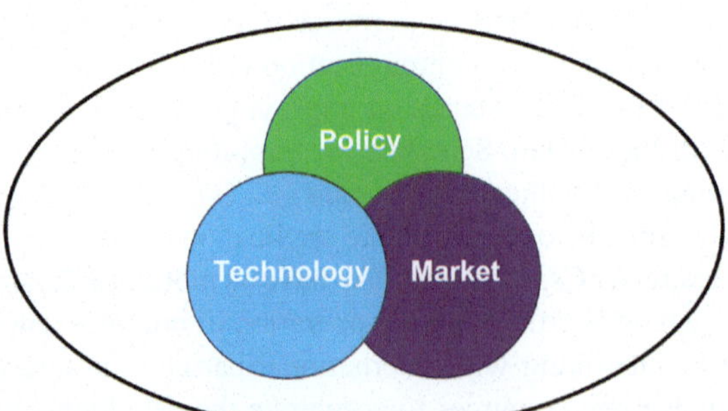

FIG. 12.9 Policy, technology, and market relationship diagram.

formation of a complete and efficient global industry chain and industrial ecosystem for AFCs.

In the field of market expansion, policy support has deepened the application of AFCs in traditional fields such as transportation and distributed power generation. At the same time, it has also shown broad application prospects in emerging fields such as marine energy and aerospace, opening up new growth paths for the industry.

Looking ahead, with the continuous growth of global demand for clean energy and the sustained promotion of policies in various countries, the AFC industry is expected to develop even more vigorously. In terms of technological innovation, it is anticipated to achieve more breakthroughs in areas such as new catalysts and membrane materials, further enhancing battery performance, significantly reducing costs, and making the battery more competitive in the energy market.

In terms of industrial cooperation, international cooperation will become more diversified and in-depth. Collaboration among countries in technology R&D, standard-setting, and market exploration will be further strengthened, promoting the globalization process of the industry. Industry-university-research cooperation will also continue to innovate in models, cultivating more professional talents and providing a continuous stream of intellectual support for industrial development.

In terms of market application, the application scale of AFCs in existing fields will continue to expand, and applications in emerging fields will gradually be implemented and commercialized. As a clean energy solution, AFCs will play an increasingly important role in the global energy transformation and sustainable development, becoming a crucial force in promoting the green energy revolution and helping the world achieve the carbon neutrality goal.

References

ACWA Power, n.d. NEOM Green Hydrogen Project. https://acwapower.com/en/what-we-do/projects/neom-green-hydrogen-project/ (Accessed 27 April 2026).

Breeze, P., 2017. Chapter 3 – The alkaline fuel cell. In: Breeze, P., (Ed.), Fuel Cells. Academic Press, pp. 23–32. doi:10.1016/B978-0-08-101039-6.00003-0.

China.org.cn, 2024. China-Europe synergy turbocharges global transition to carbon neutrality. http://www.china.org.cn/business/2024-06/09/content_117244253.htm (Accessed 27 April 2026).

European Commission, 2023. Commission Delegated Regulation (EU) 2023/1184 establishing a Union methodology for renewable fuels of non-biological origin. https://eur-lex.europa.eu/eli/reg_del/2023/1184/oj/eng (Accessed 27 April 2026).

European Hydrogen Observatory, 2024. Renewable Energy Directive. https://observatory.clean-hydrogen.europa.eu/eu-policy/renewable-energy-directive (Accessed 27 April 2026).

European Hydrogen Observatory, n.d.a Hy2Tech. https://observatory.clean-hydrogen.europa.eu/hy2tech (Accessed 27 April 2026).

European Hydrogen Observatory, n.d.b Patents. https://observatory.clean-hydrogen.europa.eu/hydrogen-landscape/research-and-innovation-activity/patents (Accessed 27 April 2026).

European Parliament and Council, 2023. Regulation (EU) 2023/956 establishing a carbon border adjustment mechanism. Off. J. Eur. Union. https://eur-lex.europa.eu/eli/reg/2023/956/oj/eng (Accessed 27 April 2026).

Hydrogen Council, 2024. Scaling up for a low-emissions hydrogen future: new Asahi Kasei plant achieves milestones. https://hydrogencouncil.com/zh/scaling-up-for-a-low-emissions-hydrogen-future-new-asahi-kasei-plant-achieves-milestones/ (Accessed 27 April 2026).

Ichiryu, M., Takano, S., 2024. Ammonia: the future fuel with sustainable potential. Nat. Energy. https://www.nature.com/articles/d42473-024-00441-4 (Accessed 27 April 2026).

International Energy Agency, 2020. New Energy Vehicle Industry Development Plan (2021-2035). https://www.iea.org/policies/15529-new-energy-vehicle-industry-development-plan-2021-2035 (Accessed 27 April 2026).

Japan Organization for Metals and Energy Security (JOGMEC), 2025. Overview of Japan's CfD Support Program. https://journal.jogmec.go.jp/content/300517125.pdf (Accessed 27 April 2026).

Lentschig, H., Patonia, A., Quitzow, R., 2025. Multilateral governance in a global hydrogen economy: an overview of main actors and institutions, key challenges and future pathways. Int. J. Hydrog. Energy 97, 76–87. doi:10.1016/j.ijhydene.2024.11.393.

Ministry of Finance, Ministry of Industry and Information Technology, Ministry of Science and Technology, National Development and Reform Commission, 2020. Notification on improving the fiscal subsidy policy for the promotion and application of new energy vehicles. https://www.gov.cn/zhengce/zhengceku/2020-04/23/content_5505502.htm (Accessed 27 April 2026).

Ministry of Trade, Industry and Energy, 2019. Hydrogen Economy Activation Roadmap. https://cn.yna.co.kr/view/ACK20190117001200881 (Accessed 27 April 2026).

National Development and Reform Commission, et al., 2024. Guiding Opinions on Vigorously Implementing the Renewable Energy Substitution Action. https://www.gov.cn/zhengce/zhengceku/202410/content_6983959.htm (Accessed 27 April 2026).

National Development and Reform Commission, National Energy Administration, 2022. 14th Five-Year Plan for the Modern Energy System. https://www.gov.cn/zhengce/zhengceku/2022-03/23/content_5680759.htm (Accessed 27 April 2026).

Piper, D.L.A., 2024. Japan's carbon neutrality strategy relies heavily on hydrogen - and North America could provide it. https://www.dlapiper.com/en-sa/insights/publications/2024/05/japans-carbon-neutrality-strategy-relies-heavily-on-hydrogen (Accessed 27 April 2026).

State Council of the People's Republic of China, 2024. Energy Conservation and Carbon Reduction Action Plan for 2024-2025. https://www.gov.cn/zhengce/zhengceku/202405/content_6954323.htm (Accessed 27 April 2026).

U.S. Department of Energy, 2024. U.S. Department of Energy Announces $52 Million for Small Business Research and Development Grants—Including $3.4 Million for Hydrogen and Fuel Cell Projects. https://www.energy.gov/eere/fuelcells/articles/us-department-energy-announces-52-million-small-business-research-and (Accessed 27 April 2026).

U.S. Department of Energy, n.d.a. Energy Frontier Research Centers. https://www.anl.gov/energy-frontier-research-centers (Accessed 27 April 2026).

U.S. Department of Energy, n.d.b. Advanced Research Projects Agency-Energy (ARPA-E). https://www.energy.gov/advanced-research-projects-agency-energy-arpa-e (Accessed 27 April 2026).

UAE Ministry of Energy and Infrastructure, 2023. National Hydrogen Strategy 2050. United Arab Emirates. https://u.ae/en/about-the-uae/strategies-initiatives-and-awards/strategies-plans-and-visions/environment-and-energy/national-hydrogen-strategy (Accessed 27 April 2026).

United States Congress, 2022. Inflation Reduction Act, H.R. 5376. https://www.irs.gov/inflation-reduction-act-of-2022 (Accessed 27 April 2026).

Vezzoni, R., 2023. Green growth for whom, how and why? The REPowerEU Plan and the inconsistencies of European Union energy policy. Energy Res. Soc. Sci. 101, 103134. doi:10.1016/j.erss.2023.103134.

Wang, H., Daoutidis, P., Zhang, Q., 2023. Ammonia-based green corridors for sustainable maritime transportation. Digit. Chem. Eng. 6, 100082. doi:10.1016/j.dche.2022.100082.

Wang, X., Fan, L.W., Zhang, H., Zhou, P., 2025. Hydrogen fuel cell technology development in China: technology evolution, city-cluster network and industry chain distribution. Energy 322, 135606. doi:10.1016/j.energy.2025.135606.

Wen, D., Aziz, M., 2024. Perspective of staged hydrogen economy in Japan: a case study based on the data-driven method. Renew. Sustain. Energy Rev. 189, 113907. doi:10.1016/j.rser.2023.113907.

White & Case LLP, 2023. Japan Hydrogen Basic Strategy. https://www.whitecase.com/insight-alert/japan-hydrogen-basic-strategy (Accessed 27 April 2026).

Zeng, R., et al., 2022. Nonprecious transition metal nitrides as efficient oxygen reduction electrocatalysts for alkaline fuel cells. Sci. Adv. 8, eabj1584. doi:10.1126/sciadv.abj1584.

Zhang, J., et al., 2025. Multimodal in situ X-ray mechanistic studies of a bimetallic oxide electrocatalyst in alkaline media. Nat. Catal. doi:10.1038/s41929-025-01289-7.

Chapter 13

Challenges and perspectives

Haozhen Li[a] **and Zhefei Pan**[a,b]
[a]*School of Energy and Power Engineering, Institute of Engineering Thermophysics, Chongqing University, Chongqing, China*
[b]*Key Laboratory of Low-grade Energy Utilization Technologies and Systems, Ministry of Education, Chongqing University, Chongqing, China*

Alkaline fuel cells (AFCs), while promising high efficiency and low-cost advantages, face several critical challenges in their development: The alkaline electrolyte tends to react with CO_2 to form carbonates, resulting in performance degradation and requiring either high-purity fuel or complex purification systems; both electrode materials and electrolytes [e.g., liquid potassium hydroxide (KOH) or solid anion exchange membranes (AEM)] exhibit corrosion susceptibility or insufficient stability in strongly alkaline environments; water management issues (cathode flooding and anode dry-out) and poor low-temperature performance further constrain their applications. Moreover, immature high-power-density technology and competition from alternative technologies like proton exchange membrane fuel cell (PEMFCs) have hindered commercialization. Future breakthroughs will depend on the development of carbonate-resistant materials, novel catalysts, and system optimization, likely limiting AFCs to specialized applications such as aerospace or stationary power generation in the near term. Despite these multifaceted challenges, continued advancements in materials, system design, and policy support will ensure AFCs maintain a crucial role in the global transition to clean energy.

1 Introduction

Amid the global energy transition, the search for efficient, clean, and sustainable energy conversion technologies has become an urgent priority. The AFC, as a highly promising energy technology, has garnered significant attention in recent years. The AFC converts chemical energy into electrical energy through electrochemical reactions, with water being the only by-product and virtually no greenhouse gas or other pollutant emissions. It is a truly green energy technology that aligns with the current pressing needs for environmental protection

Alkaline Fuel Cells. https://doi.org/10.1016/B978-0-443-36638-3.00004-8

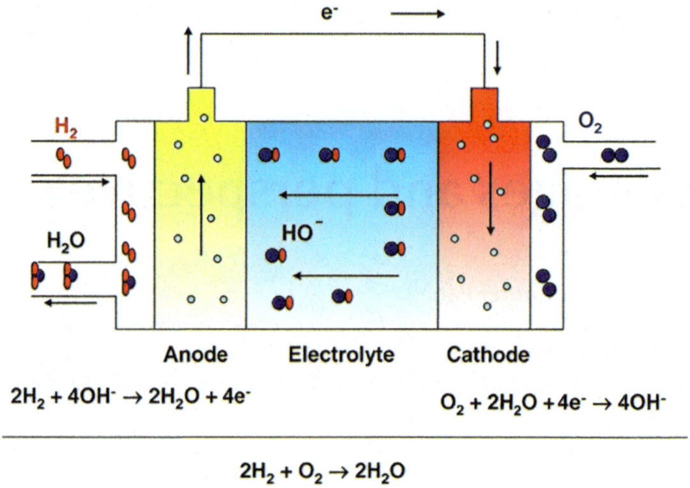

FIG. 13.1 The alkaline fuel cell. Help readers understand the principles of fuel cells and deepen their understanding of fuel cells. *(Data from Iojoiu, C., Chabert, F., Maréchal, M., Kissi, N.El., Guindet, J., Sanchez, J.-Y., 2006. From polymer chemistry to membrane elaboration. J. Power Sources, 153 (2), 198–209. https://doi.org/10.1016/j.jpowsour.2005.05.039.)*

and sustainable development. Fig. 13.1 helps readers understand the principles of fuel cells and deepen their understanding of fuel cells.

Since the concept of AFC was proposed in the early 20th century, it has undergone years of development and has been applied in specific fields such as space exploration. For example, the National Aeronautics and Space Administration of the United States used AFC as the main power source on the Apollo spacecraft. AFC has many advantages. In terms of electrolytes, the alkaline electrolyte is less corrosive than the acidic electrolyte, which allows the use of nonplatinum catalysts and metal plates. This not only reduces costs but also decreases the volume of the cell stack and increases the volumetric power density. In terms of fuel selection, it can not only use high-purity hydrogen but also high-energy-density fuels such as ammonia decomposition gas, which are incompatible with acidic electrolytes. This reduces dependence on specific fuels and lowers fuel costs.

However, on the path to broader application, AFC also faces a series of severe challenges. Its electrolyte is prone to evaporation at high temperatures, leading to a decline in cell performance; it is also highly sensitive to carbon dioxide. Carbon dioxide in the air reacts with the electrolyte to form carbonate precipitates, which affect the normal operation of the cell. Moreover, the manufacturing of AFC involves expensive catalyst materials, such as platinum and ruthenium, and requires high-purity hydrogen fuel. Combined with the costs of other components, this keeps the overall manufacturing cost high and hinders large-scale commercialization.

Despite these challenges, with continuous in-depth research in related fields such as materials science and electrochemistry, AFC has made positive progress in the development of efficient catalysts and the optimization of electrolyte materials, bringing hope for breaking through the current predicament. A thorough analysis of the challenges faced by AFC and an outlook on its future development prospects are of great significance for promoting the continuous progress of this technology and enabling it to play a greater role in the global energy structure.

2 The key challenges of alkaline fuel cells

2.1 The bottleneck of performance improvement

Kinetic limitations of the electrode reactions: The performance enhancement of AFCs is significantly constrained by the kinetics of the electrochemical reactions at the electrodes. The sluggish kinetics of the Oxygen Reduction Reaction (ORR) and the Hydrogen Oxidation Reaction (HOR) are at the core of the issue. During the oxygen reduction process occurring at the cathode, although the alkaline environment, with its unique involvement of hydroxide ions (OH^-), exhibits certain advantages over the acidic environment by reducing the reaction activation energy, the reaction rate still fails to meet the demands for efficient energy conversion.

From the perspective of catalysts, in order to reduce costs, researchers have been committed to developing nonprecious metal catalysts, such as cobalt-based and nickel-based catalysts. However, numerous studies have shown that these nonprecious metal catalysts have a significant gap in activity compared to the precious metal platinum. For example, in the case of the metal–nitrogen–carbon (M–N–C) type cobalt-based catalyst, under alkaline conditions, the onset potential and half-wave potential of its ORR are approximately 50–100 mV and 80–120 mV lower than those of platinum-based catalysts, respectively (Burchardt et al., 2002). This difference in activity directly leads to a sharp increase in cathode polarization losses when the cell operates at high current densities. In practical hydrogen–oxygen fuel cell systems, the overpotential generated by the cathodic ORR can reach as high as 300–400 mV, significantly reducing the overall output voltage of the cell and causing the energy conversion efficiency to drop by up to 15%–20% (Lin et al., 2013). Figure 13.2 enables readers to more clearly perceive the kinetic limitations of fuel cells.

Mass transfer and diffusion issues: The mass transfer process of gases within the electrode is another key barrier to the performance improvement of AFCs. AFCs typically employ a porous electrode structure, in which oxygen must sequentially pass through the diffusion layer and the catalytic layer to reach the active sites where the electrochemical reactions occur. During the diffusion process, gas molecules are impeded by the pore structures of the diffusion and catalytic layers, leading to concentration polarization phenomena.

Under high current density operating conditions, this mass transfer limitation is particularly pronounced. For example, when an AFC operates at a high

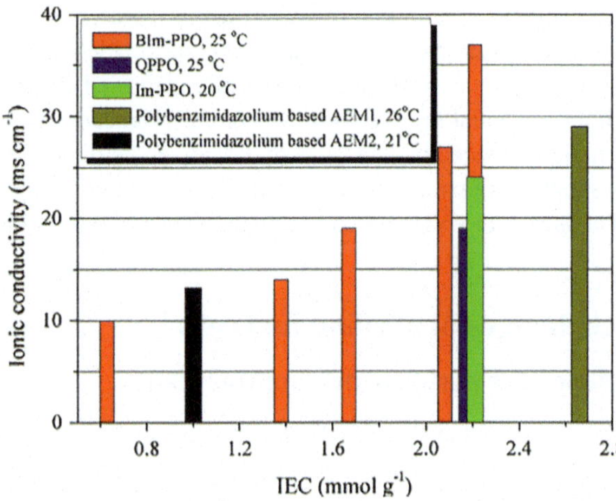

FIG. 13.2 Ionic conductivity of BIm-PPO, the benchmark QPPO and Im-PPO, well as polybenzimidazolium-based AEMs, as a function of IEC. This enables readers to more clearly perceive the kinetic limitations of fuel cells. *(Data from Lin, X, Liang, X., Poynton, S.D., Varcoe, J.R., Ong, A.L., Ran, J., Li, Y., Li, Q., Xu, T., 2013. Novel alkaline anion exchange membranes containing pendant benzimidazolium groups for alkaline fuel cells. J. Membrane Sci. 443, 193–200. https://doi.org/10.1016/j.memsci.2013.04.059.)*

current density of 1 A/cm^2, the diffusion rate of oxygen within the catalytic layer cannot meet the reaction requirements, resulting in a sharp drop in the oxygen concentration within the catalytic layer. The limiting current density is reduced by approximately 30%–40% compared to lower current densities (Li et al., 2023). This not only severely weakens the power output capability of the cell but also leads to local reaction imbalances, affecting the stability and lifespan of the cell. Moreover, the nonuniform distribution of the electrolyte within the electrode is also a problem that cannot be ignored. Nonuniform electrolyte distribution can lead to low mass transfer efficiency in some areas, forming "dry-out" regions, while other areas may be hindered in gas diffusion due to an excess of electrolyte, further exacerbating the decline in cell performance.

The impact of ohmic resistance: The internal ohmic resistance of a fuel cell significantly affects the performance of AFCs, with its sources mainly encompassing three aspects: electrode materials, electrolytes, and contact interfaces. Regarding electrode materials, commonly used carbon-based materials, despite their favorable porous structure and large specific surface area that promote gas diffusion and catalyst loading, have relatively high electrical resistivity. For instance, the resistivity of carbon paper is approximately 1–5 Ω cm, which leads to substantial energy losses during electron transport within the electrode.

The characteristics of the electrolyte are equally critical. Take KOH solution as an example; its concentration and viscosity significantly influence ionic conductivity. Although a higher concentration of KOH solution can provide more hydroxide ions to enhance ionic conduction, it also increases the solution's viscosity and exacerbates the corrosion of electrode materials. Research has shown that when the concentration of KOH solution increases from 30% to 60%, the ionic conductivity improves by about 20%, but the corrosion rate of the electrode materials accelerates by 3–5 times (Couture et al., 2011). In addition, the contact interface between the electrode and the current collector also generates contact resistance. Due to the mismatch of the microscopic surface structures of the electrode and the current collector, the actual contact area is smaller than the apparent contact area, resulting in a relatively high contact resistance. Optimizing the conductivity of the contact interface, for example, by using conductive adhesives or surface modification techniques, is an important approach to reducing ohmic resistance and enhancing cell performance.

2.2 The issue of material stability and durability

Corrosion of catalysts and electrode materials: In the operating environment of AFCs, nonprecious metal catalysts face severe challenges of corrosion and deactivation. From a chemical mechanism perspective, iron-based or cobalt-based catalysts in alkaline media are susceptible to reactions with hydroxide ions (OH^-) due to their strong nucleophilicity. These reactions can disrupt metal–metal bonds or metal–ligand bonds on the catalyst surface, leading to the dissolution of metal atoms (Verhaert et al., 2016). For example, in the case of iron–nitrogen–carbon (Fe–N–C) catalysts, after continuous operation for 500 hours, inductively coupled plasma mass spectrometry analysis revealed that the concentration of iron ions in the electrolyte increased from almost undetectable levels to 15–20 ppm. This directly resulted in a reduction of about 30% in the number of active sites of the catalyst and a decrease of 18%–22% in the cell power density.

Carbon carrier materials also struggle to maintain stability under alkaline conditions. Hydroxide ions in the alkaline environment attack the defective sites on the surface of the carbon carrier, triggering the formation of carbon–oxygen bonds, which, in turn, leads to the gradual disintegration of the carbon structure. This corrosion not only destroys the originally orderly porous structure of the electrode, making the gas diffusion path more tortuous and complex and reducing the gas diffusion efficiency by 25%–30%, but also causes the catalyst particles loaded on the carbon carrier to fall off, seriously affecting the loading stability of the catalyst (Zeng et al., 2023). In contrast, precious metal catalysts, although they exhibit good stability due to their inert electronic structure, have a high cost that makes the catalyst cost per kilowatt of power several thousand dollars. This is a key economic factor that restricts the large-scale commercial application of AFCs.

Management and stability of electrolytes: Traditional alkaline electrolytes, such as KOH solution, face multiple stability challenges in practical applications. From a chemical perspective, KOH solution is highly hygroscopic and reactive with carbon dioxide (CO_2). When exposed to air, CO_2 rapidly reacts with KOH according to the stoichiometric equation $2KOH + CO_2 \rightarrow K_2CO_3 + H_2O$, forming potassium carbonate (K_2CO_3) (Jia et al., 2020). As the reaction progresses, the generated K_2CO_3 gradually deposits within the electrode pores, forming tiny crystals that block the porous structure and reduce the effective reaction area of the electrode by 20%–25%.

The stability issues of the electrolyte are further exacerbated by temperature factors. During high-temperature operation, rapid evaporation of water significantly alters the concentration of the electrolyte. Experimental data show that when the operating temperature increases from 60°C to 80°C, the rate of water evaporation per hour increases by approximately 40%–50%, leading to a corresponding rise in electrolyte concentration. Excessive concentration can cause the KOH solution to reach a supersaturated state, triggering crystallization. These crystals not only impede ionic conduction, reducing ionic conductivity by 30%–40%, but also potentially damage the interface structure between the electrode and the electrolyte, severely threatening the normal operation of the cell (Lin et al., 2010).

Challenges of membrane materials: For alkaline anion exchange membrane (AEM) fuel cells, the stability of the membrane material directly determines the lifespan and performance of the cell. Existing AEMs suffer from multiple forms of chemical degradation in highly alkaline environments. For example, quaternary ammonium-functionalized poly(arylene ether) membranes undergo Hofmann elimination reactions under alkaline conditions, where the alkyl chains in the quaternary ammonium groups are progressively removed. This leads to a decrease in ion exchange capacity (IEC). After accelerated aging tests, in which the membrane was soaked in a 1 mol/L KOH solution at 80°C for 1000 hours, the IEC of this type of membrane decreased from an initial 2.0 meq/g to 1.2 meq/g, and the membrane's conductivity also dropped by about 40% (Varcoe and Slade, 2005; Ramaswamy and Mukerjee, 2019).

In addition to chemical degradation, the physical properties of membrane materials also require significant improvement. During cell operation, the membrane undergoes dimensional changes due to water absorption and swelling, as well as water loss and contraction. These repeated dimensional fluctuations can lead to stress concentration within the membrane, causing the formation and propagation of microcracks. Moreover, the relatively low mechanical strength of the membrane material makes it prone to damage during cell assembly and operation, making it difficult to meet the requirements for long-term stable cell operation. There is an urgent need to further optimize the membrane materials through approaches such as material modification and composite reinforcement.

2.3 System integration and cost issues

System complexity and energy loss: The efficient operation of AFC systems relies on the coordinated functioning of multiple subsystems, including fuel supply, electrolyte circulation, thermal management, and gas processing. This highly integrated design makes the system structure extremely complex and inevitably leads to energy losses. In the gas processing stage, since AFCs are highly sensitive to carbon dioxide (CO_2), any CO_2 from the air entering the system will react with the alkaline electrolyte to form carbonate precipitates, which can clog the electrode pores and degrade cell performance. Therefore, it is necessary to equip the system with specialized gas purification devices, such as chemical adsorption towers or membrane separation equipment, to remove CO_2 from the incoming air. For example, in the operation of a chemical adsorption tower, the adsorbent needs to be regenerated periodically. The regeneration process not only increases the material costs of the system but also consumes about 5%–8% of the total system energy (Zhang, 2021).

Regarding electrolyte circulation and gas supply, the operation of auxiliary equipment, such as electrolyte circulation pumps and gas compressors, also consumes a significant amount of electrical energy. The electrolyte circulation pump needs to continuously deliver the electrolyte to the electrode area to maintain the normal progress of the reaction, with its energy consumption accounting for about 12%–15% of the total system energy consumption. The gas compressor is responsible for compressing the fuel gas and oxidant gas to the appropriate pressure, with energy losses in this process reaching 10%–13% of the total system energy. The energy consumption of these auxiliary devices reduces the overall efficiency of the AFC system by 20%–30% compared to the ideal state, severely limiting its energy utilization efficiency in practical applications.

Cost constraints: Although AFCs can utilize nonprecious metal catalysts to reduce costs, the high costs of key components, such as electrode materials, electrolytes, and membrane materials, remain a major barrier to their large-scale commercialization. In terms of membrane materials, high-performance AEMs are expensive due to their complex synthesis processes. The preparation of AEMs typically involves multiple steps, including monomer synthesis, polymerization, and functionalization, with extremely strict control requirements for reaction conditions (such as temperature, pressure, and catalyst dosage). For example, the synthesis of quaternary ammonium-functionalized poly(arylene ether ketone) (PAEK) AEMs requires the use of costly quaternization reagents, and multiple purification steps are needed after the reaction, resulting in a cost of $300–500 per square meter of membrane material (Wang et al., 2011).

In addition, the manufacturing processes of fuel cell components also face cost challenges. The preparation of electrodes involves delicate operations such as catalyst coating and carbon carrier treatment. To ensure the uniformity and stability of the electrodes, high-precision coating equipment and drying devices are required. The assembly of the membrane electrode assembly (MEA) has extremely high requirements for environmental cleanliness and operational precision, typically necessitating the use of processes such as thermal pressing in a cleanroom environment. The purchase and maintenance costs of these high-precision devices are substantial. It is estimated that the costs of electrode preparation and MEA assembly account for about 40%–50% of the total cost of a fuel cell (Liu et al., 2021). The high material costs and manufacturing process costs make the cost per unit of power of AFCs much higher than that of traditional energy devices, significantly limiting their market competitiveness and the progress of large-scale commercialization.

2.4 Environmental adaptability and operational constraints

Impact of carbon dioxide: The chemical reaction between carbon dioxide (CO_2) in the air and the alkaline electrolyte in AFCs has become a key factor limiting their performance and application. From a chemical perspective, when CO_2 comes into contact with the alkaline electrolyte (such as KOH solution), it rapidly undergoes a two-step reaction: First, CO_2 reacts with water to form carbonic acid (H_2CO_3), and then H_2CO_3 neutralizes with OH^- to ultimately form carbonates (such as K_2CO_3). This reaction not only consumes the effective component OH^- in the electrolyte, gradually depleting the alkaline environment, but also produces insoluble carbonate precipitates. These precipitates can clog the porous structure of the electrodes, hindering gas diffusion and ionic conduction, leading to a significant decline in electrode performance and a reduction in cell power density by 25%–35% (Song et al., 2022).

To maintain the normal operation of AFCs in open environments, additional CO_2 removal devices must be installed, such as chemical adsorption towers and membrane separation equipment. For example, a chemical adsorption tower requires filling with active substances that have a strong adsorption capacity for CO_2 (such as amine-based adsorbents) during operation, and the adsorbent needs to be regenerated regularly. This not only increases the hardware costs of the system but also brings additional energy consumption, reducing the overall system efficiency by about 8%–12%. In urban environments, due to factors such as industrial emissions and vehicle exhaust, the concentration of CO_2 in the air is typically between 400–500 ppm, significantly higher than that in natural environments. This accelerates the degradation process of the alkaline electrolyte. Research has shown that in an environment with a CO_2 concentration of 500 ppm, the effective service life of the AFC's electrolyte is nearly 40% shorter compared to a low-concentration environment (such as 200 ppm) (Izadi et al., 2022), which greatly limits its application in urban energy scenarios.

Impact of temperature and humidity: The performance of AFCs is highly sensitive to temperature and humidity, both of which can affect the operational state of the cell in multiple ways. Under low-temperature conditions, the physical properties of the electrolyte change significantly, with its viscosity increasing sharply as the temperature decreases. For example, in the case of the commonly used KOH electrolyte, the viscosity can increase by 3–5 times when the temperature drops from 60°C to 0°C. This change directly leads to a substantial increase in the resistance to ionic conduction in the electrolyte, with the ionic conduction rate decreasing by about 50%–60%. The obstruction of ionic conduction makes the electrochemical reactions difficult to proceed smoothly, resulting in extended cell startup times, and in extreme low-temperature conditions, the cell may not start at all. In addition, the voltage drop during operation increases, and the overall performance deteriorates significantly (Babatunde et al., 2022).

While high-temperature environments can, to some extent, accelerate the kinetics of electrochemical reactions and increase reaction rates, they also bring a series of negative effects. High temperatures can intensify the chemical corrosion reactions between electrode materials, catalysts, and the electrolyte. For example, the dissolution rate of nonprecious metal catalysts increases exponentially with rising temperature. Moreover, high temperatures can also accelerate the evaporation of water from the electrolyte. Experimental data show that when the operating temperature rises from 60°C to 80°C, the evaporation rate of water in the electrolyte can increase by 2–3 times. This not only changes the concentration of the electrolyte, affecting ionic conduction, but can also lead to electrolyte dry-out, disrupting the internal material balance of the cell and severely threatening system stability (Okundamiya, 2021). Fig. 13.3 shows the performance of an AC arc furnace in terms of current density under different temperatures and constant pressure.

The impact of humidity changes on the performance of membrane materials in AFCs is also significant and cannot be ignored (Agel et al., 2001; Babatunde et al., 2022; Bib Khan and Liang, 2024; Burchardt et al., 2002; Couture et al.,

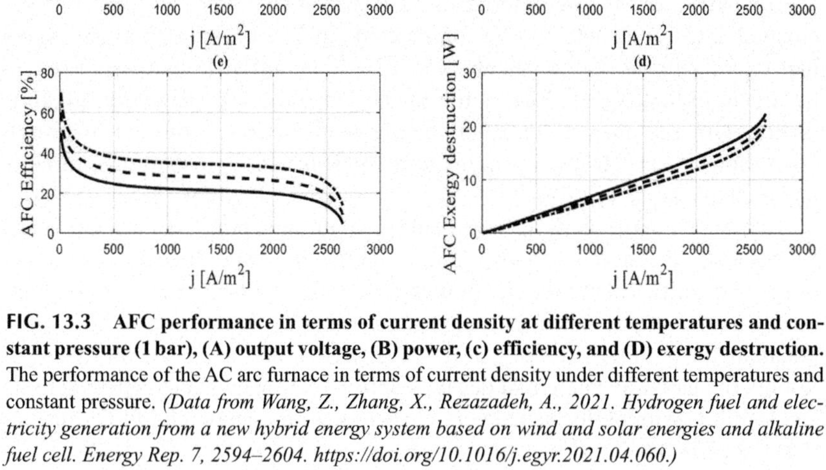

FIG. 13.3 AFC performance in terms of current density at different temperatures and constant pressure (1 bar), (A) output voltage, (B) power, (c) efficiency, and (D) exergy destruction. The performance of the AC arc furnace in terms of current density under different temperatures and constant pressure. *(Data from Wang, Z., Zhang, X., Rezazadeh, A., 2021. Hydrogen fuel and electricity generation from a new hybrid energy system based on wind and solar energies and alkaline fuel cell. Energy Rep. 7, 2594–2604. https://doi.org/10.1016/j.egyr.2021.04.060.)*

2011; Friedman et al., 2021; Hibbs et al., 2009; Iojoiu et al., 2006; Izadi et al., 2022; Li et al., 2013, 2023; Lin et al., 2010; 2013; Liu et al., 2021; Merle et al., 2011; Mustain et al., 2020; Okundamiya, 2021; Pal and Mukherjee, 2021; Pan et al., 2012; Ramaswamy and Mukerjee, 2019; Song et al., 2022; Tanaka et al., 2011; Ul Hassan et al., 2020; Varcoe and Slade, 2005; Verhaert et al., 2016, 2012; Wang et al., 2011, 2021; Zeng et al., 2023; Zhang, 2021). For AFCs that use AEMs, the hydration state of the membrane directly determines its ionic conductivity. Moderate humidity can keep the membrane well-hydrated and maintain the smoothness of ionic channels. However, when the environmental humidity is too low, the membrane may contract due to water loss, leading to reduced activity of the ion-exchange groups and decreased ionic conductivity. On the other hand, excessive humidity can cause the membrane to swell excessively, damaging its microstructure and potentially causing delamination of the MEA. Research has shown that when the relative humidity of the environment drops sharply from 50% to 20%, the ionic conductivity of the membrane can decrease by 30%–40%, severely affecting the power output of the cell (Pal and Mukherjee, 2020).

3 The development prospects of alkaline fuel cells

3.1 Material innovation and performance optimization

Development of new catalysts: The development of highly efficient non-precious metal catalysts is a core pathway to breaking through the cost and performance bottlenecks of AFCs. From the perspective of materials science, transition metal nitrides, carbides, and alloy catalysts, with their unique electronic structures and surface properties, have demonstrated excellent catalytic potential for the ORR in alkaline environments. For example, nitrogen atoms in the lattice of transition metal nitrides can regulate the d-band electronic structure of metal atoms through electron delocalization, reducing the energy barrier of the rate-determining step in the ORR and thus significantly enhancing catalytic activity (Friedman et al., 2021). Experimental data show that cobalt nitride (Co_4N) catalysts can achieve an ORR half-wave potential of up to 0.82 V (vs. Reversible Hydrogen Electrode (RHE)) in alkaline media, which is only about 50 mV lower than that of traditional platinum-on-carbon (Pt/C) catalysts, demonstrating performance close to that of noble metal catalysts.

Iron-nickel alloy catalysts have made breakthrough progress in recent years. By adjusting the atomic ratio of Fe to Ni in the alloy and the crystal structure of the nanoparticles, the distribution of active sites on the surface can be optimized. For example, when the Fe:Ni ratio in the Fe–Ni alloy is 3:7, its onset potential for the ORR reaches 0.91 V (vs. RHE), and after 10,000 cycles of testing, the half-wave potential decay is only 12 mV, showing excellent activity and stability, with performance very close to that of commercial Pt/C catalysts (Li et al., 2013).

Single-atom catalysts, due to their atomically dispersed characteristics and maximized atomic utilization, have become a cutting-edge direction in the development of AFC catalysts. For example, in the case of single-atom cobalt anchored on a nitrogen-doped carbon carrier (Co–N–C), theoretical calculations indicate that individual cobalt atoms, by forming coordination structures with surrounding nitrogen atoms, can precisely regulate the adsorption and dissociation processes of reactant molecules, thereby reducing the overpotential of the ORR reaction by about 30%. In experimental verification, this catalyst achieves a mass activity of 0.85 A/mg in an alkaline environment, which is 3–5 times higher than that of traditional nanoparticle catalysts (Tanaka et al., 2011).

Optimization of electrode structure: Rational design of electrode structures is a key strategy for improving mass transfer efficiency and reaction kinetics in AFCs. Three-dimensional porous electrode structures, achieved by constructing a hierarchical pore system, significantly optimize gas transport pathways. The micrometer-sized macropores are utilized for rapid gas diffusion, while the nanometer-sized mesopores and micropores provide high specific surface area loading spaces for catalysts. This structure can reduce gas diffusion resistance by 40%–50%, effectively alleviating concentration polarization issues under high current densities (Hibbs et al., 2009).

Gradient electrode design, by regulating the compositional and structural gradients within the electrode, achieves functional zoning and coordination. For example, from the gas diffusion layer to the catalytic layer of the electrode, a gradient distribution with gradually increasing catalyst loading and decreasing porosity can be employed. This allows for a more uniform distribution of reactant concentrations within the electrode, avoiding local "starvation effects." Experiments have shown that gradient electrode design can increase the cell voltage by 15–20 mV at a current density of 1 A/cm^2 (Pan et al., 2012).

The application of composite electrode materials integrates the advantageous properties of different materials. By combining carbon nanotubes with high electrical conductivity and metal oxide nanoparticles with high catalytic activity, both rapid electron conduction and an increased number of catalytically active sites are ensured. Utilizing a hierarchical porous structure in the composite electrode, with high-activity nanocatalyst particles loaded in the inner layer and a high-porosity diffusion layer constructed on the outer layer, this design enables the cell to achieve a maximum power density of 800 mW/cm^2, which is approximately 60% higher than that of traditional electrodes.

Improvement of electrolytes and membrane materials: Developing highly stable alkaline electrolytes and AEMs is a core task to enhance the durability of AFCs. New types of solid-state electrolytes, by combining liquid electrolytes with polymer matrices, effectively address the issues of volatility and CO_2 reactivity associated with traditional alkaline solutions. For example, aPAEK membrane doped with KOH, through hydrogen bonding and ion-dipole interactions between molecular chains, constructs stable ionic conduction channels. Its ionic conductivity reaches 0.05 S/cm at 80°C, and after continuous operation in an alkaline environment for 1000 hours, the performance degradation is <5% (Merle et al., 2011).

In the development of AEMs, the introduction of bulky groups or cross-linking structures can significantly enhance the alkaline resistance of the membranes. A poly(arylene ether) membrane functionalized with benzimidazole, by incorporating rigid benzene ring structures, increases the spatial hindrance of the polymer main chain, effectively suppressing the hydrolysis and Hofmann elimination reactions of quaternary ammonium groups in alkaline environments. Tests have shown that after soaking in a 1 mol/L KOH solution at 80°C for 2000 hours, the IEC of this membrane decreases by only 8%, while the capacity loss of traditional poly(arylene ether) quaternary ammonium membranes exceeds 40%. Moreover, by constructing interpenetrating network cross-linking structures, the mechanical strength and dimensional stability of the membrane can be further improved, allowing it to maintain good integrity under dry–wet cycling conditions and providing assurance for the long-term stable operation of AFCs.

3.2 System integration and intelligent management

Efficient system design: Optimizing the integration of AFC systems is a key approach to improving overall efficiency and reducing energy losses. In traditional AFC systems, subsystems such as fuel supply and electrolyte circulation are relatively independent, and the numerous pipe connections significantly increase fluid resistance. For example, in the case of hydrogen supply pipes, if a traditional separate design is used, the frictional resistance of the pipe's inner wall and the local resistance generated by bends and valves can cause a pressure loss of up to 0.3–0.5 bar during hydrogen transportation, resulting in an energy loss of about 5%–8%. However, by adopting an integrated fuel supply and electrolyte circulation system, which involves compactly integrating fuel storage, transportation, and electrolyte circulation equipment, the length of the pipes can be shortened and the fluid path optimized. This can reduce pipe resistance by 40%–60%, significantly decreasing energy consumption (Verhaert et al., 2012).

Thermal management is an important aspect of AFC system design. During operation, AFCs generate a significant amount of waste heat. If not effectively utilized, this not only leads to energy waste but can also affect cell performance. An integrated thermal management system can recover the waste heat generated during cell operation for multiple purposes. For example, using the waste heat to preheat the fuel gas and oxidant gas entering the cell can increase the gas temperature by 20–30°C. This accelerates the kinetics of the electrode reactions, reduces polarization losses, and thereby increases the cell's output voltage by 8–12 mV, enhancing the overall system efficiency by 3%–5%. In addition, the recovered waste heat can also be used to drive an absorption chiller to provide cooling for the battery system, creating a cycle of energy reuse and further improving the system's overall energy efficiency (Zeng et al., 2023).

Intelligent control strategies: Advanced control algorithms provide strong support for the efficient operation of AFCs. Model predictive control (MPC) establishes a dynamic model of the fuel cell system to predict its operating state over a future period and calculates the optimal control strategy based on optimization objectives. In practical applications, MPC can anticipate load change trends in advance based on real-time parameters such as the cell's voltage, current, and temperature, and precisely adjust the fuel flow rate and electrolyte concentration. For example, when the load power suddenly increases by 20%, the MPC algorithm can adjust the hydrogen flow rate within 0.5–1 s, enabling a rapid response of the cell's output power while keeping voltage fluctuations within ±5 mV. This effectively enhances the system's dynamic performance.

Adaptive control can automatically adjust control parameters based on changes in the operating parameters of the cell to accommodate different working conditions. Since the performance of AFCs is affected by various factors such as temperature, humidity, and fuel purity, adaptive control algorithms dynamically optimize control strategies by continuously monitoring these environmental variables. In environments with a large temperature range (e.g., from 20°C to 50°C), adaptive control can reduce the fluctuation amplitude of the cell's output power by 30%–40%, significantly enhancing the stability of the cell under complex working conditions.

The fuzzy logic-based energy management system also demonstrates unique advantages in AFCs. This system uses fuzzy logic rules to convert multiple input variables, such as load demand, remaining battery charge, and hydrogen storage, into clear control outputs. When the load demand is low, the fuzzy control system can reduce the output power of the fuel cell, decreasing hydrogen consumption by 15%–20%, while storing excess electrical energy in auxiliary energy storage devices. When the load demand surges, the system quickly adjusts the fuel cell output and coordinates with the energy storage devices to supply power together, ensuring stable system operation. Experimental data show that the fuzzy logic-based energy management system can increase the hydrogen utilization rate of AFCs by 12%–18%, effectively extending the fuel usage time.

3.3 Expansion of application scenarios

Transportation sector: AFCs hold significant application potential in the transportation sector, particularly in scenarios such as rail transit, shipping, and heavy-duty vehicles, where high energy demands and long-range requirements are critical. Compared to other types of fuel cells, AFCs have a higher energy density, with a theoretical energy density of up to 300–500 Wh/kg. This enables vehicles equipped with AFCs to carry more energy, meeting the needs of long-distance transportation. In addition, the rapid hydrogen refueling capability of AFCs allows for fuel replenishment in just a few minutes, significantly reducing vehicle refueling time and effectively enhancing operational efficiency.

Figure 13.4 enables readers to clearly understand the application of fuel cells in the transportation field.

In the practical application of hydrogen fuel cell buses, AFC systems can provide stable and powerful power output. Fig. 13.5 enables readers to clearly understand the application of fuel cells in the transportation field. For example, in a city where AFC buses are in operation, the AFC system installed on the buses has a rated power of up to 120 kW. Under the complex urban road conditions, whether it is frequent starting and stopping or climbing slopes, the buses can maintain stable operation with a maximum speed of up to 69 km/h. In addition, compared with traditional diesel buses, buses using AFCs achieve zero exhaust emissions, reducing carbon dioxide emissions by about 350 kg per 100 km traveled, significantly lowering carbon emissions in the transportation sector. For rail transit and shipping, the high energy density and environmental protection characteristics of AFCs can meet their long-distance and large-capacity transportation needs, helping to achieve a green transformation in the transportation industry. Fig. 13.6 shows readers the application of fuel cells in the field of aircraft. Fig. 13.7 shows readers the application in the automotive field.

Distributed energy and backup power: As a distributed energy system, AFC plays an important role in power supply in remote areas and as a backup power source for the grid. The modular design of AFCs allows them to be

FIG. 13.4 The application of alkaline fuel cells in the transportation sector enables readers to clearly understand the application of fuel cells in the transportation field.

FIG. 13.5 The application of alkaline fuel cells in the transportation sector enables readers to clearly understand the application of fuel cells in the transportation field.

FIG. 13.6 The application of alkaline fuel cells in the transportation sector shows readers the application of fuel cells in the field of aircraft.

FIG. 13.7 The application of alkaline fuel cells in the transportation sector shows readers the application in the automotive field.

flexibly combined into power generation units according to different electricity demands, ranging from small-scale household use to large-scale data center power supply. In addition, AFCs operate with very low noise levels, typically below 55 decibels, and do not cause interference to the surrounding environment. This characteristic makes them highly suitable for deployment in noise-sensitive locations such as urban residential areas, hospitals, and schools.

In remote areas such as islands or rural regions, power supply stability is poor due to their distance from the main power grid. AFCs, when combined with renewable energy sources like solar and wind power, can create off-grid energy systems that operate independently. For example, in a case study on an island, an integrated system of AFCs, solar photovoltaic panels, wind turbines, and energy storage devices was established. Fig. 13.8 shows readers the prospects of AFCs in power supply for remote areas and as a backup power source for the grid. During the day, when sunlight is abundant or wind is strong, renewable energy sources supply power to the island's users first, and any excess electricity is used for water electrolysis to produce and store hydrogen. At night or during bad weather, the AFCs generate electricity using the stored hydrogen to ensure a continuous power supply for the island's residents. In addition, in data centers where power supply stability is extremely critical, AFCs serve as backup power sources. They can start up instantaneously during a grid failure, responding to power demands at the millisecond level to prevent data loss and equipment damage caused by power outages.

FIG. 13.8 AFC plays an important role in power supply for remote areas and as a grid backup power source. Show the readers the prospects of AFC in power supply for remote areas and as a backup power source for the grid.

Coupling with renewable energy: The integration of AFCs with water electrolysis hydrogen production systems offers a feasible solution for creating an integrated "green hydrogen" energy storage and power generation system. In this system, when renewable energy sources such as solar and wind power are generating at peak levels, the excess electricity is used to drive water electrolysis devices, splitting water into hydrogen and oxygen. This process converts electrical energy into chemical energy for storage. Conversely, when renewable energy generation is insufficient or during periods of high electricity demand, the AFCs use the stored hydrogen and oxygen to generate electricity through electrochemical reactions, effectively releasing the stored energy.

This "renewable energy hydrogen production—hydrogen energy storage—AFC power generation" model effectively addresses the intermittency and variability of renewable energy generation. For example, in a project coupling a wind farm with an AFC, during periods of unstable wind resources, hydrogen storage through electrolysis can convert 15%–20% of the electrical energy that would otherwise be lost due to curtailment into hydrogen for storage. This stored hydrogen can later be used to generate electricity through the AFC, increasing the overall energy utilization efficiency to over 85%. In addition, this model helps to optimize grid stability by reducing the impact of renewable energy integration on the power grid, thereby promoting the development of the energy structure toward cleaner, more efficient, and sustainable directions. It shows the readers the combination of the AFC with the water electrolysis hydrogen production system.

3.4 Policy and market drivers

Policy support and research and experimental development (R&D) investment: Globally, the strategic deployment of clean energy technologies by governments has provided a strong impetus for the development of AFCs. Top-level policy design, through multiple means such as financial support, research and development plans, and industrial guidance, has accelerated technological breakthroughs and the commercialization of AFCs.

China explicitly listed fuel cell technology as a key focus area for energy research in its 14th Five-Year Plan and has invested billions of yuan into AFC research through national key R&D programs (such as the "Hydrogen Energy Technology" special project) and local government matching funds. For example, the "*Key Materials and System Integration of Alkaline Fuel Cells*" special project, established in 2022, focuses on core technologies such as non-precious metal catalysts and long-life electrolytes, promoting a breakthrough in the lifespan of domestic AFC stacks from the thousand-hour level to the ten-thousand-hour level. At the local level, regions such as Shanghai and Guangdong have established hydrogen energy industrial parks, offering tax exemptions and open application scenarios for AFC R&D companies, accelerating the transformation of technology from laboratory to demonstration projects.

The European Union's (EU) "Green Deal" identifies AFC as one of the core technologies for achieving the "2050 Carbon Neutrality" target and has injected over 500 million euros in R&D funding into AFC projects through the Horizon Europe program. Fig. 13.9 shows readers the emphasis that many countries around the world place on clean energy. Member states such as Germany and France have established special funds to support collaborative development of high-efficiency AFC systems between universities and enterprises. For example, the "ALK-FUEL" project, a collaboration between the Karlsruhe Institute of Technology and Thyssenkrupp in Germany, aims to increase the volumetric power density of AFCs to 2.0 kW/L while reducing system costs by 30%. Under the synergistic effects of these policies, the EU has established a full industrial chain covering hydrogen production, storage, transportation, and application, providing a mature ecosystem for the commercialization of AFCs.

Cost reduction and economies of scale: The cost optimization of AFCs is driven by multiple factors, with technological progress, production scaling, and increased supply chain maturity working in synergy. Innovations in material preparation processes serve as the core engine for cost reduction—bulk synthesis technologies for nonprecious metal catalysts (such as atomic layer deposition and spray pyrolysis) have reduced the cost of cobalt-based and iron-based catalysts by over 90% compared to platinum-based catalysts, while maintaining

FIG. 13.9 **The scene of the International Clean Energy Conference** shows readers the emphasis that many countries around the world place on clean energy.

activity close to 85% of that of commercial Pt/C catalysts. The introduction of automated production technologies (such as electrode coating robots and membrane electrode thermal pressing lines) has increased component assembly efficiency by 5–8 times, while reducing the proportion of labor costs from 40% to below 15%.

The economies of scale are particularly significant in the field of materials: The large-scale production of AEMs has broken through the early bottleneck of "laboratory-scale synthesis." A multinational chemical company has increased the annual output capacity of a single AEM production line to 100,000 m^2 through continuous flow polymerization processes, reducing the cost of membrane materials from \$500 per square meter to \$50 per square meter—a decrease of 90%. The integration of supply chains for key components, such as catalyst carriers and bipolar plates (e.g., domestic production of carbon paper and the maturation of metal bipolar plate stamping processes), further compresses costs—the price of domestic carbon paper is only one-third that of imported products, and the unit cost of metal bipolar plates has decreased from \$200 per plate to \$80 per plate. Figure 13.10 shows readers the prospects of fuel cells in terms of cost reduction and economies of scale.

According to industry forecasts, when the global annual production capacity of AFCs exceeds 10 GW, the system cost is expected to drop from the current \$5000 per kW to below \$1500 per kW, reaching the tipping point for

FIG. 13.10 AFC cost reduction and economies of scale shows the readers the prospects of fuel cells in terms of cost reduction and economies of scale.

competition with traditional fossil fuel-based power generation systems. This steep decline in the cost curve relies not only on technological iteration but also benefits significantly from the large-scale expansion of the entire hydrogen energy "production-storage-use" industry chain, creating a positive cycle of "market expansion → increased production capacity → cost reduction → widespread application."

4 Conclusion

In the process of moving toward large-scale commercial application, AFCs are facing multiple challenges in performance improvement, material stability, system cost, and environmental adaptability. In terms of performance enhancement, the sluggish kinetics of electrode reactions, especially the rate-limiting steps of the ORR and HOR, lead to significant polarization losses at high current densities, severely restricting power output. Resistance in mass transfer and diffusion processes results in insufficient gas supply within the electrodes, causing concentration polarization. Moreover, the internal ohmic resistance of the cell, which includes energy losses from electrode materials, electrolytes, and contact interfaces, further reduces the efficiency of energy conversion. Regarding material stability, nonprecious metal catalysts are prone to corrosion and deactivation, while carbon carriers are susceptible to corrosion under alkaline conditions. Although noble metal catalysts are stable, their high cost is prohibitive. Traditional alkaline electrolytes are prone to degradation through the absorption of carbon dioxide, and issues such as evaporation and concentration changes also affect their performance. AEMs are susceptible to chemical degradation in highly alkaline environments, and their dimensional stability and mechanical strength need to be improved. In terms of system cost, the integration of AFC systems is complex, with significant energy losses across multiple subsystems. The high costs of electrode materials, electrolytes, and membrane materials, coupled with the need for high-precision manufacturing equipment, further limit large-scale commercialization. Regarding environmental adaptability, the reaction of carbon dioxide with alkaline electrolytes leads to performance degradation. Variations in temperature and humidity—such as increased electrolyte viscosity at low temperatures, accelerated material corrosion at high temperatures, and the impact of humidity on membrane hydration—significantly affect cell performance.

However, through multidimensional collaborative efforts in material innovation, system optimization, and policy support, the development prospects of alkaline fuel cells are extremely broad. In terms of material innovation, the development of new catalysts has become a key breakthrough. Transition metal nitrides, carbides, alloy catalysts, and single-atom catalysts have demonstrated good oxygen reduction activity in alkaline environments, holding promise for cost reduction and performance enhancement. Optimization of electrode structures, such as three-dimensional porous electrodes, gradient electrode design, and the application of composite electrode materials, can improve mass transfer and reaction kinetics. Improvements in electrolytes and membrane materials, including the development of new solid-state electrolytes and AEMs with stronger alkaline resistance, will effectively enhance the stability of the cells. Regarding system optimization, efficient system design through integrated integration and thermal management optimization reduces energy losses from auxiliary equipment. Intelligent control strategies, such as the introduction of MPC, adaptive control, and fuzzy control algorithms, enable real-time monitoring and parameter optimization of cell operation, significantly improving operational efficiency and stability. In the realm of policy support, the emphasis on clean energy by governments worldwide has injected strong momentum into the development of alkaline fuel cells. Policy initiatives such as China's "14th Five-Year Plan" and the EU's "Green Deal" have accelerated technological development and commercialization by increasing research and development investment, providing financial support, and opening up application scenarios.

With the continuous advancement of technology, alkaline fuel cells are poised to unleash significant potential across a diverse range of application scenarios. In the transportation sector, their high energy density and rapid hydrogen refueling capabilities make them well-suited for long-distance and heavy-load applications such as rail transit, shipping, and heavy-duty vehicles. In the realm of distributed energy and backup power, their modular design and quiet operation enable them to provide electricity to remote areas and serve as a reliable backup power source for urban residential areas and data centers. Moreover, when coupled with renewable energy sources, they can form integrated "green hydrogen" energy storage and power generation systems, effectively addressing the intermittency issues of renewable energy. In the future, alkaline fuel cells are expected to hold a significant position in the clean energy domain. By continuously reducing carbon emissions and enhancing energy utilization efficiency, they will contribute crucially to the global goals of peak carbon and carbon neutrality. They are set to become one of the core technologies driving the green transformation of the energy structure.

AI disclosure

During the preparation of this work, the author(s) used Doubao, Kimi in order to polish the language using AI tools. After using this tool/service, the author(s) reviewed and edited the content as needed and take(s) full responsibility for the content of the publication.

References

Agel, E., Bouet, J., Fauvarque, J.F., 2001. Characterization and use of anionic membranes for alkaline fuel cells. J. Power Sources. 03787753. 101 (2), 267–274. doi:10.1016/s0378-7753(01)00759-5.

Babatunde, O.M., Munda, J.L., Hamam, Y., 2022. Off-grid hybrid photovoltaic – micro wind turbine renewable energy system with hydrogen and battery storage: effects of sun tracking technologies. Energy Convers. Manag. 01968904. 255, 115335. doi:10.1016/j.enconman.2022.115335.

Bib Khan, J., Liang, Y., 2024. Recent Progress in Non-Noble Metal Catalysts for Oxygen Evolution Reaction: a Focus on Transition and Rare-Earth Elements. Chem. Rec. 1527-8999. 24 (12). doi:10.1002/tcr.202400151.

Burchardt, T., Gouérec, P., Sanchez-Cortezon, E., Karichev, Z., Miners, J.H., 2002. Alkaline fuel cells: contemporary advancement and limitations☆. Fuel. 00162361. 81 (17), 2151–2155. doi:10.1016/s0016-2361(02)00163-1.

Couture, G., Alaaeddine, A., Boschet, F., Ameduri, B., 2011. Polymeric materials as anion-exchange membranes for alkaline fuel cells. Prog. Polym. Sci. 00796700. 36 (11), 1521–1557. doi:10.1016/j.progpolymsci.2011.04.004.

Friedman, A., Mizrahi, M., Levy, N., Zion, N., Zachman, M., Elbaz, L., 2021. Application of Molecular Catalysts for the Oxygen Reduction Reaction in Alkaline Fuel Cells. ACS Appl. Mater. Interfaces. 1944-8244. 13 (49), 58532–58538. doi:10.1021/acsami.1c16311.

Hibbs, M.R., Fujimoto, C.H., Cornelius, C.J., 2009. Synthesis and characterization of poly(phenylene)-based anion exchange membranes for alkaline fuel cells. Macromolecules. 00249297. 42(21), 8316–8321. Available from: http://pubs.acs.org/doi/pdfplus/10.1021/ma901538c. doi:10.1021/ma901538c

Iojoiu, C., Chabert, F., Maréchal, M., Kissi, N.E., Guindet, J., Sanchez, J.Y., 2006. France From polymer chemistry to membrane elaboration: a global approach of fuel cell polymeric electrolytes. J. Power Sources 2, 198–209. doi:10.1016/j.jpowsour.2005.05.039.

Izadi, A., Shahafve, M., Ahmadi, P., 2022. Neural network genetic algorithm optimization of a transient hybrid renewable energy system with solar/wind and hydrogen storage system for zero energy buildings at various climate conditions. Elsevier Ltd, Iran Energy Convers. Manag. 01968904. 260. Available from: https://www.journals.elsevier.com/energy-conversion-and-management. doi:10.1016/j.enconman.2022.115593.

Jia, Q., Li, Y., Zhang, X., 2020. Recent progress in non-noble metal catalysts for oxygen reduction reaction in alkaline fuel cells[J]. Chem. Soc. Rev. 49 (14), 5035–5060.

Li, N., Leng, Y., Hickner, M.A., Wang, C.Y., 2013. Highly stable, anion conductive, comb-shaped copolymers for alkaline fuel cells. J. Am. Chem. Soc. 15205126. 135 (27), 10124–10133. doi:10.1021/ja403671u.

Li, X., Wang, Z., Chen, Y., Yonggang, L., Guo, J., Zheng, J., Shenghai, L., Zhang, S., 2023. Imidazolium-based AEMs with high dimensional and alkaline-resistance stabilities for extended temperature range of alkaline fuel cells. J. Membr. Sci. 03767388. 670, 121352. doi:10.1016/j.memsci.2023.121352.

Lin, B., Qiu, L., Jianmei, L., Yan, F., 2010. Cross-Linked Alkaline Ionic Liquid-Based Polymer Electrolytes for Alkaline Fuel Cell Applications. Chem. Mater. 0897-4756. 22 (24), 6718–6725. doi:10.1021/cm102957g.

Lin, X., Liang, X., Poynton, S.D., Varcoe, J.R., Ong, A.L., Ran, J., Li, Y., Li, Q., Xu, T., 2013. Novel alkaline anion exchange membranes containing pendant benzimidazolium groups for alkaline fuel cells. J. Membr. Sci. 18733123. 443, 193–200. doi:10.1016/j.memsci.2013.04.059.

Liu, J., Cao, S., Chen, X., Yang, H., Peng, J., 2021. Energy planning of renewable applications in high-rise residential buildings integrating battery and hydrogen vehicle storage. Appl. Energy. 03062619. 281, 116038. doi:10.1016/j.apenergy.2020.116038.

Merle, G., Wessling, M., Nijmeijer, K., 2011. Anion exchange membranes for alkaline fuel cells: a review. J. Membr. Sci. 03767388. 377 (1-2), 1–35. doi:10.1016/j.memsci.2011.04.043.

Mustain, W.E., Chatenet, M., Page, M., Kim, Y.S., 2020. Durability challenges of anion exchange membrane fuel cells. Energy Environ. Sci. 17545706. 13 (9), 2805–2838. Available from: http://pubs.rsc.org/en/journals/journal/ee. doi:10.1039/d0ee01133a.

Okundamiya, M.S., 2021. Integration of photovoltaic and hydrogen fuel cell system for sustainable energy harvesting of a university ICT infrastructure with an irregular electric grid. Energy Convers. Manag. 01968904. 250, 114928. doi:10.1016/j.enconman.2021.114928.

Pal, P., Mukherjee, V., 2020. off-grid solar photovoltaic/hydrogen fuel cell system for renewable energy generation: an investigation[J]. Int. J. Hydrogen Energy 45 (44), 23491–23503.

Pal, P., Mukherjee, V., 2021. Off-grid solar photovoltaic/hydrogen fuel cell system for renewable energy generation: an investigation based on techno-economic feasibility assessment for the application of end-user load demand in North-East India. Renew. Sustain. Energy Rev. 13640321. 149, 111421. doi:10.1016/j.rser.2021.111421.

Pan, J., Chen, C., Zhuang, L., Lu, J., 2012. Designing advanced alkaline polymer electrolytes for fuel cell applications. Acc. Chem. Res. 15204898. 45 (3), 473–481. doi:10.1021/ar200201x.

Ramaswamy, N., Mukerjee, S., 2019. Alkaline Anion-Exchange Membrane Fuel Cells: challenges in Electrocatalysis and Interfacial Charge Transfer. Chem. Rev. 0009-2665. 119 (23), 11945–11979. doi:10.1021/acs.chemrev.9b00157.

Song, Y., Hailin, M., Nan, L., Shi, X., Zhao, X., Chen, C., Wang, H., 2022. Techno-economic analysis of a hybrid energy system for CCHP and hydrogen production based on solar energy. Int. J. Hydrogen Energy. 03603199. 47 (58), 24533–24547. doi:10.1016/j.ijhydene.2021.08.134.

Tanaka, M., Fukasawa, K., Nishino, E., Yamaguchi, S., Yamada, K., Tanaka, H., Bae, B., Miyatake, K., Watanabe, M., 2011. Anion Conductive Block Poly(arylene ether)s: synthesis, Properties, and Application in Alkaline Fuel Cells. J. Am. Chem. Soc. 0002-7863. 133 (27), 10646–10654. doi:10.1021/ja204166e.

Ul Hassan, N., Mandal, M., Huang, G., Firouzjaie, H.A., Kohl, P.A., Mustain, W.E., 2020. Achieving High-Performance and 2000 h Stability in Anion Exchange Membrane Fuel Cells by Manipulating Ionomer Properties and Electrode Optimization. Adv. Energy Mater. 1614-6832. 10 (40). doi:10.1002/aenm.202001986.

Varcoe, J.R., Slade, R.C.T., 2005. Prospects for Alkaline Anion-Exchange Membranes in Low Temperature Fuel Cells. Fuel Cells. 1615-6846. 5 (2), 187–200. doi:10.1002/fuce.200400045.

Verhaert, I., Verhelst, S., Huisseune, H., Poels, I., Janssen, G., Mulder, G., De Paepe, M., 2012. Thermal and electrical performance of an alkaline fuel cell. Appl. Therm. Eng. 13594311. 40, 227–235. doi:10.1016/j.applthermaleng.2012.01.042.

Verhaert, I., Mulder, G., De Paepe, M., 2016. Evaluation of an alkaline fuel cell system as a micro-CHP. Energy Convers. Manag. 01968904. 126, 434–445. doi:10.1016/j.enconman.2016.07.058.

Wang, J., Wang, J., Shenghai, L., Zhang, S., 2011. Poly(arylene ether sulfone)s ionomers with pendant quaternary ammonium groups for alkaline anion exchange membranes: preparation and stability issues. J. Membr. Sci. 03767388. 368 (1-2), 246–253. doi:10.1016/j.memsci.2010.11.058.

Wang, Z., Zhang, X., Rezazadeh, A., 2021. Hydrogen fuel and electricity generation from a new hybrid energy system based on wind and solar energies and alkaline fuel cell. Energy Rep. 23524847. 7, 2594–2604. doi:10.1016/j.egyr.2021.04.060.

Zeng, R., Tang, X., Deng, Y., Zhang, X., Hongqiang, L., Yin, W., Zhang, G., 2023. Design and optimization of solar energy system with hydrogen energy storage and alkaline fuel cell. Energy Convers. Manag. 01968904. 295, 117628. doi:10.1016/j.enconman.2023.117628.

Zhang, L., 2021. Recent advances in alkaline fuel cells: from materials to system applications. Energy Environ. Sci. 14 (12), 7351–7384.

Index

Note: Page numbers followed by "*f*" indicate figures, "*t*" indicate tables.

B

E

F

G

H

Q

R

S

T

Printed and bound by CPI Group (UK) Ltd, Croydon, CR0 4YY
14/09/2026
02216408-0003